STUDY GUIDE

SECOND EDITION

LINEAR ALGEBRA AND ITS APPLICATIONS

DAVID C. LAY

UNIVERSITY OF MARYLAND

ADDISON-WESLEY

An imprint of Addison Wesley Longman, Inc.

Reading, Massachusetts • Menlo Park, California • New York • Harlow, England
Don Mills, Ontario • Sydney • Mexico City • Madrid • Amsterdam

Reproduced by Addison-Wesley from camera-ready copy supplied by the author.

ISBN 0-201-82477-9

2 3 4 5 6 7 8 9 10 VG 00999897

TABLE OF CONTENTS

INFORMATION ABOUT MATRIX PROGRAMS

The *Study Guide* contains valuable information for using several matrix programs with the text.

MATLAB

MATLAB boxes in many *Study Guide* sections discuss MATLAB commands as they are needed for the exercises. An index of these commands is at the back of the *Study Guide*. I encourage you to try MATLAB—it is so easy to learn. If MATLAB is already installed on computers at your school, about all you need to know to get started is how to enter and exit the MATLAB program. To save time on homework, obtain the MATLAB Toolbox that has the data for over 800 exercises, together with special programs that reinforce basic concepts in the course. The Toolbox is available via "ftp" on the Internet:

//ftp.aw.com/math/authors/lay/la2e

TI-85 and HP-48G Calculators

Notes for the TI-85 and HP-48G graphing calculators are included in the last two sections of this *Study Guide*. The notes correspond to the MATLAB boxes and translate MATLAB commands into commands appropriate for the graphing calculators. The notes provide the specific information you need to use your calculator effectively and they describe special programs that implement algorithms from the text. You can obtain these programs via ftp from the Internet address listed above. The programs can be downloaded to your TI-85, TI-86, or HP-48G calculator.

The TI-85 notes and programs were written by Professor Luz Maria DeAlba, of Drake University (Des Moines, Iowa) The HP-48G notes and programs were written by Professor Thomas W. Polaski, of Winthrop University (Rock Hill, South Carolina), who has used these notes in his classes several times. Both professors have been using technology successfully with *Linear Algebra and Its Applications* for several years. I greatly appreciate their contributions to this *Study Guide*.

HOW TO STUDY LINEAR ALGEBRA

A first course in linear algebra is dramatically different from most mathematics courses that precede it. The focus shifts from learning computational procedures to digesting and mastering basic concepts that underlie the computations. To survive, you may need to learn a new way to study mathematics. That's why I wrote this *Study Guide*—to show you how to succeed in the course and to give you tools to do this.

Because you are likely to use linear algebra later in your career, you need to learn the material at a level that will carry you far beyond the final exam. I believe that the strategies below are crucial to success.

STRATEGIES FOR SUCCESS IN LINEAR ALGEBRA

1. **Read each section thoroughly before you begin the exercises.** Most students are not used to doing this in courses that precede linear algebra. They could survive by looking at the examples only when they were unable to work an exercise. That simply will not work in linear algebra. If you "copy" an example (with necessary modifications), you may think you are "understanding" the problem, but very little true learning will take place. (You'll find that out on your first exam.) For example, in addition to knowing *how* to carry out a certain procedure, you must learn *when* that procedure is appropriate and (most importantly) *why* it works.

   After reading the text, use the "Key Ideas" and "Study Notes" in this *Guide* to help you work through the text again. *Then* start on the exercises! In the long run, this strategy will improve your performance and save you time.

2. **Prepare for each class as you would for a language class.** Mastery of the subject requires that you learn a rich vocabulary. Your goal now is to become so familiar with concepts that you can use them easily (and correctly) in conversation and in writing. In homework, try to write complete sentences, such as you'll find in the *Study Guide* solutions. Pay attention, too, to the warnings here about misuse of terminology.

   This course resembles a language class because of the preparation needed between class meetings to avoid falling behind. Most sections in the text build on preceding sections, and once you are behind, catching up with the class is often difficult. The fact that concepts may seem "simple" does not mean you can afford to postpone your study until the weekend. The homework may be harder than you expect. The most valuable advice I can give you is to keep up with the course.

3. **Concentrate more on learning definitions, facts, and concepts, than on practicing routine computations or algorithms.** Seek connections *between* concepts. Many theorems and boxed "facts" describe such connections. For instance, see Theorem 2 in Section 1.2, and Theorems 3 and 4 in Section 1.4. Your goal is to think in general terms, to *imagine* typical computations without performing any arithmetic, and to focus on the principles behind the computations.

4. **Review frequently**. Review and reflection are key ingredients for success in learning the material. At strategic points in this *Guide*, I have inserted special subsections labeled "Mastering Linear Algebra Concepts," to provide specific help for your review of each main concept. I urge you to prepare the sheets as you reach each review box, thinking carefully about the material as you write. Later, you may choose to add further notes and, of course, use the sheets to review for exams. A Glossary Checklist at the end of each chapter may help you learn important definitions.

> **CAUTION** Because you can find complete solutions here to many exercises, you will be tempted to read the explanations before you really try to write out the solutions yourself. Don't do it! If you merely think a bit about a problem and then check to see if your idea is basically correct, you are likely to overestimate your understanding. Some of my students have done this and miserably failed the first exam. By then the damage was done, and they had great difficulty catching up with the class. Proper use of the *Study Guide*, however, will help you to succeed and enjoy the course at the same time.

A Personal Note

Students who have used this material have told me how much it helped them learn linear algebra and prepare for tests. The first time my students used these notes, they had already taken one exam. Grades on the next exam were substantially higher. For some students, the improvement was dramatic. I hope this *Study Guide* will encourage you to master linear algebra and to perform at a level higher than you ever dreamed possible.

David C. Lay

1 LINEAR EQUATIONS IN LINEAR ALGEBRA

As you work through this chapter and the next, your experience may resemble several walks through a village at different seasons of the year. The surroundings will be familiar, but the landscape will change. You will examine various mathematical concepts from several points of view, and a major problem will be to learn all the new terminology and the many connections between the concepts. In Chapter 4, you will see these ideas in a more abstract setting. Diligent work now will make the trip through Chapter 4 just another walk through the same village.

1.1 SYSTEMS OF LINEAR EQUATIONS

The fundamental concepts presented in this section and the next must be mastered, for they will be used throughout the course.

STUDY NOTES

Please read **How to Study Linear Algebra**, on the preceding two pages, before you continue.

The text uses boldface type to identify important terms the first time they appear. You need to learn them; some students write selected terms on 3×5 cards, for review. At the end of each chapter in this Study Guide, a glossary checklist may help you learn definitions.

The text defines the **size** of a matrix. Don't use the term *dimension*, even though that appears in some computer programming languages, because in linear algebra, *dimension* refers to another concept (in Section 4.5).

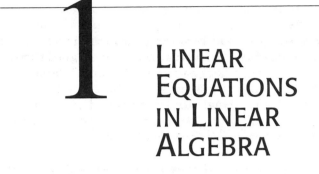

The first few examples are so simple that they could be solved by a variety of techniques. But it is important to learn the systematic method presented here, because it easily handles more complicated linear systems, and it works in all cases.

The calculations in this section are based on the following important fact:

> *When elementary row operations are applied to a linear system, the new system has exactly the same solution set.*

(See the text.) The steps in the summary below will be modified slightly in Section 1.2.

Summary of the Elimination Method (for This Section)

1. The first equation must contain an x_1. Interchange equations, if necessary. This will create a nonzero entry in the first row, first column, of the augmented matrix.
2. Eliminate x_1 terms in the other equations. That is, use replacement operations to create zeros in the first column of the matrix below the first row.
3. Obtain an x_2 term in the second equation. (Interchange the second equation with one below, if needed, but don't touch the first equation.) You may scale the second equation, if desired, to create a 1 in the second column and second row of the matrix.
4. Eliminate x_2 terms in equations below the second equation, using replacement operations.
5. Continue with x_3 in the third equation, x_4 in the fourth equation, etc., eliminating these variables in the equations below. This will produce a "triangular" system (at least for systems in this section).
6. Check if the system in triangular form is consistent. If it is, a solution is found by starting with the last nonzero equation and working back up to the first equation. Each variable on the "diagonal" is used to eliminate the terms in that variable above it. The solution to the system becomes apparent when the system is finally transformed into "diagonal" form.
7. Check any solutions you find by substituting them into the original system.

The *solution set* of a system of linear equations either is empty, or contains one solution, or contains infinitely many solutions. When asked to "solve" a system, you may write "inconsistent" if the system has no solution.

As you will see later, determining the number of solutions in the solution set is sometimes more important than actually computing the solution or solutions. For that reason, pay close attention to the subsection on existence and uniqueness questions. Key Exercises: 21—26.

SOLUTIONS TO EXERCISES

Get into the habit *now* of working the Practice Problems before you start the exercises. Probably, you should attempt all the Practice Problems before checking the solutions at the end of the exercise set, because once you start reading the first solution, you might tend to read on through the other solutions and spoil your chance to benefit from those problems.

1. $\begin{aligned} x_1 + 7x_2 &= 4 \\ -2x_1 - 9x_2 &= 2 \end{aligned}$ $\begin{bmatrix} 1 & 7 & 4 \\ -2 & -9 & 2 \end{bmatrix}$

Replace row 2 by row 2 plus 2·row 1: $\begin{aligned} x_1 + 7x_2 &= 4 \\ 5x_2 &= 10 \end{aligned}$ $\begin{bmatrix} 1 & 7 & 4 \\ 0 & 5 & 10 \end{bmatrix}$

Scale row 2 (multiply by 1/5): $\begin{aligned} x_1 + 7x_2 &= 4 \\ x_2 &= 2 \end{aligned}$ $\begin{bmatrix} 1 & 7 & 4 \\ 0 & 1 & 2 \end{bmatrix}$

Replace row 1 by row 1 + (-7)·row 2: $\begin{aligned} x_1 \quad &= -10 \\ x_2 &= 2 \end{aligned}$ $\begin{bmatrix} 1 & 0 & -10 \\ 0 & 1 & 2 \end{bmatrix}$

The solution is (-10,2). Check:

$$\begin{aligned} (-10) + 7(2) &= -10 + 14 = 4 \\ -2(-10) - 9(2) &= 20 - 18 = 2 \end{aligned}$$

7. $\begin{bmatrix} 1 & 3 & 0 & 5 & -5 \\ 0 & 1 & -6 & 9 & 0 \\ 0 & 0 & 2 & 7 & 1 \\ 0 & 0 & 1 & 4 & -2 \end{bmatrix}$. To simplify hand computation, the next step is either to interchange rows 3 and 4 or to multiply row 3 by 1/2. Another possibility is to replace row 4 by row 4 + (-1/2)·row 3.

13. Work with the augmented matrix: $\begin{bmatrix} 0 & 1 & 5 & -4 \\ 1 & 4 & 3 & -2 \\ 2 & 7 & 1 & -1 \end{bmatrix}$.

Interchange rows 1 and 2: $\begin{bmatrix} 1 & 4 & 3 & -2 \\ 0 & 1 & 5 & -4 \\ 2 & 7 & 1 & -1 \end{bmatrix}$

Add (-2)·row 1 to row 3: $\begin{bmatrix} 1 & 4 & 3 & -2 \\ 0 & 1 & 5 & -4 \\ 0 & -1 & -5 & 3 \end{bmatrix}$

Add row 2 to row 3: $\begin{bmatrix} 1 & 4 & 3 & -2 \\ 0 & 1 & 5 & -4 \\ 0 & 0 & 0 & -1 \end{bmatrix}$ $\quad \begin{aligned} x_1 + 4x_2 + 3x_3 &= -2 \\ x_2 + 5x_3 &= -4 \\ 0x_3 &= -1 \end{aligned}$

The equation $0x_3 = -1$ or $(0x_1 + 0x_2 + 0x_3 = -1)$ has no solution, so the solution set of this triangular system is empty. Since the original system has the same solution set, the system has no solution.

Study Tip: When writing a coefficient matrix or augmented matrix for a system of linear equations, be sure that the variables appear in all equations *in the same order*. Arrange the variables in columns, as in the text; place zeros in the matrix whenever a variable is missing from an equation.

19. Work with the augmented matrix: $\begin{bmatrix} 0 & 2 & 2 & 0 & 0 \\ 1 & 0 & 0 & -2 & -3 \\ 0 & 0 & 1 & 3 & -4 \\ -2 & 3 & 2 & 1 & 5 \end{bmatrix}$.

Interchange rows 1 and 2: $\begin{bmatrix} 1 & 0 & 0 & -2 & -3 \\ 0 & 2 & 2 & 0 & 0 \\ 0 & 0 & 1 & 3 & -4 \\ -2 & 3 & 2 & 1 & 5 \end{bmatrix}$

Add 2·row 1 to row 4: $\begin{bmatrix} 1 & 0 & 0 & -2 & -3 \\ 0 & 2 & 2 & 0 & 0 \\ 0 & 0 & 1 & 3 & -4 \\ 0 & 3 & 2 & -3 & -1 \end{bmatrix}$

Add (-3/2)·row 2 to row 4: $\begin{bmatrix} 1 & 0 & 0 & -2 & -3 \\ 0 & 2 & 2 & 0 & 0 \\ 0 & 0 & 1 & 3 & -4 \\ 0 & 0 & -1 & -3 & -1 \end{bmatrix}$

Add 1·row 3 to row 4: $\begin{bmatrix} 1 & 0 & 0 & -2 & -3 \\ 0 & 2 & 2 & 0 & 0 \\ 0 & 0 & 1 & 3 & -4 \\ 0 & 0 & 0 & 0 & -5 \end{bmatrix}$

The system is inconsistent, because $0x_4 = -5$ has no solution.

Study Tip: Pay attention to how a problem is worded. If you are instructed only to determine existence or uniqueness of a solution, as in Exercise 19, stop row operations when you reach a "triangular" form.

25. This is an important problem. Try to work it out completely yourself before you read further in this *Study Guide.* Row reduce the matrix, treating g, h, and k as unspecified numbers:

$$\begin{bmatrix} 1 & -4 & 7 & g \\ 0 & 3 & -5 & h \\ -2 & 5 & -9 & k \end{bmatrix} \sim \begin{bmatrix} 1 & -4 & 7 & g \\ 0 & 3 & -5 & h \\ 0 & -3 & 5 & k+2g \end{bmatrix} \sim \begin{bmatrix} 1 & -4 & 7 & g \\ 0 & 3 & -5 & h \\ 0 & 0 & 0 & k+2g+h \end{bmatrix} \qquad (*)$$

The original system has a solution if and only if $k + 2g + h = 0$. That is, $2g + h + k = 0$.

To see why this conclusion is true, think of $k+2g+h$ as some number b. Then the third equation has the form $0x_1 + 0x_2 + 0x_3 = b$, which has no solution when $b \neq 0$. But if b *is* zero, the system

$$\begin{aligned} x_1 - 4x_2 + 7x_3 &= g \\ 3x_2 - 5x_3 &= h \\ 0 &= 0 \end{aligned}$$

has a solution no matter what the values of g and h.

31. $\begin{bmatrix} 1 & 3 & -1 & 5 \\ 0 & 1 & -4 & 2 \\ 0 & 2 & -5 & -1 \end{bmatrix}$, $\begin{bmatrix} 1 & 3 & -1 & 5 \\ 0 & 1 & -4 & 2 \\ 0 & 0 & 3 & -5 \end{bmatrix}$. Only the third row changes. Replace row 3 of the left matrix by its sum with -2 times row 2. To reverse the process, replace row 3 of the right matrix by its sum with +2 times row 2.

33. My own students have recommended that I never give you the complete answers to the true/false questions. They felt that the temptation to read the answers is too great. After working both with and without answers, they realized how much doing the true/false work on their own helped them. So all you will see here are the places where you can find the answers.

a. See the remarks following the box *Elementary Row Operations.*

b. See page 5.

c. See the middle of page 3.

d. See the box before Example 2.

A Mathematical Note: "If ..., then ...."

Many important facts and theorems in the text are written as *implication statements*, in the form "If *P*, then *Q*", where *P* and *Q* represent complete sentences. For instance the statement in the box at the end of page 7 has the form

$$\text{If } \left\{\begin{matrix}\text{the augmented matrices}\\ \text{of two linear systems}\\ \text{are row equivalent}\end{matrix}\right\}, \text{ then } \left\{\begin{matrix}\text{the two systems}\\ \text{have the same}\\ \text{solution set}\end{matrix}\right\} \qquad (1)$$

An implication statement "If *P*, then *Q*" is itself true provided that statement *Q* is true *whenever* statement *P* is true. In mathematical terminology, we say that "*P* implies *Q*," and we write $P \Rightarrow Q$.

Be careful to distinguish between an implication statement "*P* implies *Q*" and the **converse** or "opposite" implication, "*Q* implies *P*". The converse may or may not be true when the original implication is true. For instance, the converse of (1) above is *not* true, because there exist two linear systems with the same solution set but whose augmented matrices are *not* row equivalent. For example:

$$\begin{aligned} x_1 + x_2 &= 1 \\ 2x_1 + 2x_2 &= 2 \end{aligned} \qquad\qquad \begin{aligned} x_1 + x_2 &= 1 \\ 2x_1 + 2x_2 &= 2 \\ 3x_1 + 3x_2 &= 3 \end{aligned}$$

MATLAB Row Operations

To obtain the data for the exercises in this section, install the m-files from the Toolbox data disk, available from your instructor. Then, while you are running MATLAB, type **c1s1** (chapter 1 section 1), and press <Enter>. You should see a list of exercises for which data are available. Type the number of the appropriate exercise (and press <Enter>).

In this exercise set, the data for each exercise are stored in a matrix *M*. Row operations on *M* are performed by the following commands (from the Toolbox data disk):

replace(M,r,m,s)	Replaces row r of M by row r + m·row s
swap(M,r,s)	Interchanges rows r and s of M
scale(M,r,c)	Multiplies row r of M by a nonzero scalar c

(Press <Enter> after each MATLAB command, displayed in boldface type.) The name of any matrix in your MATLAB workspace can be inserted in place of *M*; the letters *r*, *m*, *s*, and *c* stand for numbers you choose.

If you enter one of these commands, say, **swap(M,1,3)**, then the new matrix, produced from *M*, is stored in the matrix "ans" (for "answer"). If, instead, you type **M1 = swap(M,1,3)**, then the answer is stored in a new matrix *M*1. If the next operation is **M2 = replace(M1,2,5,1)**, then the result of changing *M*1 is placed in *M*2, and so on.

The advantage of giving a new name to each new matrix is that you can easily go back a step if you don't like what you just did to a matrix. If, instead, you type **M = replace(M,2,5,1)** , then the result is placed back in *M* and the "old" *M* is lost. Of course, the "reverse" operation, **M = replace(M,2,-5,1)** will bring back the old *M*.

Note: For the simple problems in this section and the next, the multiple *m* you need in the command **replace(M,r,m,s)** will usually be a small integer or fraction that you can compute in your head. In general, *m* may not be so easy to compute mentally. The next two paragraphs describe how to handle such a case.

The entry in row *r* and column *c* of a matrix *M* is denoted by *M*(*r*,*c*). If the number stored in *M*(*r*,*c*) is displayed with a decimal point, then the displayed value may be accurate to only about five digits. In this case, use the *symbol* *M*(*r*,*c*) instead of the displayed value in calculations.

For instance, if you want to use the entry *M*(*s*,*c*) to change *M*(*r*,*c*) to 0, enter the commands

```
m = -M(r,c)/M(s,c)        The multiple of row s to be added to row r
M = replace(M,r,m,s)      Adds m times row s to row r
```

Or, you can use just one command: **M = replace(M,r,-M(r,c)/M(s,c),s)**.

Finally, the command **format compact** will eliminate extra space between displays, so you can see more data on the screen. The command **format** will return the screen to the normal display.

Warning: Using a matrix program such as MATLAB is fun and will save you time, but make sure you can perform row operations rapidly and accurately with pencil and paper. Probably, you should work all the exercises in Section 1.1 by hand and only use your matrix program to check your work.

1.2 ROW REDUCTION AND ECHELON FORMS

Our interest in the row reduction algorithm lies mostly in the echelon forms that are created by the algorithm. For practical work, a computer should perform the calculations. However, you need to understand the algorithm so you can learn how to use it for various tasks. Also, unless you take your exams at a computer or with a matrix programmable calculator, you must be able to perform row reduction quickly and accurately by hand.

STUDY NOTES

The row reduction algorithm applies to any matrix, not just an augmented matrix for a linear system. In many cases, all you need is an echelon form. The reduced echelon form is mainly used when it comes from an augmented matrix and you have to find all the solutions of a linear system.

Strategies for faster and more accurate row reduction:

- Avoid subtraction in a row replacement. It leads to mistakes in arithmetic. Instead, add a negative multiple of one row to another.
- Always enclose each matrix with brackets or large parentheses.
- To save time, combine all row replacement operations that use the *same* pivot position, and write just one new matrix. Never "clean out" more than one column at a time. (You can combine several scaling operations, or combine several interchanges, if you are careful. But that seldom will be necessary.)
- *Never* combine an interchange with a replacement. In general, don't combine different types of row operations. This will be particularly important when you evaluate determinants, in Chapters 3 and 5.

How to avoid copying errors:

- Practice neat writing, not too small. Develop proper habits in homework so your work on tests will be accurate and complete.
- Write a matrix row by row. Your eye may less likely to read from the wrong row if you place the new matrix beside the old one. Arrange your sequence of matrices across the page, rather than down the page.
- Don't let your work flow from one side of a paper to the reverse side.

Study Tips: Theorem 2 is a key result for future work. Also, study the procedure in the box following Theorem 2. Failure to write out the system of equations (step 4) is a common source of errors.

SOLUTIONS TO EXERCISES

1. To check whether a matrix is in echelon form ask the questions:

   (i) *Is every nonzero row above the all-zero rows (if any)?*

   The matrix (d) fails this test, so it is *not* in echelon form.

   (ii) *Are the leading entries in a stair-step pattern, with zeros below each leading entry?*

The matrices (a), (b), (c) all pass tests (i) and (ii), so they are in echelon form.

To check whether a matrix in echelon form is actually in *reduced* echelon form, ask two more questions:

(iii) *Is there a 1 in every pivot position?*

The matrices (a), (b), (c) all pass this test, so ask:

(iv) *Is each leading 1 the only nonzero entry in its column?*

Matrices (a) and (b) pass all four tests, so they are in reduced echelon; but matrix (c) is not in reduced echelon form.

7. $\begin{bmatrix} 1 & 0 & 2 & 5 \\ 2 & 0 & 3 & 6 \end{bmatrix} \sim \begin{bmatrix} 1 & 0 & 2 & 5 \\ 0 & 0 & -1 & -4 \end{bmatrix} \sim \begin{bmatrix} 1 & 0 & 2 & 5 \\ 0 & 0 & 1 & 4 \end{bmatrix} \sim \begin{bmatrix} \textcircled{1} & 0 & 0 & -3 \\ 0 & 0 & \textcircled{1} & 4 \end{bmatrix}$

The corresponding system of equations is: $\begin{aligned} x_1 \quad &= -3 \\ x_3 &= 4 \end{aligned}$

The basic variables are x_1 and x_3, corresponding to the pivot columns 1 and 3 in the matrix. Any other variables are free, so the general solution is $\begin{cases} x_1 = -3 \\ x_2 \text{ is free.} \\ x_3 = 4 \end{cases}$ Two incorrect answers are $\begin{cases} x_1 = -3 \\ x_2 = 0 \\ x_3 = 4 \end{cases}$ and $\begin{cases} x_1 = -3 \\ x_3 = 4 \end{cases}$ (with nothing said about x_2).

Warning: The remarks for Exercise 7 are important, because students often have difficulty with a problem in which a variable does not appear in the equation.

13. Combine two replacement operations in the first step:

$$\begin{bmatrix} 2 & -4 & 3 \\ -6 & 12 & -9 \\ 4 & -8 & 6 \end{bmatrix} \sim \begin{bmatrix} 2 & -4 & 3 \\ 0 & 0 & 0 \\ 0 & 0 & 0 \end{bmatrix} \sim \begin{bmatrix} \textcircled{1} & -2 & 3/2 \\ 0 & 0 & 0 \\ 0 & 0 & 0 \end{bmatrix} \qquad \begin{aligned} \textcircled{x_1} - 2x_2 &= 3/2 \\ 0 &= 0 \\ 0 &= 0 \end{aligned}$$

Thus $x_1 = 3/2 + 2x_2$, and the general solution is $\begin{cases} x_1 = 3/2 + 2x_2 \\ x_2 \text{ is free} \end{cases}$.

Study Tip: Be sure to work on Exercises 17—22. The experience will help you later. These exercises make nice quiz questions, too.

19. $\begin{bmatrix} 1 & h & 3 \\ 2 & 8 & 1 \end{bmatrix} \sim \begin{bmatrix} 1 & h & 3 \\ 0 & 8-2h & -5 \end{bmatrix}$. There are only two possibilities:

(1) $8 - 2h = 0$. In this case $\begin{bmatrix} 1 & h & 3 \\ 0 & 0 & -5 \end{bmatrix}$ represents an inconsistent system. So the system has no solution when $2h = 8$, and $h = 4$.

(2) $8 - 2h \neq 0$. In this case the augmented matrix corresponds to a consistent system, and there are no free variables. So the system has a unique solution when $h \neq 4$.

23. **a.** See Theorem 1.

b. See the beginning of the section.

c. Basic variables are defined after equation (4).

d. See the subsection *Parametric Descriptions of Solution Sets.* Actually, this question does not take the case of an inconsistent system into account. A better true/false statement would be: "If a linear system is consistent, then finding a parametric description of the solution set is the same as *solving* the system."

e. The row shown corresponds to the equation $5x_4 = 0$.

29. *If a linear system is consistent, then the solution is unique if and only if every column in the coefficient matrix is a pivot column; otherwise, there are infinitely many solutions.*

This statement is true because the free variables correspond to *nonpivot* columns of the coefficient matrix. The columns are all pivot columns if and only if there are no free variables. And there are no free variables if and only if the solution is unique, by Theorem 2.

Study Tip: Notice from Exercise 29 that the question of uniqueness of the solution of a linear system is not influenced by the numbers in the rightmost column of the augmented matrix.

31. The full solution is in the text.

A Mathematical Note: "If and only if"

You need to know what the phrase "if and only if" means. It was used above in Exercise 29, and you will see it again in theorems and boxed facts. The phrase "if and only if" always appears between two complete statements. Look at Theorem 2, for instance:

$$\left\{\begin{matrix}\text{A specific}\\ \text{linear system}\\ \text{is consistent}\end{matrix}\right\} \text{ if and only if } \left\{\begin{matrix}\text{the rightmost column of the}\\ \text{augmented matrix of the linear}\\ \text{system is not a pivot column.}\end{matrix}\right\} \quad (1)$$

The entire sentence means that the two statements in parentheses are either both true or both false.

Sentence (1) has the general form

P if and only if Q (2)

where P denotes the first statement and Q denotes the second statement. The sentence (2) says two things:

If statement P is true, then statement Q is true, too. (3)
If statement Q is true, then statement P is also true. (4)

A mathematical shorthand for (2) is "$P \Leftrightarrow Q$".

1.3 VECTOR EQUATIONS

Do not be deceived by the rather simple beginning of Section 1.3. The important material on Span$\{\mathbf{v}_1, \ldots, \mathbf{v}_p\}$ will take time to digest. Exercises 11 – 14 and 21 – 26 are important. Each exercise involves an *existence* question about whether a certain vector equation has a solution. (You don't have to find the solution.) Note how the same basic question can be asked in several different ways.

STUDY NOTES

Develop the habit of reading the section carefully once or twice before looking at the *Study Guide* and before starting the exercises. (Don't just look at the pictures and examples! Important comments lurk in between.)

In nearly all of the text, a *scalar* is just a real number. By convention, scalars usually are written to the left of vectors, such as $5\mathbf{v}$ or $c\mathbf{v}$, rather than $\mathbf{v}5$ or $\mathbf{v}c$. To identify vectors in your lecture notes and homework, you can write underlined letters for vectors. (Some students write arrows above the letters, but that takes longer.)

Vectors must be the same size to be added or used in a linear combination. For instance, a vector in $\mathbb{R}^3$ cannot be added to a vector in $\mathbb{R}^2$.

SOLUTIONS TO EXERCISES

1. $\mathbf{u} = \begin{bmatrix} 3 \\ 2 \end{bmatrix}$, $\mathbf{v} = \begin{bmatrix} 2 \\ -1 \end{bmatrix}$, $\mathbf{u} + \mathbf{v} = \begin{bmatrix} 3 + 2 \\ 2 + (-1) \end{bmatrix} = \begin{bmatrix} 5 \\ 1 \end{bmatrix}$

$$\mathbf{u} - 2\mathbf{v} = \begin{bmatrix}3\\2\end{bmatrix} - 2\begin{bmatrix}2\\-1\end{bmatrix} = \begin{bmatrix}3\\2\end{bmatrix} - \begin{bmatrix}4\\-2\end{bmatrix} = \underbrace{\begin{bmatrix}3-4\\2-(-2)\end{bmatrix}}_{\text{usually not written}} = \begin{bmatrix}-1\\4\end{bmatrix}$$

7. To write **a** as a linear combination of **u** and **v**, imagine walking from the origin to **a** along the grid "streets" and keep track of how many "blocks" you travel in the **u**-direction and how many in the **v**-direction. For instance, starting at the origin, you can reach **a** if you travel 2 units in the **u**-direction and then -1 units in the **v**-direction. Hence,

$$\mathbf{a} = 2\cdot\mathbf{u} + (-1)\cdot\mathbf{v} = 2\mathbf{u} - \mathbf{v}$$

Similarly, $\mathbf{b} = 2\mathbf{u} - 2\mathbf{v}$. For **c**, it might be easier to visualize -2 units in the **v**-direction and then 3.5 units in the **u**-direction, so that $\mathbf{c} = -2\mathbf{v} + 3.5\mathbf{u}$ or $\mathbf{c} = 3.5\mathbf{u} - 2\mathbf{v}$.

To get to **d** by a path that stays in the figure, note that to get from **b** to **d** you could travel 2 units in the **u**-direction and -1 unit in the **v**-direction. So $\mathbf{d} = \mathbf{b} + 2\mathbf{u} - \mathbf{v} = (2\mathbf{u} - 2\mathbf{v}) + 2\mathbf{u} - \mathbf{v} = 4\mathbf{u} - 3\mathbf{v}$.

13. Denote the columns of A by $\mathbf{a}_1$, $\mathbf{a}_2$, $\mathbf{a}_3$. To determine if **b** is a linear combination of these columns, use the boxed fact on page 34. Row reduce the augmented matrix:

$$[\mathbf{a}_1 \quad \mathbf{a}_2 \quad \mathbf{a}_3 \quad \mathbf{b}] = \begin{bmatrix}1 & 0 & 2 & -5\\-2 & 5 & 0 & 11\\2 & 5 & 8 & -7\end{bmatrix} \sim \begin{bmatrix}1 & 0 & 2 & -5\\0 & 5 & 4 & 1\\0 & 5 & 4 & 3\end{bmatrix} \sim \begin{bmatrix}\textcircled{1} & 0 & 2 & -5\\0 & \textcircled{5} & 4 & 1\\0 & 0 & 0 & \textcircled{2}\end{bmatrix}$$

The system for this augmented matrix is inconsistent, so **b** is *not* a linear combination of the columns of A.

19. By inspection, $\mathbf{v}_2 = (-3/2)\mathbf{v}_1$. Any linear combination of $\mathbf{v}_1$ and $\mathbf{v}_2$ is actually just a multiple of $\mathbf{v}_1$. For instance,

$$a\mathbf{v}_1 + b\mathbf{v}_2 = a\mathbf{v}_1 + b(-3/2)\mathbf{v}_1 = (a - 3b/2)\mathbf{v}_1$$

So Span$\{\mathbf{v}_1, \mathbf{v}_2\}$ is the set of points on the line through $\mathbf{v}_1$ and **0**.

23. The vector $\mathbf{y} = \begin{bmatrix}h\\k\end{bmatrix}$ is in Span$\{\mathbf{v}_1, \mathbf{v}_2\}$ if and only if the vector equation corresponding to augmented matrix $[\mathbf{v}_1 \quad \mathbf{v}_2 \quad \mathbf{y}]$ has a solution. Compute:

$$\begin{bmatrix}2 & 2 & h\\-1 & 1 & k\end{bmatrix} \sim \begin{bmatrix}\textcircled{2} & 2 & h\\0 & \textcircled{2} & k + h/2\end{bmatrix}$$

The augmented column cannot be a pivot column, so this matrix corresponds to a consistent system for all h and k. Hence each **y** in $\mathbb{R}^2$ is in Span{$\mathbf{v}_1, \mathbf{v}_2$}. In other words, Span{$\mathbf{v}_1, \mathbf{v}_2$} is all of $\mathbb{R}^2$.

25. **a.** There are only three vectors in the set {$\mathbf{a}_1, \mathbf{a}_2, \mathbf{a}_3$}, and **b** is not one of them.

b. There are infinitely many vectors in W = Span{$\mathbf{a}_1, \mathbf{a}_2, \mathbf{a}_3$}. To determine if **b** is in W, use the method of Exercise 13.

$$\begin{bmatrix} 1 & 0 & -4 & 4 \\ 0 & 3 & -2 & 1 \\ -2 & 6 & 3 & -4 \end{bmatrix} \sim \begin{bmatrix} 1 & 0 & -4 & 4 \\ 0 & 3 & -2 & 1 \\ 0 & 6 & -5 & 4 \end{bmatrix} \sim \begin{bmatrix} \textcircled{1} & 0 & -4 & 4 \\ 0 & \textcircled{3} & -2 & 1 \\ 0 & 0 & \textcircled{-1} & 2 \end{bmatrix}$$

$$\begin{matrix} \uparrow & \uparrow & \uparrow & \uparrow \\ \mathbf{a}_1 & \mathbf{a}_2 & \mathbf{a}_3 & \mathbf{b} \end{matrix}$$

The system for this augmented matrix is consistent, so **b** is in W.

c. $\mathbf{a}_1 = 1\mathbf{a}_1 + 0\mathbf{a}_2 + 0\mathbf{a}_3$. See the last line of page 34.

31. **a.** See the remarks following Example 1.

b. See the statement preceding Example 3.

c. See the line displayed just before Example 4.

d. See the box that discusses the matrix in (5).

e. Read the geometric description of Span{**u**, **v**} carefully.

Study Tip: I urge my students to work by themselves on the true/false questions and then meet together in groups of two or three, to compare and discuss their answers.

MATLAB Constructing a Matrix

To access the data for Section 1.3, give the command **c1s3**. The data for Exercise 25, for example, consists of a matrix A, its columns **a1**, **a2**, **a3**, and the vector **b**. The command **M = [a1 a2 a3 b]** creates a matrix using the vectors as its columns. The same matrix is created by the command **M = [A b]**.

Each time you want data for a new exercise in Section 1.3, you need the command **c1s3**. After the first exercise, you can use the up-arrow (↑). This will make MATLAB scroll back through your old commands. You may be able to find "c2s1" faster than you can retype it. Press <Enter> to reuse the command.

Exercises 11–14 and 25–28 can be solved using the commands **replace**, **swap**, and (occasionally) **scale**, described on page 1-6.

1.4 THE MATRIX EQUATION Ax = b

The ideas, boxed statements, and theorems in this section are absolutely fundamental for the rest of the text, so you should read the section extremely carefully.

KEY IDEAS

The definition of $A\mathbf{x}$ as a linear combination of the columns of A will be used often. You should learn the definition in *words* as well as symbols. *Note*: It is not wrong to write a scalar on the *right* side of a vector and write $A\mathbf{x}$ as $\mathbf{a}_1 x_1 + \cdots + \mathbf{a}_n x_n$, but the text follows the usual practice of writing a scalar on the *left* side of a vector.

You need to understand *why* Theorem 4 is true. That may take some time and effort. Example 3 should help, along with the proof. Theorem 4(c) can be restated as "The reduced echelon form of A has no row of zeros."

The phrase *logically equivalent* is explained in the statement of Theorem 4. This phrase is used with several statements in the same way that *if and only if*, or the symbol ⇔, is used between two statements. (See the Mathematical Note on page 1-11 in this Study Guide.) Saying that statements (a), (b), and (c) are logically equivalent means the same thing as saying that (a) <=> (b) and (a) <=> (c).

Key exercises are 17-28, 37, 38. Think about 37 and 38, even if they are not assigned, because they introduce ideas you will need soon. (Don't check the solution of 37 until you have written your own answer.)

Checkpoint 1: True or False? If an augmented matrix $[A \quad \mathbf{b}]$ has a pivot position in every row, then the equation $A\mathbf{x} = \mathbf{b}$ is consistent.

Note: You should work a checkpoint problem when you see it, provided that you have already read the text at least once. Always *write* your answer before comparing it with the one I have written. The checkpoint answer will be at the end of the solutions to the exercises.

SOLUTIONS TO EXERCISES

1. The text has the solution. Exercises 1-6 are designed to help you learn the definition of $A\mathbf{x}$.

7. The solution is in the text. The goal of Exercises 7-16 is to help you learn Theorem 3. If a problem involves vectors, say $\mathbf{v}_1, \mathbf{v}_2, \mathbf{v}_3$, you can place the vectors into a matrix $[\mathbf{v}_1 \quad \mathbf{v}_2 \quad \mathbf{v}_3]$, if that is helpful. If a problem involves a matrix A, you can give names to the columns of A, say $\mathbf{a}_1, \mathbf{a}_2, \mathbf{a}_3$, and reformulate a matrix equation as a vector equation.

If a problem leads to a system of linear equations, you may regard it as either a vector equation or a matrix equation, whichever is most useful.

Warning: Be careful to distinguish between the *matrix equation* $A\mathbf{x} = \mathbf{b}$ and the *augmented matrix* $[\mathbf{a}_1 \ \cdots \ \mathbf{a}_p \ \ \mathbf{b}]$, which is used in Theorem 3 to refer to a system of linear equations having this augmented matrix. Thus, the answer to Exercise 7 is *not* the augmented matrix $\begin{bmatrix} 3 & -1 & 4 & 1 \\ -4 & 1 & -5 & 0 \\ 0 & 1 & -3 & 6 \end{bmatrix}$.

13. The given equation is in the form $A\mathbf{x} = \mathbf{b}$. The equivalent vector equation is in the text's answer. Note how the entries in the vector $\mathbf{x}$ are used as the weights in the linear combination of the columns of A.

17. To justify the answer, make an appropriate calculation, and then write a sentence that mentions an important fact described in this section.

19. The augmented matrix for $A\mathbf{x} = \mathbf{b}$ is $\begin{bmatrix} -3 & 1 & b_1 \\ 6 & -2 & b_2 \end{bmatrix}$. One row operation produces $\begin{bmatrix} -3 & 1 & b_1 \\ 0 & 0 & b_2 + 2b_1 \end{bmatrix}$, which shows that the equation $A\mathbf{x} = \mathbf{b}$ is not consistent when $2b_1 + b_2 \neq 0$. The set of $\mathbf{b}$ for which the equation *is* consistent is a line through the origin—the set of all points (b_1, b_2) satisfying $b_2 = -2b_1$.

25. $A = \begin{bmatrix} 0 & 0 & 2 \\ 0 & -5 & 1 \\ 4 & 6 & -3 \end{bmatrix} \sim \begin{bmatrix} \textcircled{4} & 6 & -3 \\ 0 & \textcircled{-5} & 1 \\ 0 & 0 & \textcircled{2} \end{bmatrix}$, so the matrix has a pivot in each row. By Theorem 4, the columns of A span $\mathbb{R}^3$.

Study Tip: The answer shown here for Exercise 25 is the sort of answer expected for Exercises 23—28. A simple calculation is not enough. The phrase "*so the matrix has a pivot in each row*" is needed because it explains how Theorem 4 is used. On a test, you probably would not have to know the theorem number. It might be enough to say "*By a theorem*", instead of "*By Theorem 4*". (Check with your instructor.)

Checkpoint 2: Given $\mathbf{v}_1, \mathbf{v}_2, \mathbf{v}_3$ as in Exercise 27, find a specific vector in $\mathbb{R}^4$ that is *not* in Span$\{\mathbf{v}_1, \mathbf{v}_2, \mathbf{v}_3\}$. (If necessary, reread Example 3.)

29. **a.** See the paragraph following equation (3).

b. See the box before Example 3.

c. See the warning following Theorem 4.

d. See Example 4. **e.** See Theorem 4. **f.** See Theorem 4.

31. Whenever you encounter a matrix-vector product $A\mathbf{x}$, as in this exercise, you should immediately recognize it as a linear combination of the columns of A. More often than not, you will find this point of view helpful. In this exercise, the scalars you need are the entries in the vector $\mathbf{x}$, namely, $c_1 = -1$, $c_2 = 4$, and $c_3 = 2$.

37. If A is $m \times n$ with more rows than columns, then A cannot possibly have a pivot in every row, so the three statements in Theorem 4 are all false. Thus, the equation $A\mathbf{x} = \mathbf{b}$ cannot be consistent for all $\mathbf{b}$ in $\mathbb{R}^m$. (Be sure to try Exercise 38. It contains a valuable idea.)

43. **[M]** Use a matrix program to obtain an echelon form of the matrix. Try deleting (or simply covering) various columns of this echelon form, one column at a time, and ask yourself if the columns of the resulting matrix span $\mathbb{R}^4$. If you can delete one column, can you delete a second column? Why or why not? A good answer would include a discussion of your reasoning.

Answers to Checkpoints:

1. False. See page 42. If you missed this, you are not studying the text properly. You should read the text thoroughly *before* you look at the *Study Guide* and before you work on the exercises.

2. Let $A = [\mathbf{v}_1 \quad \mathbf{v}_2 \quad \mathbf{v}_3] = \begin{bmatrix} 1 & 0 & 1 \\ 0 & 1 & 0 \\ -1 & 0 & 0 \\ 0 & -1 & -1 \end{bmatrix}$ and $\mathbf{b} = \begin{bmatrix} b_1 \\ b_2 \\ b_3 \\ b_4 \end{bmatrix}$. Row reduce the augmented matrix for $A\mathbf{x} = \mathbf{b}$ to determine values of $b_1, \ldots, b_4$ that make the equation *in*consistent.

$$\begin{bmatrix} 1 & 0 & 1 & b_1 \\ 0 & 1 & 0 & b_2 \\ -1 & 0 & 0 & b_3 \\ 0 & -1 & -1 & b_4 \end{bmatrix} \sim \begin{bmatrix} 1 & 0 & 1 & b_1 \\ 0 & 1 & 0 & b_2 \\ 0 & 0 & 1 & b_3 + b_1 \\ 0 & -1 & -1 & b_4 \end{bmatrix} \sim \begin{bmatrix} 1 & 0 & 1 & b_1 \\ 0 & 1 & 0 & b_2 \\ 0 & 0 & 1 & b_3 + b_1 \\ 0 & 0 & -1 & b_4 + b_2 \end{bmatrix}$$

$$\sim \begin{bmatrix} 1 & 0 & 1 & b_1 \\ 0 & 1 & 0 & b_2 \\ 0 & 0 & 1 & b_3 + b_1 \\ 0 & 0 & 0 & b_4 + b_2 + b_3 + b_1 \end{bmatrix}$$

Take $\mathbf{b} = (1,1,0,0)$, for example, or any other choice of $b_1, \ldots, b_4$ whose sum is *not* zero.

Mastering Linear Algebra Concepts: Span

Please begin by reviewing "How to Study Linear Algebra", at the beginning of the *Study Guide*.

To really understand a key concept, you need to form an image in your mind that consists of the basic definition(s) together with many related ideas. Your goal at this point is to collect various ideas associated with the set Span$\{\mathbf{v}_1, \ldots, \mathbf{v}_p\}$ and the concept of a set that "spans" $\mathbb{R}^n$. Here are specific things to do now to prepare a sheet (or sheets) for review and reference.

- Write the **definition** of Span$\{\mathbf{v}_1, \ldots, \mathbf{v}_p\}$. (Learn it word-for-word.)
- Write the **definition** of the phrase: "$\{\mathbf{v}_1, \ldots, \mathbf{v}_p\}$ spans $\mathbb{R}^n$." (See page 42.) Here *span* is a verb rather than a noun as in Span$\{\mathbf{v}_1, \ldots, \mathbf{v}_p\}$.
- Add the **equivalent description** of what is meant for a vector $\mathbf{b}$ to be in Span$\{\mathbf{v}_1, \ldots, \mathbf{v}_p\}$. (See page 34.)
- Copy **Theorem** 4 word-for-word. (If you try to rephrase or summarize it in your own words, you are likely to change the meaning.)
- Sketch some **geometric interpretations** of Span$\{\mathbf{v}_1, \ldots, \mathbf{v}_p\}$. (Select some of Figs. 10, 11, 12 in Sec. 1.3 or Fig. 1 in Sec. 1.4.)
- Identify **special cases**. (Describe Span$\{\mathbf{u}\}$ and Span$\{\mathbf{u}, \mathbf{v}\}$ in words.)
- Summarize **algorithms or typical computations** (such as Example 6 and Exercises 11—14 and 21—26 in Sec. 1.3, or Example 3 and Exercises 17—20 and 23—26 in Sec. 1.4.)
- Describe **connections with other concepts**. (See pages 41—42.)

Whenever you encounter new examples or situations that help you understand the concept of a spanning set, add them to this review sheet.

MATLAB gauss and bgauss

To solve $A\mathbf{x} = \mathbf{b}$, row reduce the matrix **M = [A b]**. The command **x = [5;3;-7]** creates a column vector **x** with entries 5,3,-7. Matrix-vector multiplication is **A*x**.

To speed up row reduction of $M = [A \;\; \mathbf{b}]$, the command **gauss(M,r)** will use the leading entry in row r of M as a pivot, and use row replacements to create zeros in the pivot column below this pivot entry. The result is stored in the default matrix "ans", unless you assign the result to some other variable, such as M itself.

For the backward phase of row reduction, use **bgauss(M,r)**, which selects the leading entry in row r of M as the pivot, and creates zeros in the column *above* the pivot. Use **scale** to create leading 1's in the pivot positions. The commands **gauss** , **bgauss** , and **scale** are in the Toolbox, available from your instructor.

1.5 SOLUTION SETS OF LINEAR SYSTEMS

Many of the concepts and computations in linear algebra involve sets of vectors which are visualized geometrically as lines and planes. The most important examples of such sets are the solution sets of linear systems.

KEY IDEAS

Visualize the solution set of a homogeneous equation $A\mathbf{x} = \mathbf{0}$ as:

- the single point $\mathbf{0}$, when $A\mathbf{x} = \mathbf{0}$ has only the trivial solution,
- a line through $\mathbf{0}$, when $A\mathbf{x} = \mathbf{0}$ has one free variable,
- a plane through $\mathbf{0}$, when $A\mathbf{x} = \mathbf{0}$ has two free variables. (For more than two free variables, also use a plane through $\mathbf{0}$.)

For $\mathbf{b} \neq \mathbf{0}$, visualize the solution set of $A\mathbf{x} = \mathbf{b}$ as:

- empty, if $\mathbf{b}$ is not a linear combination of the columns of A,
- one nonzero point (vector), when $A\mathbf{x} = \mathbf{b}$ has a unique solution,
- a line not through $\mathbf{0}$, when $A\mathbf{x} = \mathbf{b}$ is consistent and has one free variable,
- a plane not through $\mathbf{0}$, when $A\mathbf{x} = \mathbf{b}$ is consistent and has two or more free variables.

The solution set of $A\mathbf{x} = \mathbf{b}$ is said to be described *implicitly*, because the equation is a condition an $\mathbf{x}$ must satisfy in order to be in the set, yet the equation does not show how to find such an $\mathbf{x}$. When the solution set of $A\mathbf{x} = \mathbf{0}$ is written as Span$\{\mathbf{v}_1, \ldots, \mathbf{v}_p\}$, the set is said to be described *explicitly*; each element in the set is produced by forming a linear combination of $\mathbf{v}_1, \ldots, \mathbf{v}_p$.

A common explicit description of a set is an equation in *parametric vector form*. Examples are:

$\mathbf{x} = t\mathbf{v}$,	a line through $\mathbf{0}$ in the direction of $\mathbf{v}$,
$\mathbf{x} = \mathbf{p} + t\mathbf{v}$,	a line through $\mathbf{p}$ in the direction of $\mathbf{v}$,
$\mathbf{x} = x_2\mathbf{u} + x_3\mathbf{v}$,	a plane through $\mathbf{0}$, $\mathbf{u}$ and $\mathbf{v}$,
$\mathbf{x} = \mathbf{p} + x_2\mathbf{u} + x_3\mathbf{v}$,	a plane through $\mathbf{p}$ parallel to the plane whose equation is $\mathbf{x} = x_2\mathbf{u} + x_3\mathbf{v}$.

An equation in parametric vector form describes a set explicitly because the equation shows how to produce each $\mathbf{x}$ in the set.

Solving an equation $A\mathbf{x} = \mathbf{b}$ means to find an explicit description of the solution set. This description can be written in parametric vector form in which the parameters are the free variables from the system. *Important:* The number of free variables in $A\mathbf{x} = \mathbf{b}$ depends only on A, not on $\mathbf{b}$.

Theorem 6 and the paragraph following it are important. They describe how the solutions of $A\mathbf{x} = \mathbf{0}$ and $A\mathbf{x} = \mathbf{b}$ are related. See Figs. 5 and 6.

SOLUTIONS TO EXERCISES

1. Row reduce the augmented matrix until the presence or absence of a free variable is evident:

$$\begin{bmatrix} 1 & -5 & 9 & 0 \\ -1 & 4 & -3 & 0 \\ 2 & -8 & 9 & 0 \end{bmatrix} \sim \begin{bmatrix} 1 & -5 & 9 & 0 \\ 0 & -1 & 6 & 0 \\ 0 & 2 & -9 & 0 \end{bmatrix} \sim \begin{bmatrix} \textcircled{1} & -5 & 9 & 0 \\ 0 & \textcircled{-1} & 6 & 0 \\ 0 & 0 & \textcircled{3} & 0 \end{bmatrix}$$

There are no free variables; the system has only the trivial solution.

7. Always use the *reduced* echelon form of an augmented matrix to find solutions of a system. See the text's discussion of back substitution on pages 22—23.

$$\begin{bmatrix} 1 & -5 & 0 & 2 & 0 & -4 \\ 0 & 0 & 0 & 1 & 0 & -3 \\ 0 & 0 & 0 & 0 & 1 & 5 \\ 0 & 0 & 0 & 0 & 0 & 0 \end{bmatrix} \sim \begin{bmatrix} \textcircled{1} & -5 & 0 & 0 & 0 & 2 \\ 0 & 0 & 0 & \textcircled{1} & 0 & -3 \\ 0 & 0 & 0 & 0 & \textcircled{1} & 5 \\ 0 & 0 & 0 & 0 & 0 & 0 \end{bmatrix}, \quad \begin{aligned} x_1 - 5x_2 \quad &= 2 \\ x_4 &= -3 \\ x_5 &= 5 \end{aligned}$$

If you wrote something like the system above, then you made a common mistake. The matrix in the problem is a coefficient matrix, not an augmented matrix. You should row reduce $[A \;\; \mathbf{0}]$. The correct system of equations is

$$\begin{aligned} \textcircled{x_1} - 5x_2 \qquad\qquad + 2x_6 &= 0 \\ \textcircled{x_4} \quad - 3x_6 &= 0 \\ \textcircled{x_5} + 5x_6 &= 0 \end{aligned}$$

Some students are not sure what to do with x_3. Some ignore it, others set it equal to zero. In fact, x_3 is free; there is no constraint on x_3 at all. So $x_1 = 5x_2 - 2x_6$, $x_4 = 3x_6$, $x_5 = -5x_6$, with x_2, x_3, x_6 free. The general solution is

$$\mathbf{x} = \begin{bmatrix} 5x_2 - 2x_6 \\ x_2 \\ x_3 \\ 3x_6 \\ -5x_6 \\ x_6 \end{bmatrix} = \begin{bmatrix} 5x_2 \\ x_2 \\ 0 \\ 0 \\ 0 \\ 0 \end{bmatrix} + \begin{bmatrix} 0 \\ 0 \\ x_3 \\ 0 \\ 0 \\ 0 \end{bmatrix} + \begin{bmatrix} -2x_6 \\ 0 \\ 0 \\ 3x_6 \\ -5x_6 \\ x_6 \end{bmatrix} = x_2 \underset{\uparrow \atop \mathbf{u}}{\begin{bmatrix} 5 \\ 1 \\ 0 \\ 0 \\ 0 \\ 0 \end{bmatrix}} + x_3 \underset{\uparrow \atop \mathbf{v}}{\begin{bmatrix} 0 \\ 0 \\ 1 \\ 0 \\ 0 \\ 0 \end{bmatrix}} + x_6 \underset{\uparrow \atop \mathbf{w}}{\begin{bmatrix} -2 \\ 0 \\ 0 \\ 3 \\ -5 \\ 1 \end{bmatrix}}$$

The solution set is the same as Span$\{\mathbf{u}, \mathbf{v}, \mathbf{w}\}$. Originally, the solution set was described implicitly, by a set of equations. Now the solution set is described explicitly, in parametric vector form.

Study Tip: When solving a system, identify (and perhaps circle) the basic variables. All other variables are free.

13. Row reduce the augmented matrix:

$$\begin{bmatrix} 1 & -3 & -2 & -5 \\ 0 & 1 & -1 & 4 \\ -2 & 3 & 7 & -2 \end{bmatrix} \sim \cdots \sim \begin{bmatrix} \textcircled{1} & 0 & -5 & 7 \\ 0 & \textcircled{1} & -1 & 4 \\ 0 & 0 & 0 & 0 \end{bmatrix}, \quad \begin{aligned} \textcircled{x_1} \quad - 5x_3 &= 7 \\ \textcircled{x_2} - x_3 &= 4 \\ 0 &= 0 \end{aligned}$$

So $x_1 = 7 + 5x_3$, $x_2 = 4 + x_3$, with x_3 free, and $\mathbf{x} = \begin{bmatrix} 7 \\ 4 \\ 0 \end{bmatrix} + x_3\begin{bmatrix} 5 \\ 1 \\ 1 \end{bmatrix}$. The solution set is the line through $\begin{bmatrix} 7 \\ 4 \\ 0 \end{bmatrix}$, parallel to the solution set (a line) determined by the homogeneous system in Exercise 5.

Checkpoint: Let A be a 2×2 matrix. Answer True or False: If the solution set of $A\mathbf{x} = \mathbf{0}$ is a line through the origin in $\mathbb{R}^2$ and if $\mathbf{b} \neq \mathbf{0}$, then the solution set of $A\mathbf{x} = \mathbf{b}$ is a line not through the origin.

19. $\mathbf{p} = \begin{bmatrix} -1 \\ 4 \end{bmatrix}$, $\mathbf{q} = \begin{bmatrix} 0 \\ 7 \end{bmatrix}$, $\mathbf{q} - \mathbf{p} = \begin{bmatrix} 1 \\ 3 \end{bmatrix}$. The line through $\mathbf{p}$ and $\mathbf{q}$ is parallel to the line through $\mathbf{q} - \mathbf{p}$ and the origin. A parametric representation of the line through $\mathbf{p}$ and $\mathbf{q}$ has the form $\mathbf{x} = \mathbf{a} + t\mathbf{b}$, where $\mathbf{a}$ is some point on the line (such as $\mathbf{p}$ or $\mathbf{q}$) and $\mathbf{b}$ is any nonzero multiple of $\mathbf{q} - \mathbf{p}$. So one parametric equation of the line is $\mathbf{x} = \mathbf{p} + t(\mathbf{q} - \mathbf{p})$, or

$$\begin{bmatrix} x_1 \\ x_2 \end{bmatrix} = \begin{bmatrix} -1 \\ 4 \end{bmatrix} + t\begin{bmatrix} 1 \\ 3 \end{bmatrix}$$

Another choice is $\mathbf{x} = \mathbf{p} + s(\mathbf{p} - \mathbf{q})$. Other answers are possible. The first answer given is often written in the form $\mathbf{x} = (1 - t)\mathbf{p} + t\mathbf{q}$. As t varies from 0 to 1, $\mathbf{x}$ varies from the point $\mathbf{p}$ to the point $\mathbf{q}$.

21. **a.** See the first paragraph of the subsection *Homogeneous Linear Systems.*

b. See the first two sentences of the subsection *Parametric Vector Form.*

c. See the box before Example 1.

d. See the paragraph that precedes Fig. 5.

e. See Theorem 6.

25. Suppose **p** satisfies $A\mathbf{x} = \mathbf{b}$. Then $A\mathbf{p} = \mathbf{b}$. Theorem 6 says that the solution set of $A\mathbf{x} = \mathbf{b}$ equals the set $S = \{\mathbf{w} : \mathbf{w} = \mathbf{p} + \mathbf{v}_h$ for some $\mathbf{v}_h$ such that $A\mathbf{v}_h = \mathbf{0}\}$. There are two things to prove: (a) every vector in S satisfies $A\mathbf{x} = \mathbf{b}$, (b) every vector that satisfies $A\mathbf{x} = \mathbf{b}$ is in S.

a. Let **w** have the form $\mathbf{w} = \mathbf{p} + \mathbf{v}_h$, where $A\mathbf{v}_h = \mathbf{0}$. Then

$$A\mathbf{w} = A(\mathbf{p} + \mathbf{v}_h) = A\mathbf{p} + A\mathbf{v}_h \quad \text{By Theorem 5(a) in Section 1.4}$$
$$= \mathbf{b} + \mathbf{0} = \mathbf{b}$$

So every vector of the form $\mathbf{p} + \mathbf{v}_h$ satisfies $A\mathbf{x} = \mathbf{b}$.

b. Now let **w** be any solution of $A\mathbf{x} = \mathbf{b}$, and set $\mathbf{v}_h = \mathbf{w} - \mathbf{p}$. Then

$$A\mathbf{v}_h = A(\mathbf{w} - \mathbf{p}) = A\mathbf{w} - A\mathbf{p} = \mathbf{b} - \mathbf{b} = \mathbf{0}$$

So $\mathbf{v}_h$ satisfies $A\mathbf{x} = \mathbf{0}$. Thus every solution of $A\mathbf{x} = \mathbf{b}$ has the form $\mathbf{w} = \mathbf{p} + \mathbf{v}_h$.

31. A is a 3×2 matrix with two pivot positions. (a) Since A has a pivot position in each column, each variable in $A\mathbf{x} = \mathbf{0}$ is a basic variable. So there are no free variables, and $A\mathbf{x} = \mathbf{0}$ has no nontrivial solution. (b) A cannot have a pivot position in each of its three rows, so the equation $A\mathbf{x} = \mathbf{b}$ does not have a solution for *every possible* **b**, by Theorem 4 in Section 1.4.

Note: The term *possible* here means that of course we only consider **b** in $\mathbb{R}^3$, because A has 3 rows. The context of the problem determines what vectors are possible candidates for **b**.

37. **a.** Fill in the exchange table one column at a time. The entries in a column describe where a sector's output goes. The decimal fractions in each column sum to 1.

Distribution of Output From:				
Chemicals	Fuel	Machinery		Purchased by:
output ↓	↓	↓		
.2	.8	.4	input ⟶	Chemicals
.3	.1	.4	⟶	Fuels
.5	.1	.2	⟶	Machinery

b. Denote the total annual output (in dollars) of the sectors by p_C, p_F, and p_M. From the first row of the table, the total input to the Chemical & Metals sector is $.2p_C + .8p_F + .4p_M$. So the equilibrium prices must satisfy

$$\underset{\text{income}}{p_C} = \underset{\text{expenses}}{.2p_C + .8p_F + .4p_M}$$

From the second and third rows of the table, the income/expense requirements for the Fuels & Power sector and the Machinery sector are, respectively,

$$p_F = .3p_C + .1p_F + .4p_M$$
$$p_M = .5p_C + .1p_F + .2p_M$$

Move all variables to the left side and combine like terms:

$$\begin{aligned} .8p_C - .8p_F - .4p_M &= 0 \\ -.3p_C + .9p_F - .4p_M &= 0 \\ -.5p_C - .1p_F + .8p_M &= 0 \end{aligned}$$

c. **[M]** You can obtain the reduced echelon form with a matrix program. Actually, hand calculations are not too messy. To simplify the calculations, first scale each row of the augmented matrix by 10, then continue as usual.

$$\begin{bmatrix} 8 & -8 & -4 & 0 \\ -3 & 9 & -4 & 0 \\ -5 & -1 & 8 & 0 \end{bmatrix} \sim \begin{bmatrix} 1 & -1 & -.5 & 0 \\ -3 & 9 & -4 & 0 \\ -5 & -1 & 8 & 0 \end{bmatrix} \sim \begin{bmatrix} 1 & -1 & -.5 & 0 \\ 0 & 6 & -5.5 & 0 \\ 0 & -6 & 5.5 & 0 \end{bmatrix} \sim$$

$$\begin{bmatrix} 1 & -1 & -.5 & 0 \\ 0 & 1 & -.917 & 0 \\ 0 & 0 & 0 & 0 \end{bmatrix} \sim \begin{bmatrix} 1 & 0 & -1.417 & 0 \\ 0 & 1 & -.917 & 0 \\ 0 & 0 & 0 & 0 \end{bmatrix}$$

The number of decimal places displayed is somewhat arbitrary.

The general solution is $p_C = 1.417p_M$, $p_F = .917p_M$, with p_M free. If p_M is assigned the value 100, then $p_C = 141.7$ and $p_F = 91.7$. Note that only the *ratios* of the prices are determined. This makes sense, for if the prices were converted from, say, dollars to yen or Deutschmarks, the inputs and outputs of each sector would still balance. The economic equilibrium is not affected by a proportional change in prices.

Answer to Checkpoint: False. The solution set could be empty. (See the paragraph in the text preceding Theorem 6.) Suppose $A = \begin{bmatrix} 1 & 2 \\ 1 & 2 \end{bmatrix}$, $\mathbf{v} = \begin{bmatrix} -2 \\ 1 \end{bmatrix}$, and $\mathbf{b} = \begin{bmatrix} 5 \\ h \end{bmatrix}$. Then the general solution of $A\mathbf{x} = \mathbf{0}$ is the line $\mathbf{x} = t\mathbf{v}$, but if $h \neq 5$, the solution set of $A\mathbf{x} = \mathbf{b}$ is empty. (You are not expected to furnish such an example at this point in the course.)

MATLAB The command **zeros(m,n)** creates an $m \times n$ matrix of zeros. When solving an equation $A\mathbf{x} = \mathbf{0}$, create an augmented matrix:

M = [A zeros(m,1)] m is the number of rows in A.

Then use **gauss** , **bgauss** , and **scale** to row reduce M completely.

1.6 LINEAR INDEPENDENCE

This section is as important as Section 1.4 and should be studied just as carefully. Full understanding of the concepts will take time, so get started on the section now.

KEY IDEAS

Figures 1 and 2, along with Theorem 7, will help you understand the nature of a linearly dependent set. (Fig. 2 applies only when **u** and **v** are independent.) But you must also learn the *definitions* of linear dependence and linear independence, word for word! Many theoretical problems involving a linearly dependent set are treated by the definition, because it provides an equation (the dependence equation) with which to work. (See the proof of Theorem 7.)

The box before Example 2 contains a very useful fact. Any time you need to study the linear independence of a set of p vectors in $\mathbb{R}^n$, you can always form an $n \times p$ matrix A with those vectors as columns and then study the matrix equation $A\mathbf{x} = \mathbf{0}$. This is not the only method, however. Stay alert for three special situations:

- A set of two vectors. Always check this by inspection; don't waste time on row reduction of $[A \;\; \mathbf{0}]$. The set is linearly independent if neither of the vectors is a multiple of the other. (For brevity, I sometimes say that "the vectors are not multiples.") See Example 3.
- A set that contains too many vectors, that is, more vectors than entries in the vectors; the columns of a short, fat matrix. Theorem 8.
- A set that contains the zero vector. Theorem 9.

The most common mistake students make when checking a set of three or more vectors for independence is to think they only have to verify that no vector is a multiple of one of the other vectors. That's wrong! Study Example 5 and Figure 4.

SOLUTIONS TO EXERCISES

1. Let $\mathbf{u} = \begin{bmatrix} 3 \\ 0 \\ 0 \end{bmatrix}$, $\mathbf{v} = \begin{bmatrix} -3 \\ 2 \\ 3 \end{bmatrix}$, $\mathbf{w} = \begin{bmatrix} 6 \\ 4 \\ 0 \end{bmatrix}$. To test the linear independence of $\{\mathbf{u}, \mathbf{v}, \mathbf{w}\}$, use an augmented matrix to study the solution set of

$$x_1\mathbf{u} + x_2\mathbf{v} + x_3\mathbf{w} = \mathbf{0} \qquad (*)$$

Since $\begin{bmatrix} 3 & -3 & 6 & 0 \\ 0 & 2 & 4 & 0 \\ 0 & 3 & 0 & 0 \end{bmatrix} \sim \begin{bmatrix} \textcircled{3} & -3 & 6 & 0 \\ 0 & \textcircled{2} & 4 & 0 \\ 0 & 0 & \textcircled{-6} & 0 \end{bmatrix}$, there are three basic variables, no free variables, and so (*) has *only* the trivial solution. The vectors are linearly independent.

Warning: Whenever you study a homogeneous equation, you may be tempted to omit the augmented column of zeros because it never changes under row operations. I urge you to keep the zeros, to avoid possibly misinterpreting your own calculations. In Exercise 1, if you wrote

$$\begin{bmatrix} 3 & -3 & 6 \\ 0 & 2 & 4 \\ 0 & 3 & 0 \end{bmatrix} \sim \begin{bmatrix} 3 & -3 & 6 \\ 0 & 2 & 4 \\ 0 & 0 & -6 \end{bmatrix}$$

you might conclude that "the system is inconsistent" and then go on to make some crazy statement about linear dependence or independence. Don't laugh. I have seen this happen on exams. A more common error occurs in a problem like Exercise 11. In that exercise, if you write

$$\begin{bmatrix} 1 & 1 & 0 & 4 \\ -1 & 0 & 3 & -1 \\ 0 & -2 & 1 & 1 \\ 1 & 0 & -1 & 3 \end{bmatrix} \sim \cdots \sim \begin{bmatrix} 1 & 1 & 0 & 4 \\ 0 & 1 & 3 & 3 \\ 0 & 0 & 7 & 7 \\ 0 & 0 & 0 & 0 \end{bmatrix}$$

you might conclude that "the system has a unique solution" and the vectors are linearly independent. However, the four columns of the matrix are actually linearly dependent. In both cases, the error is to misinterpret your matrix as an augmented matrix.

7. Study the equation $A\mathbf{x} = \mathbf{0}$. You could row reduce $\begin{bmatrix} 1 & 3 & -2 & 0 & 0 \\ 3 & 10 & -7 & 1 & 0 \\ -5 & -5 & 3 & 7 & 0 \end{bmatrix}$,

but that would be a waste of time. There are only three rows, so there are at most three pivot positions. At least one of the four variables must be free. So the equation $A\mathbf{x} = \mathbf{0}$ has a nontrivial solution and the columns of A are linearly dependent. (If you know Theorem 8, you can even omit most of this discussion.)

Checkpoint: What is wrong with the following statement?

The vectors $\begin{bmatrix} 3 \\ -1 \end{bmatrix}, \begin{bmatrix} 2 \\ 8 \end{bmatrix}, \begin{bmatrix} -5 \\ 3 \end{bmatrix}, \begin{bmatrix} 7 \\ -4 \end{bmatrix}$ are linearly dependent "because there is a free variable", or "because there are more variables than equations".

13. $\mathbf{v}_1 = \begin{bmatrix} 1 \\ 3 \\ -2 \end{bmatrix}, \mathbf{v}_2 = \begin{bmatrix} -2 \\ -6 \\ 4 \end{bmatrix}, \mathbf{v}_3 = \begin{bmatrix} 1 \\ 2 \\ h \end{bmatrix}$

a. For $\mathbf{v}_3$ to be in Span$\{\mathbf{v}_1, \mathbf{v}_2\}$, the system $x_1\mathbf{v}_1 + x_2\mathbf{v}_2 = \mathbf{v}_3$ must be consistent. To find out if this is true, row reduce $[\mathbf{v}_1 \ \ \mathbf{v}_2 \ \ \mathbf{v}_3]$, considered as an augmented matrix.

$$\begin{bmatrix} 1 & -2 & 1 \\ 3 & -6 & 2 \\ -2 & 4 & h \end{bmatrix} \sim \begin{bmatrix} 1 & -2 & 1 \\ 0 & 0 & -1 \\ -2 & 4 & h \end{bmatrix}$$

No further work is needed. The second equation, $0x_1 + 0x_2 = -1$, is impossible, so the original system is inconsistent. Thus $\mathbf{v}_3$ is not in Span$\{\mathbf{v}_1, \mathbf{v}_2\}$ for any value of h.

b. For $\{\mathbf{v}_1, \mathbf{v}_2, \mathbf{v}_3\}$ to be linearly independent, the system $x_1\mathbf{v}_1 + x_2\mathbf{v}_2 + x_3\mathbf{v}_3 = \mathbf{0}$ should have only the trivial solution. Row reduce the associated augmented matrix $[\mathbf{v}_1 \ \ \mathbf{v}_2 \ \ \mathbf{v}_3 \ \ \mathbf{0}]$:

$$\begin{bmatrix} 1 & -2 & 1 & 0 \\ 3 & -6 & 2 & 0 \\ -2 & 4 & h & 0 \end{bmatrix} \sim \begin{bmatrix} 1 & -2 & 1 & 0 \\ 0 & 0 & -1 & 0 \\ 0 & 0 & h+2 & 0 \end{bmatrix} \sim \begin{bmatrix} \textcircled{1} & -2 & 1 & 0 \\ 0 & 0 & \textcircled{-1} & 0 \\ 0 & 0 & 0 & 0 \end{bmatrix}$$

For every value of h, x_2 is a free variable, and so the homogeneous system has a nontrivial solution. Thus $\{\mathbf{v}_1, \mathbf{v}_2, \mathbf{v}_3\}$ is linearly dependent for all h. (Note: You can avoid all calculation if you happen to notice that $\mathbf{v}_2 = -2\mathbf{v}_1$, and then mention Theorem 7. In general, however, you should not spend much time looking for special cases such as this.)

Warning: Exercise 13 and Practice Problem 3 emphasize that to check whether a set such as $\{\mathbf{v}_1, \mathbf{v}_2, \mathbf{v}_3\}$ is linearly independent, it is *not* wise to check instead whether $\mathbf{v}_3$ is a linear combination of $\mathbf{v}_1$ and $\mathbf{v}_2$.

19. The set $\begin{bmatrix} 5 \\ 5 \end{bmatrix}, \begin{bmatrix} 6 \\ 1 \end{bmatrix}, \begin{bmatrix} 2 \\ 4 \end{bmatrix}, \begin{bmatrix} 3 \\ -6 \end{bmatrix}$ is obviously linearly dependent, by Theorem 8, because there are more vectors (4) than entries in the vectors. (On a test, you would probably not have to know the theorem number.)

25. The answer in the text is correct because $A\mathbf{x} = \mathbf{0}$ has only the trivial solution if and only if the columns of A are linearly independent. If A has only two nonzero columns, they are linearly independent if and only if they are *not* multiples (more precisely, neither column is a multiple of the other).

Examples: $A = \begin{bmatrix} 1 & 0 \\ 0 & 1 \\ 0 & 1 \end{bmatrix}$, $B = \begin{bmatrix} 2 & 1 \\ 2 & 1 \\ 2 & 1 \end{bmatrix}$.

27. **a.** See the box before Example 2.

b. See the warning after Theorem 7.

c. See Fig. 3, after Theorem 8.

d. See the remark following Example 4.

31. The text uses Theorem 7 to conclude that $\{\mathbf{v}_1, \mathbf{v}_2, \mathbf{v}_3, \mathbf{v}_4\}$ is linearly dependent. Another argument is to rewrite the equation $\mathbf{v}_3 = 2\mathbf{v}_1 + \mathbf{v}_2$ as $2\mathbf{v}_1 + 1\mathbf{v}_2 + (-1)\mathbf{v}_3 + 0\mathbf{v}_4 = \mathbf{0}$, which is a linear dependence relation.

33. The first printing of the text has a misprint. The first phrase should read, "If $\mathbf{v}_1$ and $\mathbf{v}_2$ are in $\mathbb{R}^4$ and $\mathbf{v}_2$ is not a scalar multiple of $\mathbf{v}_1$,".

37. If one of the five columns of the 7×5 matrix A were *not* a pivot column, then that column would correspond to a free variable in the equation $A\mathbf{x} = \mathbf{0}$, in which case the columns of A would be linearly dependent. So, for the columns to be linearly independent, all five columns must be pivot columns.

39. If for all $\mathbf{b}$, the equation $A\mathbf{x} = \mathbf{b}$ has at most one solution, then take $\mathbf{b} = \mathbf{0}$, and conclude that the equation $A\mathbf{x} = \mathbf{0}$ has at most one solution. Of course the trivial solution is a solution, so it is the only solution. Thus the columns of A are linearly independent.

Answer to Checkpoint: The set of four vectors contains only vectors, no variables of any kind, and no equations. It makes no sense to talk about the variables in a set of vectors. Variables appear in an equation. One cannot assume that the writer of the statement has any idea of the appropriate equation. If you want to give an explanation involving variables, then you must specify the equation. One correct answer is: the vectors are linearly dependent because the equation $x_1\begin{bmatrix} 3 \\ -1 \end{bmatrix} + x_2\begin{bmatrix} 2 \\ 8 \end{bmatrix} + x_3\begin{bmatrix} -5 \\ 3 \end{bmatrix} + x_4\begin{bmatrix} 7 \\ -4 \end{bmatrix} = \begin{bmatrix} 0 \\ 0 \end{bmatrix}$ necessarily has a free variable.

Mastering Linear Algebra Concepts: Linear Independence

In Section 1.4 of this *Guide*, I described how to begin forming a mental image of the concept of a spanning set. The same technique works for linear independence. The goal is to merge all the ideas you find regarding linear independence into a single mental image, with each part immediately available in your mind for use as needed. Start now to organize on paper your understanding of linear independence/dependence, using the following list as a guide. In each case, write information that you think will be helpful. (Definitions and theorems should be copied word-for-word.)

- definitions of linear independence and dependence
- equivalent descriptions — Theorem 7
- geometric interpretations, — Figs. 1,2,4
- special cases, — Theorems 8,9, box on p. 61, Examples 3,5,6
- examples and "counterexamples", — Figs. 1,2,3,4, Exercises 13-18, 31-36
- algorithms or typical computations, — Examples 1,2, Exercises 1-12
- connections with other concepts. — Box on p. 60, Examples 2,4, Exercises 37-39

As you work on your notes, be careful to use terminology correctly. For instance, the term "linearly independent" may be applied to a set of vectors, but it *never* is applied to a matrix or to an equation. The *columns* of a matrix may be linearly independent, but it is meaningless to refer to a linearly independent matrix. Similarly, *solutions* of a system of linear equation may be linearly independent, but the term "linearly independent equations" has never been defined. Finally, a set of vectors or a matrix cannot have a "nontrivial solution". Only equations have solutions.

1.7 INTRODUCTION TO LINEAR TRANSFORMATIONS

Linear transformations are important for both the theory and the applications of linear algebra. You will see both uses in a variety of settings throughout the text. The graphical descriptions in this section will be augmented in Section 1.8 and in a later section on computer graphics.

STUDY NOTES

Viewing the correspondence from a vector $\mathbf{x}$ to a vector $A\mathbf{x}$ as a mapping provides a dynamic interpretation of matrix-vector multiplication and a new way to understand the equation $A\mathbf{x} = \mathbf{b}$. Using the language of computer science, we can describe a matrix in two ways—as a data structure (a rectangular array of numbers) and as a program (a prescription for transforming vectors). Strictly speaking, however, the actual linear transformation is the function or mapping $\mathbf{x} \mapsto A\mathbf{x}$ rather than just A itself.

Here is a way to visualize a matrix acting as a linear transformation. The entries in the input vector $\mathbf{x}$ are assigned as weights that multiply the corresponding columns of A, then the resulting weighted columns are added together to produce the output vector $\mathbf{b}$.

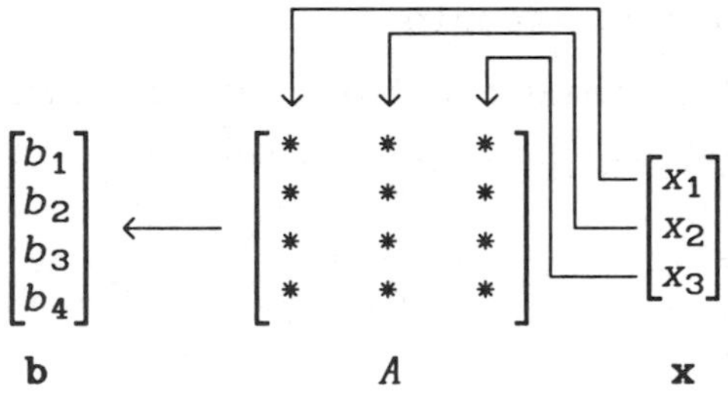

As you learn the definition of a linear transformation T, don't forget the crucial phrases "for all $\mathbf{u}$ and $\mathbf{v}$ in the domain of T" and "for all $\mathbf{u}$ and all scalars c." The mapping T defined by $T(x_1, x_2) = (|x_2|, |x_1|)$ is *not* a linear mapping, and yet T satisfies the linearity properties for *some* vectors in its domain and *some* scalars.

SOLUTIONS TO EXERCISES

1. $A = \begin{bmatrix} 3 & 0 \\ 0 & 3 \end{bmatrix}$. $T\mathbf{u} = \begin{bmatrix} 3 & 0 \\ 0 & 3 \end{bmatrix}\begin{bmatrix} 1 \\ 5 \end{bmatrix} = \begin{bmatrix} 3 \\ 15 \end{bmatrix}$, $T\mathbf{v} = \begin{bmatrix} 3 & 0 \\ 0 & 3 \end{bmatrix}\begin{bmatrix} -4 \\ -1 \end{bmatrix} = \begin{bmatrix} -12 \\ -3 \end{bmatrix}$.

Can you show that T is a dilation transformation?

7. If A is 7×5, then $\mathbf{x}$ must have exactly 5 entries (that is, be in $\mathbb{R}^5$) for $A\mathbf{x}$ to be defined. Since $A\mathbf{x}$ is a linear combination of the columns of A and each column has 7 entries, $A\mathbf{x}$ is in $\mathbb{R}^7$. If $T(\mathbf{x}) = A\mathbf{x}$, then T maps $\mathbb{R}^a$ into $\mathbb{R}^b$, where $a = 5$ and $b = 7$.

13. a.

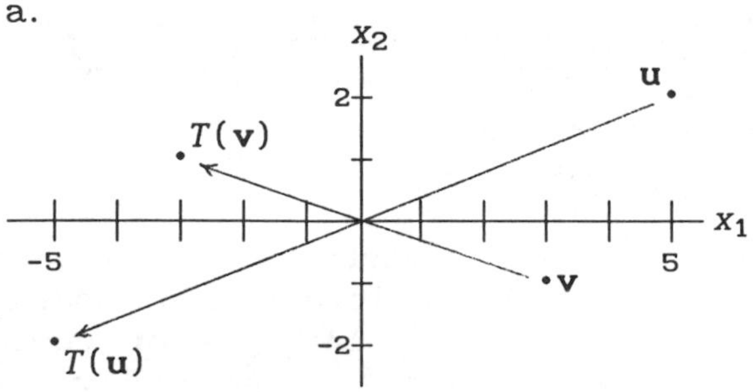

b. A reflection through the origin; two other good answers are: a a rotation of $\pm \pi$ radians about the origin.

19. By linearity, since **u** maps onto $\begin{bmatrix} 2 \\ 0 \end{bmatrix}$, 2**u** maps onto $2\begin{bmatrix} 2 \\ 0 \end{bmatrix} = \begin{bmatrix} 4 \\ 0 \end{bmatrix}$. Likewise, since **v** maps onto $\begin{bmatrix} 1 \\ -4 \end{bmatrix}$, 3**v** maps onto $3\begin{bmatrix} 1 \\ -4 \end{bmatrix} = \begin{bmatrix} 3 \\ -12 \end{bmatrix}$. Finally, $T(2\mathbf{u} + 3\mathbf{v})$ is the sum of the images of 2**u** and 3**v**, namely, $\begin{bmatrix} 7 \\ -12 \end{bmatrix}$.

23. **a.** A function is another word for transformation or mapping.

b. See the paragraph before Example 1.

c. See the paragraph following the box that contains equation (4).

d. See the paragraph following equation (5).

25. A point **x** on the line through **p** in the direction of **v** satisfies the parametric equation $\mathbf{x} = \mathbf{p} + t\mathbf{v}$. By linearity, the image $T(\mathbf{x})$ satisfies the parametric equation

$$T(\mathbf{x}) = T(\mathbf{p} + t\mathbf{v}) = T(\mathbf{p}) + tT(\mathbf{v}) \qquad (*)$$

If $T(\mathbf{v}) = \mathbf{0}$, then $T(\mathbf{x}) = T(\mathbf{p})$ for all values of t, and the image of the original line is just a single point. Otherwise, (*) is the parametric equation of a line through $T(\mathbf{p})$ in the direction of $T(\mathbf{v})$.

Study Tip: Exercise 31 is important, because it will help you to connect the concepts of linear dependence and linear transformation. Be sure to *try* the exercise first, before looking in the answer section of the text. Don't feel badly if you need to peek at the hint there. Only my best students can do this problem unaided. Once you have seen the hint, try hard to construct the desired explanation without consulting the solution I have written below. Don't give up too soon. Reread the definitions of linear dependence and linear transformation, if necessary.

After you have written your best attempt at an explanation, check it against the *Study Guide* solution. Also, study the strategy there of how I found the solution. Even if your attempt is quite unsatisfactory, the time spent on this problem is worthwhile, because you will learn more from the solution here.

31. To help you use this *Study Guide* properly, I have hidden the solution at the end of the solutions for Section 1.8. *Do not look there until you have followed the instructions above.* (I will not "hide" a solution again, but I wanted this one time to emphasize the importance of working seriously on a problem before checking the solution.)

Mastering Linear Algebra Concepts: Linear Transformation

Start to form a robust mental image of a linear transformation by preparing a review sheet that covers the following categories:

▶ definition	Page 70
▶ equivalent descriptions	Equations (4) and (5)
▶ geometric interpretations	Figs. 1 or 2
▶ special cases	Matrix transformation: page 70
▶ examples and "counterexamples"	Superposition, Examples 2-6
	Paragraph before Exercise 1, in this Guide
	Exercises 29,30,33
▶ connections with other concepts	Existence and uniqueness: page 69
	Linear dependence: Exercise 31

Note: Exercise 31 should enrich your mental image of linear dependence, so add a note about it to your list for "linear independence". If your course does not emphasize the next section, turn now to the end of the *Study Guide* material for Section 1.8 and read the box on *Existence and Uniqueness*.

1.8 THE MATRIX OF A LINEAR TRANSFORMATION

Every matrix transformation is a linear transformation. This section shows that every linear transformation from $\mathbb{R}^n$ to $\mathbb{R}^m$ is a matrix transformation. Chapters 4 and 5 will discuss other examples of linear transformations.

KEY IDEAS

A linear transformation $T:\mathbb{R}^n \rightarrow \mathbb{R}^m$ is completely determined by what it does to the columns of the identity matrix I_n. The jth column of the standard matrix for T is $T(\mathbf{e}_j)$, where $\mathbf{e}_j$ is the jth column of I_n.

There are two ways to compute the standard matrix A. Either compute $T(\mathbf{e}_1),\ldots,T(\mathbf{e}_n)$, which is easy to do when T is described geometrically, as in Exercises 1-14, or fill in the entries of A by inspection, which is easy to do when T is described by a formula, as in Exercises 15-22.

Existence and uniqueness questions about the mapping $\mathbf{x} \mapsto A\mathbf{x}$ are determined by properties of A. You should know how this works. The proof of Theorem 11 also applies to linear transformations on the general vector spaces in Chapter 4. Here is a shorter proof that applies only to matrix transformations.

Let A be the standard matrix of T. Then T is one-to-one if and only if the equation $A\mathbf{x} = \mathbf{b}$ has at most one solution for each $\mathbf{b}$. This happens if and only if every column of A is a pivot column which, in turn, happens if and only if $A\mathbf{x} = \mathbf{0}$ has only the trivial solution.

The "if and only if" phrase in Theorem 11 (and in the proof above) was discussed in a *Mathematical Note*, in Section 1.2 of this *Guide*.

SOLUTIONS TO EXERCISES

1. The columns of the standard matrix A of T are the images of $\mathbf{e}_1$ and $\mathbf{e}_2$. Since $T(\mathbf{e}_1) = \begin{bmatrix} 4 \\ -1 \\ 2 \end{bmatrix}$ and $T(\mathbf{e}_2) = \begin{bmatrix} -5 \\ 3 \\ -6 \end{bmatrix}$, we have $A = \begin{bmatrix} 4 & -5 \\ -1 & 3 \\ 2 & -6 \end{bmatrix}$.

7. $T(\mathbf{e}_1) = \mathbf{e}_1 + 2\mathbf{e}_2 = \begin{bmatrix} 1 \\ 2 \end{bmatrix}$, $T(\mathbf{e}_2) = \mathbf{e}_2 = \begin{bmatrix} 0 \\ 1 \end{bmatrix}$; so $A = \begin{bmatrix} 1 & 0 \\ 2 & 1 \end{bmatrix}$.

Checkpoint: Use an idea of this section to explain why the linear transformation T that reflects points in $\mathbb{R}^2$ through the origin, $T(x_1, x_2) = (-x_1, -x_2)$, is the same as the linear transformation R that rotates points about the origin in $\mathbb{R}^2$ through π radians.

13. The vector $\mathbf{e}_1$ rotates into $\begin{bmatrix} \cos \pi/4 \\ \sin \pi/4 \end{bmatrix} = \begin{bmatrix} 1/\sqrt{2} \\ 1/\sqrt{2} \end{bmatrix}$ and then reflects through the x_2-axis into $\begin{bmatrix} -1/\sqrt{2} \\ 1/\sqrt{2} \end{bmatrix}$. This image of $\mathbf{e}_1$ is the first column of the standard matrix A for T. The vector $\mathbf{e}_2$ rotates into $\begin{bmatrix} -1/\sqrt{2} \\ 1/\sqrt{2} \end{bmatrix}$ and then reflects into $\begin{bmatrix} 1/\sqrt{2} \\ 1/\sqrt{2} \end{bmatrix}$. Thus, $A = \begin{bmatrix} -1/\sqrt{2} & 1/\sqrt{2} \\ 1/\sqrt{2} & 1/\sqrt{2} \end{bmatrix}$.

19. The matrix A that changes (x_1, x_2, x_3) into $(3x_2 - x_3,\ x_1 + 4x_2 + x_3)$ can be found by inspection when vectors are written in column format. Since the equation

$$\begin{bmatrix} ? & ? & ? \\ ? & ? & ? \end{bmatrix} \begin{bmatrix} x_1 \\ x_2 \\ x_3 \end{bmatrix} = \begin{bmatrix} 3x_2 - x_3 \\ x_1 + 4x_2 + x_3 \end{bmatrix}$$

must hold for all (x_1, x_2, x_3), you can see that $A = \begin{bmatrix} 0 & 3 & -1 \\ 1 & 4 & 1 \end{bmatrix}$.

Study Tip: When T is described by a formula as in Exercises 15–22, you can use the method of Exercise 19 to find an A such that $T(\mathbf{x}) = A\mathbf{x}$, *provided* that T is a linear transformation. (Finding A *proves* that T is linear.) If you can't find the matrix, T is probably *not* a linear transformation. To show that such a T is not linear, you have to find either two vectors $\mathbf{u}$ and $\mathbf{v}$ such that $T(\mathbf{u} + \mathbf{v}) \neq T(\mathbf{u}) + T(\mathbf{v})$ or a vector $\mathbf{u}$ and a scalar c such that $T(c\mathbf{u}) \neq cT(\mathbf{u})$.

Note: The text does not give you practice determining whether a transformation is linear because the time needed to develop this skill would have to be taken away from some other topic. If you are expected to have this skill, you will need some exercises. Check with your instructor.

23. **a.** See Theorem 10.

b. See Example 3.

c. See the definition of "onto" on page 81.

25. Row reduce the standard matrix A of the transformation T in Exercise 3:

$$A = \begin{bmatrix} 1 & -2 & 3 \\ 4 & 9 & -8 \end{bmatrix} \sim \begin{bmatrix} 1 & -2 & 3 \\ 0 & 17 & -20 \end{bmatrix}$$

Since the third column of A is not a pivot column, any equation of the form $A\mathbf{x} = \mathbf{b}$ will have x_3 as a free variable. So T is not one-to-one.

Another way to show that T is not one-to-one is to look at A and see that its columns are obviously linearly dependent (why?), and then use Theorem 12(b).

31. *T is one-to-one if and only if A has n pivot columns.* This statement follows by combining Theorem 12(b) with the statement in Exercise 38 of Section 1.6.

A Mathematical Note: One-to-one

Many students have difficulty with the concept of a one-to-one mapping. Figure 4 should help. The transformation T on the left appears to map *three* (or even more) points to one image point. In contrast, the transformation T on the right maps three points to three points. You could say that T is three-to-three (or six-to-six), but the standard terminology is one-to-one.

33. Define $T: \mathbb{R}^n \rightarrow \mathbb{R}^m$ by $T(\mathbf{x}) = B\mathbf{x}$ for some $m \times n$ matrix B, and let A be the standard matrix for T. By definition, $A = [T(\mathbf{e}_1) \ \cdots \ T(\mathbf{e}_n)]$, where $\mathbf{e}_j$ is the jth column of I_n. However, by matrix-vector multiplication, $T(\mathbf{e}_j) = B\mathbf{e}_j = \mathbf{b}_j$, the jth column of B. So $A = [\mathbf{b}_1 \ \cdots \ \mathbf{b}_n] = B$.

35. If $T:\mathbb{R}^n \to \mathbb{R}^m$ maps $\mathbb{R}^n$ *onto* $\mathbb{R}^m$, then its standard matrix A has a pivot in each row, by Theorem 12 and by Theorem 4 in Section 1.4. So A must have at least as many columns as rows, so $m \le n$.

When T is one-to-one, A must have a pivot in each column, by Theorem 12, so $m \ge n$.

37. **[M]** There is no pivot in the fourth column, so the columns of the matrix are not linearly independent and hence the linear transformation is not one-to-one (Theorem 12).

39. **[M]** Row reduction of the matrix shows that columns 1,2,3, and 5 contain pivots, but there is no pivot in the fifth row, so the columns of the matrix do not span $\mathbb{R}^5$. By Theorem 12, the linear transformation is not onto.

31. *(This solution is for Section 1.7.) To construct the proof, first write in mathematical terms what is given.*

Since $\{\mathbf{v}_1, \mathbf{v}_2, \mathbf{v}_3\}$ is linearly dependent, there exist scalars c_1, c_2, c_3, not all zero, such that

$$c_1\mathbf{v}_1 + c_2\mathbf{v}_2 + c_3\mathbf{v}_3 = \mathbf{0} \qquad (*)$$

Next, think about what you must prove. In this problem, to prove that the image points are linearly dependent, you need a dependence relation among $T(\mathbf{v}_1)$, $T(\mathbf{v}_2)$, and $T(\mathbf{v}_3)$. That fact suggests the next step.

Apply T to both sides of (*) and use linearity of T, obtaining

$$T(c_1\mathbf{v}_1 + c_2\mathbf{v}_2 + c_3\mathbf{v}_3) = T(\mathbf{0})$$

and

$$c_1T(\mathbf{v}_1) + c_2T(\mathbf{v}_2) + c_3T(\mathbf{v}_3) = \mathbf{0}$$

Since not all the weights are zero, $\{T(\mathbf{v}_1), T(\mathbf{v}_2), T(\mathbf{v}_3)\}$ is a linearly dependent set. This completes the proof.

Study Tip: Analyze the strategy above for solving Exercise 31 (in Section 1.7). This approach will work later in a variety of situations.

Answer to Checkpoint: The reflection T has the property that $T(\mathbf{e}_1) = -\mathbf{e}_1$ and $T(\mathbf{e}_2) = -\mathbf{e}_2$, while the rotation R has the property that $R(\mathbf{e}_1) = -\mathbf{e}_1$ and $R(\mathbf{e}_2) = -\mathbf{e}_2$. Since a linear transformation is completely determined by what it does to the columns $\mathbf{e}_1$ and $\mathbf{e}_2$ of the identity matrix, T and R must be the same transformation. (You could also explain this by observing that T and R have the same standard matrix, namely, $[-\mathbf{e}_1 \ \ -\mathbf{e}_2]$.)

Mastering Linear Algebra Concepts: Existence and Uniqueness

It's time to review and organize what you have learned about existence and uniqueness concepts, if you have not already done so. The review will help to prepare you for an exam on the chapter material.

Search through the chapter and collect all the various ways to express existence and uniqueness statements. Most of them can be found in boxes (and theorems) with an "if and only if" statement. Also, check the exercises. For existence, make two lists—one that concerns the equation $A\mathbf{x} = \mathbf{b}$ for some fixed $\mathbf{b}$ (but not always phrased as a matrix equation), and one that concerns the existence of solutions of $A\mathbf{x} = \mathbf{b}$ for all $\mathbf{b}$.

1.9 LINEAR MODELS IN BUSINESS, SCIENCE, AND ENGINEERING

This is the first of eleven sections devoted to uses of linear algebra. The applications in the text were selected to give you an impression of the power of linear algebra. You are likely to encounter some of these topics again—in school or in your career—and the discussions in your text will be valuable references.

The main point of this first section is to present several interesting applications in which "linearity" arises naturally.

STUDY NOTES

Nutrition Problem: In some applied problems such as the nutrition problem considered here, the data are already organized naturally in a manner that leads to a vector equation of the type we have discussed. The steps to the solution in this case may be diagrammed as follows:

Applied Problem → Vector Equation → Augmented Matrix → Solution Set → Answer to Problem

The nutrition model is linear because the nutrients supplied by each foodstuff are *proportional* to the amount of the foodstuff added to the diet mixture, and each nutrient in the mixture is the *sum* of the amounts from each foodstuff. Study equations (1) and (2) on page 87.

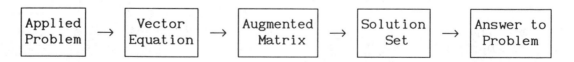

The nutrition problem leads naturally into linear programming, a subject that uses linear algebra and has applications in agriculture, business, engineering, and other areas. In the 1950's and 1960's, one of the most common applications of linear algebra (measured in millions of dollars per year for computer time) was to linear programming problems. Such problems are still of great importance in operations research and management science. The following reference gives an entertaining introduction to linear programming. Matrix notation is used in the appendix (pp. 127-152).

Saul I. Gass, *An Illustrated Guide to Linear Programming*, New York: McGraw-Hill, 1970. Republished by Dover Publications, 1990.

Electrical Networks: The linearity of this model, which is evident from the matrix equation $R\mathbf{i} = \mathbf{v}$, comes from the linearity of Ohm's law and Kirchhoff's voltage law. (Kirchhoff's current law, which is also linear, is needed when studying another model that involves branch currents.)

Population Movement: The entries in each *column* of the migration matrix must sum to one because the (decimal) fractions in a column account for the entire population in one region. A certain fraction of the population in a region remains in (or moves within) that region, and other fractions move elsewhere.

Warning: The order of the entries in a column of a migration matrix must match the order of the columns. For instance, if the first column concerns the population in the city, then the first entry in *each* column must be the fraction of a population that moves to (or remains in) the city.

SOLUTIONS TO EXERCISES

1. **a.** If x_1 is the number of servings of Cheerios and x_2 is the number of servings of 100% Natural Cereal, then x_1 and x_2 should satisfy

$$x_1\begin{bmatrix}\text{nutrients}\\ \text{per serving}\\ \text{of Cheerios}\end{bmatrix} + x_2\begin{bmatrix}\text{nutrients}\\ \text{per serving of}\\ \text{100\% Natural}\end{bmatrix} = \begin{bmatrix}\text{quantities}\\ \text{of nutrients}\\ \text{required}\end{bmatrix}$$

That is,

$$x_1\begin{bmatrix}110\\4\\20\\2\end{bmatrix} + x_2\begin{bmatrix}130\\3\\18\\5\end{bmatrix} = \begin{bmatrix}295\\9\\48\\8\end{bmatrix}$$

b. The equivalent matrix equation is $\begin{bmatrix}110 & 130\\4 & 3\\20 & 18\\2 & 5\end{bmatrix}\begin{bmatrix}x_1\\x_2\end{bmatrix} = \begin{bmatrix}295\\9\\48\\8\end{bmatrix}$. To solve this, row reduce the augmented matrix for this equation.

$$\begin{bmatrix} 110 & 130 & 295 \\ 4 & 3 & 9 \\ 20 & 18 & 48 \\ 2 & 5 & 8 \end{bmatrix} \sim \begin{bmatrix} 2 & 5 & 8 \\ 4 & 3 & 9 \\ 20 & 18 & 48 \\ 110 & 130 & 295 \end{bmatrix} \sim \begin{bmatrix} 1 & 2.5 & 4 \\ 4 & 3 & 9 \\ 10 & 9 & 24 \\ 110 & 130 & 295 \end{bmatrix}$$

$$\sim \begin{bmatrix} 1 & 2.5 & 4 \\ 0 & -7 & -7 \\ 0 & -16 & -16 \\ 0 & -145 & -145 \end{bmatrix} \sim \begin{bmatrix} 1 & 2.5 & 4 \\ 0 & 1 & 1 \\ 0 & 0 & 0 \\ 0 & 0 & 0 \end{bmatrix} \sim \begin{bmatrix} 1 & 0 & 1.5 \\ 0 & 1 & 1 \\ 0 & 0 & 0 \\ 0 & 0 & 0 \end{bmatrix}$$

The desired nutrients are provided by 1.5 servings of Cheerios together with 1 serving of 100% Natural Cereal.

Study Tip: Be sure to distinguish between (i) the vector equation, (ii) the matrix equation (which has the form $A\mathbf{x} = \mathbf{b}$), and (iii) the augmented matrix (which has the form $[A \quad \mathbf{b}]$) that represents a system of linear equations.

7. Loop 1: The resistance vector is

$$\mathbf{r}_1 = \begin{bmatrix} 11 \\ -5 \\ 0 \\ -5 \end{bmatrix} \begin{matrix} \text{Total of three RI voltage drops for current } I_1 \\ \text{Voltage drop for } I_2 \text{ is negative, } I_2 \text{ flows in opposite direction} \\ \text{Current } I_3 \text{ does not flow in loop 1} \\ \text{Voltage drop for } I_4 \text{ in loop 1} \end{matrix}$$

Loop 2:

$$\mathbf{r}_2 = \begin{bmatrix} -5 \\ 12 \\ -5 \\ 0 \end{bmatrix} \begin{matrix} \text{Voltage drop for } I_1 \text{ is negative} \\ \text{Total of three RI voltage drops for current } I_2 \\ \text{Voltage drop for } I_3 \text{ in loop 2} \\ \text{Current } I_4 \text{ does not flow in loop 2} \end{matrix}$$

Constructing $\mathbf{r}_3$ and $\mathbf{r}_4$ similarly, and listing the four loop voltages in $\mathbf{v}$, we obtain

$$\mathbf{r}_3 = \begin{bmatrix} 0 \\ -5 \\ 13 \\ -5 \end{bmatrix}, \quad \mathbf{r}_4 = \begin{bmatrix} -5 \\ 0 \\ -5 \\ 14 \end{bmatrix}, \quad R = [\mathbf{r}_1 \ \cdots \ \mathbf{r}_4] = \begin{bmatrix} 11 & -5 & 0 & -5 \\ -5 & 12 & -5 & 0 \\ 0 & -5 & 13 & -5 \\ -5 & 0 & -5 & 14 \end{bmatrix}, \quad \mathbf{v} = \begin{bmatrix} 40 \\ 30 \\ 20 \\ 10 \end{bmatrix}$$

Note that the off-diagonal entries of R are all negative because the loop current directions are all in the same direction on the figure. (For each loop j, this forces the currents in the other loops to flow in the direction opposite to current I_j.) Also, remember to check whether the voltages are positive or negative. (See the statement before the box on Kirchhoff's voltage law.)

The loop current vector **i** satisfies $R\mathbf{i} = \mathbf{v}$. Row reduction of $[R \ \ \mathbf{v}]$ produces $\mathbf{i} = (12.99,\ 11.85,\ 9.46,\ 8.73)$, to two decimal places.

13. **[M]** Set $M = \begin{bmatrix} .95 & .03 \\ .05 & .97 \end{bmatrix}$ and $\mathbf{x}_0 = \begin{bmatrix} 600{,}000 \\ 400{,}000 \end{bmatrix}$.

a. Some of the population vectors are

$$\mathbf{x}_5 = \begin{bmatrix} 523{,}293 \\ 476{,}067 \end{bmatrix},\quad x_{10} = \begin{bmatrix} 472{,}737 \\ 527{,}263 \end{bmatrix},\quad x_{15} = \begin{bmatrix} 445{,}018 \\ 554{,}982 \end{bmatrix},\quad x_{20} = \begin{bmatrix} 425{,}161 \\ 574{,}839 \end{bmatrix}$$

The data here shows that the city population is declining and the suburban populations is increasing, but the changes in population each year seem to grow smaller.

b. When $\mathbf{x}_0 = \begin{bmatrix} 350{,}000 \\ 650{,}000 \end{bmatrix}$, the situation is different. Now

$$\mathbf{x}_5 = \begin{bmatrix} 358{,}523 \\ 641{,}477 \end{bmatrix},\quad x_{10} = \begin{bmatrix} 364{,}140 \\ 635{,}860 \end{bmatrix},\quad x_{15} = \begin{bmatrix} 367{,}843 \\ 632{,}157 \end{bmatrix},\quad x_{20} = \begin{bmatrix} 370{,}283 \\ 629{,}717 \end{bmatrix}$$

The city population is increasing slowly and the suburban population is decreasing. No other conclusions are expected. If you find the results somewhat interesting, you might explore further. (This example will be analyzed in greater detail later in the text.)

MATLAB Generating a Sequence

The m-file (in the Toolbox) for Exercises 9—13 in Section 1.9 stores initial vectors in **x0**. Set **x = x0** to put the initial data into **x**. Then use the command **x = M*x** repeatedly to generate the sequence $\mathbf{x}_1$, $\mathbf{x}_2$,.... You only type the command once. After that, use the up-arrow (↑) key to recall the command, and press <Enter>.

In Exercise 11, you need 6 decimal places to get four significant figures in M(1,2). Use the command **format long** and then **M** to see more decimal places in *M*. The command **format short** will return MATLAB to the standard four decimal place display. (The display format does not affect MATLAB's accuracy in computations.)

Numbers are entered in MATLAB without commas. The number 600,000 in MATLAB scientific notation is 6e5. A small number such as .00000012 is 1.2e-7.

CHAPTER 1 SUPPLEMENTARY EXERCISES

Justifications for the answers to the True/False questions are given below. Other justifications are possible. A solution to Exercise 11 is also supplied, because the text's answer is only a hint.

1. a. False. Counterexample: Let

$$A = \begin{bmatrix} 1 & 2 \\ 3 & 4 \end{bmatrix},\ B = \begin{bmatrix} 1 & 2 \\ 0 & -2 \end{bmatrix},\ C = \begin{bmatrix} 1 & 2 \\ 0 & 1 \end{bmatrix}$$

Then B and C are in echelon form, and A is row equivalent to B and C.

b. False. Counterexample: Let A be any $n \times n$ matrix with fewer than n pivot columns. Then the equation $A\mathbf{x} = \mathbf{0}$ has infinitely many solutions.

c. True. If a linear system has more than one solution, it is a consistent system and has a free variable. By the Existence and Uniqueness Theorem in Section 1.2, the system has infinitely many solutions.

d. False. Counterexample: The following system has no free variables and has no solution

$$\begin{aligned} x_1 + x_2 &= 1 \\ x_2 &= 5 \\ x_1 + x_2 &= 2 \end{aligned}$$

e. True. See the box at the bottom of page 7. If $[A \ \ \mathbf{b}]$ is transformed into $[C \ \ \mathbf{d}]$ by elementary row operations, then the two augmented matrices are row equivalent.

f. True. Theorem 6 in Section 1.5 essentially says that when $A\mathbf{x} = \mathbf{b}$ is consistent, the solutions sets of the nonhomogeneous equation and the homogeneous equation are translates of each other; in this case, the two equations have the same number of solutions.

g. False. For the columns of A to span $\mathbb{R}^m$, the equation $A\mathbf{x} = \mathbf{b}$ must be consistent for *all* $\mathbf{b}$ in $\mathbb{R}^m$, not for just one vector $\mathbf{b}$ in $\mathbb{R}^m$.

h. False. *Any* matrix can be transformed by elementary row operations into reduced echelon form.

i. False. Every equation $A\mathbf{x} = \mathbf{0}$ has the trivial solution whether or not some variables are free.

j. True, by Theorem 4 in Section 1.4. If the equation $A\mathbf{x} = \mathbf{b}$ is consistent for every $\mathbf{b}$ in $\mathbb{R}^m$, then A must have a pivot position in every one of its m rows. If A has m pivot positions, A must have at least m columns.

k. False. Counterexample. Let A be any matrix with m pivot columns but more than m columns altogether. Then the equation $A\mathbf{x} = \mathbf{0}$ has m basic variables and at least one free variable, so the equation $A\mathbf{x} = \mathbf{0}$ does not have a unique solution.

l. False. See Example 5 in Section 1.6.

m. False. If a set $\{\mathbf{v}_1, \ldots, \mathbf{v}_4\}$ were to span $\mathbb{R}^5$, then the matrix $A = [\mathbf{v}_1 \cdots \mathbf{v}_4]$ would have a pivot position in each of its five rows, which is impossible since A has only four columns.

n. True. Any set of three vectors in $\mathbb{R}^2$ would have to be linearly dependent, by Theorem 8, in Section 1.6.

o. True, by Theorem 7 in Section 1.6, because $\mathbf{u}$ is nonzero.

p. True. A transformation is a function (page 67), and a linear transformation is a special type of transformation.

q. True. For the transformation $\mathbf{x} \mapsto A\mathbf{x}$ to map $\mathbb{R}^4$ onto $\mathbb{R}^5$, the matrix A would have to have a pivot in every row and hence have five pivot columns. This is impossible, because A has only four columns.

11. Let M be the line through the origin that is parallel to the line through $\mathbf{v}_1, \mathbf{v}_2$, and $\mathbf{v}_3$. Then $\mathbf{v}_2 - \mathbf{v}_1$ and $\mathbf{v}_3 - \mathbf{v}_1$ are both on M. So one of these two vectors is a multiple of the other, say $\mathbf{v}_2 - \mathbf{v}_1 = k(\mathbf{v}_3 - \mathbf{v}_1)$. This equation produces the linear dependence relation $(k - 1)\mathbf{v}_1 + \mathbf{v}_2 - k\mathbf{v}_3 = \mathbf{0}$.

A second solution: A parametric equation of the line is $\mathbf{x} = \mathbf{v}_1 + t(\mathbf{v}_2 - \mathbf{v}_1)$. Since $\mathbf{v}_3$ is on the line, there is some t_0 such that $\mathbf{v}_3 = \mathbf{v}_1 + t_0(\mathbf{v}_2 - \mathbf{v}_1) = (1 - t_0)\mathbf{v}_1 + t_0\mathbf{v}_2$. So $\mathbf{v}_3$ is a linear combination of $\mathbf{v}_1$ and $\mathbf{v}_2$, and $\{\mathbf{v}_1, \mathbf{v}_2, \mathbf{v}_3\}$ is linearly dependent.

CHAPTER 1 GLOSSARY CHECKLIST

Check your knowledge by attempting to write definitions of the terms below. Then compare your work with the definitions given in the text's Glossary. Ask your instructor which definitions, if any, might appear on a test.

affine transformation: A mapping $T:\mathbb{R}^n \to \mathbb{R}^m$ of the form $T(\mathbf{x})$ = ...

augmented matrix: A matrix made up of a ....

back-substitution (with matrix notation): The ... phase of row reduction of an ....

basic variable: A variable in a linear system that ....

codomain (of $T:\mathbb{R}^n \to \mathbb{R}^m$): The set ...that contains ....

coefficient matrix: A matrix whose entries are ....

consistent linear system: A linear system with ....

contraction: A mapping $\mathbf{x} \mapsto$ ....

difference equation (or **linear recurrence relation**): An equation of the form ... whose solution is ....

dilation: A mapping $\mathbf{x} \mapsto$ ....

domain (of a transformation T): The set of ....

echelon form (or **row echelon form**, of a matrix): An echelon matrix that ...

echelon matrix (or **row echelon matrix**): A rectangular matrix that has three properties: (1) ... (2) ... (3) ....

elementary row operations: (1) ... (2) ... (3) ....

equal vectors: Vectors in $\mathbb{R}^n$ whose ....

equivalent (linear) systems: Linear systems with the property that ....

existence question: Asks, "Does ... exist?" or "Is ...?" Also, "Does ... exist for ...?"

floating point arithmetic: Arithmetic with numbers represented as ....

flop: One arithmetic operation ....

free variable: Any variable in a linear system that ....

Gaussian elimination: *See* row reduction algorithm.

general solution (of a linear system): A ... description of a solution set that expresses ....

homogeneous equation: An equation of the form ....

identity matrix (denoted by I or I_n): A square matrix with ....

image (of a vector **x** under a transformation T): The vector .... (Use symbols)

inconsistent linear system: A linear system with ....

leading entry: The ... entry in a row of a matrix.

linear combination: A sum of ....

linear dependence relation: A ... equation where ....

linear equation (in the variables $x_1, \ldots, x_n$): An equation that can be written in the form ....

linearly dependent (vectors): A set $\{\mathbf{v}_1, \ldots, \mathbf{v}_p\}$ with the property that ....

linearly independent (vectors): A set $\{\mathbf{v}_1, \ldots, \mathbf{v}_p\}$ with the property ....

linear system: A collection of one or more ... equations involving ....

linear transformation: A transformation $T: \mathbb{R}^n \to \mathbb{R}^m$ for which (i) ..., and (ii) ....

line through p parallel to v: The set .... (Use symbols)

matrix: A rectangular ....

matrix equation: An equation that ....

matrix transformation: A mapping $\mathbf{x} \mapsto$ ....

migration matrix: A matrix that gives the ... movement between different locations, from ....

$m \times n$ matrix: A matrix with ....

nontrivial solution: A nonzero solution of ....

one-to-one (mapping): A mapping $T: \mathbb{R}^n \to \mathbb{R}^m$ such that ....

onto (mapping): A mapping $T: \mathbb{R}^n \to \mathbb{R}^m$ such that ....

overdetermined system: A system of equations with ....

parallelogram rule for addition: A geometric interpretation of ....

parametric equation of a line: An equation of the form ....

parametric equation of a plane: An equation of the form ....

pivot: A ... number that either is used ... or is ....

pivot column: A column that ....

pivot position: A position in a matrix that corresponds ....

plane through u, v, and the origin: A set whose parametric equation is....

product $A\mathbf{x}$: ....

range (of a linear transformation T): The set of ....

reduced echelon form (or **reduced row echelon form**, of a matrix): A ... matrix that is ....

reduced echelon matrix: A rectangular matrix in echelon form that has these additional properties: ....

roundoff error: Error in floating point arithmetic caused when ....

row equivalent (matrices): Two matrices for which there exists ....

row reduced (matrix): A matrix that has been transformed ....

row reduction algorithm: A systematic method using ....

row replacement: An elementary row operation that ....

rule for computing $A\mathbf{x}$: ....

scalar: ....

scalar multiple of $\mathbf{u}$ by c: The vector ....

set spanned by $\{\mathbf{v}_1, \ldots, \mathbf{v}_p\}$: ....

size (of a matrix): Two numbers ....

solution (of a linear system): ....

solution set: The set of ....

Span $\{\mathbf{v}_1, \ldots, \mathbf{v}_p\}$: The set ....

standard matrix (for a linear transformation T): The matrix ....

system of linear equations (or a **linear system**): A collection of ....

transformation (or **function** or **mapping**) T from $\mathbb{R}^n$ to $\mathbb{R}^m$: A rule that assigns to each vector $\mathbf{x}$ in $\mathbb{R}^n$ a .... Notation: $T: \mathbb{R}^n \longrightarrow \mathbb{R}^m$.

translation (by a vector $\mathbf{p}$): The operation of ....

trivial solution: The solution ... of a ....

underdetermined system: A system of equations with ....

uniqueness question: Asks, "If a solution of a system ...?"

vector: ....

vector equation: An equation involving ....

weights: ....

2 MATRIX ALGEBRA

2.1 MATRIX OPERATIONS

Most of this chapter is an outgrowth of the idea in Section 1.7 that a matrix can transform data. This dynamic role of matrices suggests that we study the *combined effect* of several matrices on data (that is, on a vector or a set of vectors). Sections 2.1 to 2.5 describe this *matrix algebra.*

KEY IDEA

Matrix multiplication corresponds to composition of linear transformations. The definition of AB, using the columns of B, is critical for the development of both the theory and some of the applications in the text.

STUDY NOTES

Double-subscript notation: The subscripts tell the location of an entry in the matrix—the first subscript identifies the row and the second subscript the column. (Remember: *Row* is shorter than *column*, so *row* goes first.)

In the product AB, left multiplication (that is, multiplication on the left) by A acts on the columns of B, by definition, while right multiplication by B acts on the rows of A (see page 104). That is,

$$\begin{bmatrix}\text{column } j\\ \text{of } AB\end{bmatrix} = A\begin{bmatrix}\text{column } j\\ \text{of } B\end{bmatrix} \quad \text{and} \quad [\text{row } i \text{ of } AB] = [\text{row } i \text{ of } A]B$$

To compute a specific matrix product by hand, use the Row-Column Rule. If A is $m \times n$, then the (i, j)-entry of AB is written with sigma notation as

$$(AB)_{ij} = \sum_{k=1}^{n} a_{ik}b_{kj}$$

Remember that if you change the *order* (position) of the factors in a matrix product, the new product may be different, or it may not even be defined. For instance, $(A + C)B$ and $AB + BC$ are probably *not* equal! Also, see the warning box on page 106.

SOLUTIONS TO EXERCISES

1. $-2A = (-2)\begin{bmatrix} 7 & 0 & -1 \\ -1 & 5 & 2 \end{bmatrix} = \begin{bmatrix} -14 & 0 & 2 \\ 2 & -10 & -4 \end{bmatrix}$. Next, use $B - 2A = B + (-2A)$:

$$B - 2A = \begin{bmatrix} -1 & 4 & 1 \\ 5 & -3 & 0 \end{bmatrix} + \begin{bmatrix} -14 & 0 & 2 \\ 2 & -10 & -4 \end{bmatrix} = \begin{bmatrix} -15 & 4 & 3 \\ 7 & -13 & -4 \end{bmatrix}$$

The product AC is not defined because the number of columns of A does not match the number of rows of C.

$CD = \begin{bmatrix} 1 & 4 \\ -4 & 0 \end{bmatrix}\begin{bmatrix} 1 & 0 \\ -2 & 1 \end{bmatrix} = \begin{bmatrix} -7 & 4 \\ -4 & 0 \end{bmatrix}$. For hand computation, the row-column rule is faster to use than the definition.

7. Since A has 5 columns, B must have 5 rows. Otherwise, AB is not defined. Since AB has 7 columns, so does B. Thus B is 5×7.

13. If you had difficulty with this problem, read the definition of AB from *right to left*. Here is the definition, written in reverse order:

$$[A\mathbf{b}_1 \;\; \cdots \;\; A\mathbf{b}_p] = A[\mathbf{b}_1 \;\; \cdots \;\; \mathbf{b}_p] = AB, \quad \text{when } B = [\mathbf{b}_1 \;\; \cdots \;\; \mathbf{b}_p]$$

Thus $[Q\mathbf{r}_1 \;\; \cdots \;\; Q\mathbf{r}_p] = QR$, when $R = [\mathbf{r}_1 \;\; \cdots \;\; \mathbf{r}_p]$.

15. **a.** See the definition of AB.

b. See the box after Example 3.

c. See Theorem 2(b), read right to left.

d. See Theorem 3(b), read right to left.

e. See the box after Theorem 3.

19. A solution is in the text. The main point of the exercise is that the columns of AB are $A\mathbf{b}_1, \ldots, A\mathbf{b}_p$.

21. Let $\mathbf{b}_p$ be the last column of B. Then $A\mathbf{b}_p = \mathbf{0}$, but $\mathbf{b}_p$ is not the zero vector. Thus the equation $A\mathbf{b}_p = \mathbf{0}$ is a linear dependence relation, and the columns of A are linearly dependent.

23. Suppose that $A\mathbf{x} = \mathbf{0}$ for some $\mathbf{x}$. Then $C(A\mathbf{x}) = C\mathbf{0} = \mathbf{0}$, and so $(CA)\mathbf{x} = \mathbf{0}$, by associativity. But $CA = I$, which implies that $\mathbf{x}$ must be zero. So the equation $A\mathbf{x} = \mathbf{0}$ has only the trivial (zero) solution.

Supplementary Exercises: The following two problems will help you review concepts from Chapter 1. Try them now. The answers are at the end of this section. In each problem, assume that the product AB is defined.

a. Show that if $\mathbf{y}$ is a linear combination of the columns of AB, then $\mathbf{y}$ is a linear combination of the columns of A.

b. Show that if the columns of B are linearly dependent, then so are the columns of AB.

25. The product $\mathbf{u}^T\mathbf{v}$ is a 1×1 matrix, which usually is identified with a real number and is written without the matrix brackets. The text shows that $\mathbf{u}^T\mathbf{v}$ and $\mathbf{v}^T\mathbf{u}$ both equal $-2a + 3b - 4c$. Since $\mathbf{u}$ is 3×1 and $\mathbf{v}^T$ is 1×3, the outer product $\mathbf{u}\mathbf{v}^T$ is 3×3, and so is $\mathbf{v}\mathbf{u}^T$. Look at the text's answers for $\mathbf{v}\mathbf{u}^T$ and $\mathbf{u}\mathbf{v}^T$ to see how they are related.

Study Tip: *Inner* products ($\mathbf{u}^T\mathbf{v}$ and $\mathbf{v}^T\mathbf{u}$) have the transpose symbol in the middle. *Outer* products ($\mathbf{u}\mathbf{v}^T$ and $\mathbf{v}\mathbf{u}^T$) have the transpose symbol on the outside.

27. The (i,j)-entry of $A(B + C)$ equals the (i,j)-entry of $AB + AC$, because

$$\sum_{k=1}^{n} a_{ik}(b_{kj} + c_{kj}) = \sum_{k=1}^{n} a_{ik}b_{kj} + \sum_{k=1}^{n} a_{ik}c_{kj}$$

The (i,j)-entry of $(B + C)A$ equals the (i,j)-entry of $BA + CA$, because

$$\sum_{k=1}^{n} (b_{ik} + c_{ik})a_{kj} = \sum_{k=1}^{n} b_{ik}a_{kj} + \sum_{k=1}^{n} c_{ik}a_{kj}$$

31. The (i,j)-entry in $(AB)^T$ is the (j,i)-entry in AB, which is

$$a_{j1}b_{1i} + \cdots + a_{jn}b_{ni}$$

The entries in row i of B^T are $b_{1i}, \cdots, b_{ni}$, because they come from the ith column of B. Likewise, the entries in column j of A^T are

$a_{j1}, \ldots, a_{jn}$, because they come from the jth row of A. Thus the (i,j)-entry in B^TA^T is $b_{1i}a_{j1} + \cdots + b_{ni}a_{jn}$. This sum equals the other sum displayed, so the (i,j)-entries of $(AB)^T$ and B^TA^T are equal. This proves Theorem 3(d).

Solutions to Supplementary Exercises:

a. If **d** is a linear combination of the columns of AB, then the equation $AB\mathbf{x} = \mathbf{d}$ has a solution. (See the box on page 41 in the text.) If **x** is such a solution, then associativity of matrix multiplication shows that $A(B\mathbf{x}) = \mathbf{d}$. That is, the equation $A\mathbf{u} = \mathbf{d}$ has a solution. So **d** is a linear combination of the columns of A.

b. If the columns of B are linearly dependent, then there is a nonzero **x** such that $B\mathbf{x} = \mathbf{0}$. Hence $A(B\mathbf{x}) = A\mathbf{0} = \mathbf{0}$, and $(AB)\mathbf{x} = \mathbf{0}$, by associativity, which shows that the columns of AB are linearly dependent, because **x** is not zero.

MATLAB Matrix Notation and Operations

To create a matrix, enter the data row-by-row, with a space between entries and a semicolon between rows. For instance, the command

A = [1 2 3;4 5 -6] Use brackets around the data.

creates a 2×3 matrix A. If A is $m\times n$, then **size(A)** is the row vector $[m\ \ n]$. The (i,j)-entry in A is **A(i,j)**. If i or j is replaced by a colon, the result is a column or row of A, respectively. Examples:

A(:,3) is column 3 of A
A(2,:) is row 2 of A

To specify columns 3, 4 and 5 of A, you can use

A(:,[3 4 5]) or **A(:,3:5)**

The symbols 3:5 (read "3 to 5") stand for the vector [3 4 5]. Similar notation works for selected rows of A.

MATLAB uses +, -, and * to denote matrix addition, subtraction, and multiplication, respectively. If A is square and k is a positive integer, **A^k** denotes the kth power of A. The transpose of A is **A'** (with an apostrophe for the prime symbol). Note: when A has complex entries, the (i,j)-entry of **A'** is the complex conjugate of the (j,i)-entry of **A**.

Use a single column (or row) matrix for a vector. If **u** and **v** are column vectors of the same size, then **u'*v** is their inner product, and **u*v'** is an outer product.

2.2 THE INVERSE OF A MATRIX

Matrix inverses are essential for many discussions in linear algebra. This section and the next describe the main properties of invertible matrices.

STUDY NOTES

The inverse formula for a 2×2 matrix will be used frequently in exercises later in the text. (See Theorem 4.) To invert a 2×2 matrix, interchange the diagonal entries, reverse the signs of the off-diagonal entries, and divide each entry by the determinant (assuming $ad - bc \neq 0$).

Theorem 5 and its proof are important. The phrase "has a unique solution" includes the assertion that a solution exists, so the proof has two parts. The equation $AA^{-1} = I$ is used to prove that a solution exists, and the equation $A^{-1}A = I$ is used to show that the solution is unique.

Except when A is 2×2, Theorem 5 is practically never used to solve $A\mathbf{x} = \mathbf{b}$. Row reduction of $[A \quad \mathbf{b}]$ is faster. Actually, in practical work, you will seldom need to compute A^{-1}. (However, Example 3 illustrates a case in which the entries of A^{-1} could be useful.)

When using an inverse in matrix algebra, remember that matrix multiplication is not commutative. The phrase "left-multiply B by A^{-1}" means to multiply B on its left side by A^{-1}. *Never* write $\frac{B}{A}$ (or B/A) because it could stand for $A^{-1}B$ or BA^{-1}.

Elementary matrices are used in this text mainly to link row reduction to matrix multiplication. Each elementary row operation amounts to left-multiplication by an elementary matrix. So, if A can be row reduced to U, then there is a product F of elementary matrices such that $FA = U$.

Theorem 7 includes an *if and only if* statement, which was discussed in the Appendix to Section 1.2 in this *Study Guide*. The proof of this statement in Theorem 7 has two parts: (1) assume that A is invertible and prove that $A \sim I_n$; and (2) assume that $A \sim I_n$ and prove that A is invertible.

SOLUTIONS TO EXERCISES

1. $$\begin{bmatrix} -4 & -5 \\ 5 & 6 \end{bmatrix}^{-1} = \frac{1}{(-4)\cdot 6 - (-5)\cdot 5}\begin{bmatrix} 6 & -(-5) \\ -5 & -4 \end{bmatrix} = \begin{bmatrix} 6 & 5 \\ -5 & -4 \end{bmatrix}$$

7. a. $$A^{-1} = \frac{1}{1\cdot 8 - 2\cdot 3}\begin{bmatrix} 8 & -2 \\ -3 & 1 \end{bmatrix} = \frac{1}{2}\begin{bmatrix} 8 & -2 \\ -3 & 1 \end{bmatrix} \text{ or } \begin{bmatrix} 4 & -1 \\ -3/2 & 1/2 \end{bmatrix}$$

$$\mathbf{x} = A^{-1}\mathbf{b}_1 = \frac{1}{2}\begin{bmatrix} 8 & -2 \\ -3 & 1 \end{bmatrix}\begin{bmatrix} 5 \\ 7 \end{bmatrix} = \frac{1}{2}\begin{bmatrix} 26 \\ -8 \end{bmatrix} = \begin{bmatrix} 13 \\ -4 \end{bmatrix}$$

Similar calculations give $A^{-1}\mathbf{b}_2 = \begin{bmatrix} -6 \\ 2 \end{bmatrix}$, $A^{-1}\mathbf{b}_3 = \begin{bmatrix} 7 \\ -2 \end{bmatrix}$, $A^{-1}\mathbf{b}_4 = \begin{bmatrix} -3 \\ 2 \end{bmatrix}$.

b. $$[A \;\; \mathbf{b}_1 \;\; \mathbf{b}_2 \;\; \mathbf{b}_3 \;\; \mathbf{b}_4] = \begin{bmatrix} 1 & 2 & 5 & -2 & 3 & 1 \\ 3 & 8 & 7 & -2 & 5 & 7 \end{bmatrix}$$

$$\sim \begin{bmatrix} 1 & 2 & 5 & -2 & 3 & 1 \\ 0 & 2 & -8 & 4 & -4 & 4 \end{bmatrix} \sim \begin{bmatrix} 1 & 2 & 5 & -2 & 3 & 1 \\ 0 & 1 & -4 & 2 & -2 & 2 \end{bmatrix}$$

$$\sim \begin{bmatrix} 1 & 0 & 13 & -6 & 7 & -3 \\ 0 & 1 & -4 & 2 & -2 & 2 \end{bmatrix}$$

The solutions are $\begin{bmatrix} 13 \\ -4 \end{bmatrix}$, $\begin{bmatrix} -6 \\ 2 \end{bmatrix}$, $\begin{bmatrix} 7 \\ -2 \end{bmatrix}$, $\begin{bmatrix} -3 \\ 2 \end{bmatrix}$, the same as in part (a).

Note: This exercise was designed to make the arithmetic simple for both methods, but (a) requires more arithmetic than (b). In fact, (a) requires 22 multiplications or divisions and 9 additions or subtractions, but (b) only takes 14 multiplications or divisions and 9 additions or subtractions. In general, the arithmetic for method (b) can be unpleasant for hand calculation. However, when A is larger than 2×2, method (b) is *much* faster than (a).

Study Tip: Notice in Exercise 7(a) how the 1/2 in the formula for A^{-1} was kept outside the matrix $\begin{bmatrix} 8 & -2 \\ -3 & 1 \end{bmatrix}$ when computing $A^{-1}\mathbf{x}$. This trick often simplifies hand calculation by postponing the arithmetic with fractions (or decimals) until the end.

9. a. See the definition of *invertible*. **b.** See Theorem 6(b).

c. If $A = \begin{bmatrix} 1 & 1 \\ 0 & 0 \end{bmatrix}$, then $ab - cd = 1 \cdot 1 - 0 \cdot 0 = 1 \neq 0$, but Theorem 4 shows that this matrix is not invertible, because $ad - bc = 0$.

d. See Theorem 5. **e.** See the box just before Example 6.

11. a. Given: A is $n \times n$ and invertible, and B is $n \times p$. Suppose that X_1 satisfies $AX_1 = B$. Then, left-multiplying each side by A^{-1},

$$A^{-1}(AX_1) = A^{-1}B, \quad IX_1 = A^{-1}B, \quad \text{and} \quad X_1 = A^{-1}B$$

If there is a solution of $AX = B$, then the solution must be $A^{-1}B$. To show that $A^{-1}B$ *is* a solution, substitute it for X and compute

$$A(A^{-1}B) = AA^{-1}B = IB = B$$

b. Write $B = [\mathbf{b}_1 \cdots \mathbf{b}_p]$ and $X = [\mathbf{u}_1 \cdots \mathbf{u}_p]$. By definition of matrix multiplication, $AX = [A\mathbf{u}_1 \cdots A\mathbf{u}_p]$. So the equation $AX = B$ is equivalent to the p systems:

$$A\mathbf{u}_1 = \mathbf{b}_1, \quad \ldots, \quad A\mathbf{u}_p = \mathbf{b}_p \qquad (1)$$

Since A is the coefficient matrix in each system, these systems may be solved simultaneously, placing the augmented columns of these systems next to A to form $[A \;\; \mathbf{b}_1 \;\cdots\; \mathbf{b}_p] = [A \;\; B]$. Since the solutions $\mathbf{u}_1,\ldots,\mathbf{u}_p$ in (1) are uniquely determined, we know that $[A \;\; \mathbf{b}_1 \;\cdots\; \mathbf{b}_p]$ reduces to $[I \;\; \mathbf{u}_1 \;\cdots\; \mathbf{u}_p]$, which is $[I \;\; X]$, where X is the solution of $AX = B$.

Another solution: Since A is invertible, it can be row reduced to I. Thus $[A \;\; B]$ can be row reduced to a matrix of the form $[I \;\; X]$ for some X. Let $E_1,\ldots,E_k$ be elementary matrices that implement this row reduction. Then

$$(E_k\cdots E_1)A = I \quad \text{and} \quad (E_k\cdots E_1)B = X \qquad (2)$$

Right-multiplying each side of the first equation by A^{-1}, we find that $(E_k\cdots E_1)AA^{-1} = IA^{-1}$, $(E_k\cdots E_1)I = A^{-1}$, and $E_k\cdots E_1 = A^{-1}$. The second equation in (2) then shows that $A^{-1}B = X$.

Study Tip: Whenever you are told "A is invertible", you know that A^{-1} exists, and you may use A^{-1} to solve an equation or to make appropriate calculations.

13. If $AB = AC$ and A is invertible, then we can left-multiply each side of the equation by A^{-1}: $A^{-1}AB = A^{-1}AC$, $IB = IC$, and $B = C$. To show that the conclusion $B = C$ does not always follow when A is singular, find a counterexample: If A is the zero matrix, then $AB = AC$ for any $n \times p$ matrices B and C. For another example, let

$$A = \begin{bmatrix} 1 & 0 \\ 0 & 0 \end{bmatrix}, \quad B = \begin{bmatrix} 1 & 2 \\ 3 & 4 \end{bmatrix}, \quad C = \begin{bmatrix} 1 & 2 \\ 5 & 6 \end{bmatrix}$$

Then $AB = AC = \begin{bmatrix} 1 & 2 \\ 0 & 0 \end{bmatrix}$, because B and C have the same first row.

Warning: A common mistake in Exercise 16 is to try to use the formula $(AB)^{-1} = B^{-1}A^{-1}$. But this formula is available only when you know, in advance, that A and B are invertible. In Exercise 16, you must *prove* that A is invertible.

19. Suppose that X satisfies the equation $C^{-1}(A + X)B^{-1} = I$. Then

$$C^{-1}(A + X)B^{-1} = I \quad \Rightarrow \quad CC^{-1}(A + X)B^{-1} = CI \qquad \text{Left-multiplication by C}$$

$$\Rightarrow I(A + X)B^{-1} = C \quad \Rightarrow \quad (A + X)B^{-1}B = CB \qquad \text{Right-multiplication by B}$$

$$\Rightarrow \quad (A + X)I = CB \quad \Rightarrow \quad A + X = CB \quad \Rightarrow \quad X = CB - A$$

(Be careful above when you multiply by B. The right side must be CB, not BC.) To show that $CB - A$ really *is* a solution, substitute it for X:

$$C^{-1}[A + (CB - A)]B^{-1} = C^{-1}[CB]B^{-1} = C^{-1}CBB^{-1} = II = I$$

After this section, your instructor may permit you to include fewer details in your calculations (check on this). For instance, the argument above could be shortened to:

$$C^{-1}(A + X)B^{-1} = I \quad \Leftrightarrow \quad (A + X)B^{-1} = C \qquad \text{Left-multiplication by C or } C^{-1}$$

$$\Leftrightarrow \quad (A + X) = CB \qquad \text{Right-multiplication by B or } B^{-1}$$

$$\Leftrightarrow \quad X = CB - A$$

21. Suppose A is invertible. By Theorem 5, the equation $A\mathbf{x} = \mathbf{0}$ has only one solution, namely, the zero solution. This means that the columns of A are linearly independent, by a remark in Section 1.6.

23. Suppose A is $n \times n$ and the equation $A\mathbf{x} = \mathbf{0}$ has only the trivial solution. Then there are no free variables in the equation $A\mathbf{x} = \mathbf{0}$, and so A has n pivot columns. Since A is *square* and the n pivot positions must be in different rows, the pivots in an echelon form of A must be on the main diagonal. Hence A is row equivalent to the $n \times n$ identity matrix.

25. Suppose that $ad - bc = 0$. If a, b, c, d are all zero, then A is the zero matrix and the equation $A\mathbf{x} = \mathbf{0}$ is true for *all* $\mathbf{x}$. Otherwise, at least one of the vectors $\begin{bmatrix} -b \\ a \end{bmatrix}$ and $\begin{bmatrix} d \\ -c \end{bmatrix}$ is nonzero, and both of them satisfy $A\mathbf{x} = \mathbf{0}$. For instance, $\begin{bmatrix} a & b \\ c & d \end{bmatrix}\begin{bmatrix} -b \\ a \end{bmatrix} = \begin{bmatrix} -ab + ba \\ -cb + da \end{bmatrix} = \begin{bmatrix} 0 \\ 0 \end{bmatrix}$, because $ad - bc = 0$. When $A\mathbf{x} = \mathbf{0}$ has more than one solution, A cannot be invertible, by Theorem 5.

31.
$$\begin{bmatrix} 1 & 0 & 5 & 1 & 0 & 0 \\ 1 & 1 & 0 & 0 & 1 & 0 \\ 3 & 2 & 6 & 0 & 0 & 1 \end{bmatrix} \sim \begin{bmatrix} 1 & 0 & 5 & 1 & 0 & 0 \\ 0 & 1 & -5 & -1 & 1 & 0 \\ 0 & 2 & -9 & -3 & 0 & 1 \end{bmatrix} \sim$$

$$\begin{bmatrix} 1 & 0 & 5 & 1 & 0 & 0 \\ 0 & 1 & -5 & -1 & 1 & 0 \\ 0 & 0 & 1 & -1 & -2 & 1 \end{bmatrix} \sim \begin{bmatrix} 1 & 0 & 0 & 6 & 10 & -5 \\ 0 & 1 & 0 & -6 & -9 & 5 \\ 0 & 0 & 1 & -1 & -2 & 1 \end{bmatrix}, \qquad A^{-1} = \begin{bmatrix} 6 & 10 & -5 \\ -6 & -9 & 5 \\ -1 & -2 & 1 \end{bmatrix}$$

Study Tip: In Exercise 40, the arithmetic to invert D by hand is not bad. It may help to multiply D by 1000 first. Then row operations will produce the inverse of $1000D$. But $(1000D)^{-1} = 1000^{-1}D^{-1}$. So multiply the inverse matrix you find by 1000 to obtain D^{-1}. Of course, using MATLAB is easier.

MATLAB The $n \times n$ identity matrix is denoted by eye(n). If A is 5×5, then the command **M = [A eye(5)]** creates the augmented matrix $[A \;\; I]$. Use **gauss**, **bgauss**, and **scale** to reduce $[A \;\; I]$. See page 1-17.

There are other MATLAB commands that row reduce matrices, invert matrices, and solve equations $A\mathbf{x} = \mathbf{b}$, but they are not discussed here because they will not help you learn the concepts in this section.

2.3 CHARACTERIZATIONS OF INVERTIBLE MATRICES

In many linear algebra texts, the equivalent of Chapter 4 is the "killer" chapter. But you won't have difficulties if you prepare well now. Review the major concepts in the previous sections, and spend more study time on this short section than you ordinarily spend on one section.

KEY IDEAS

The Invertible Matrix Theorem (IMT) only applies to square matrices. However, some groups of these statements in the IMT are also equivalent for rectangular matrices. See Theorem 4 in Section 1.4, for example.

STATEMENTS FROM THE INVERTIBLE MATRIX THEOREM

Equivalent statements, for any $n \times p$ matrix A.	Equivalent only for an $n \times n$ square matrix A.	Equivalent statements, for any $n \times p$ matrix A.
k. There is a matrix D such that $AD = I$.	a. A is an invertible matrix.	j. There is a matrix C such that $CA = I$.
*. A has a pivot posi- in every row.	c. A has n pivot positions.	*. A has a pivot position in every column.
h. The columns of A span $\mathbb{R}^n$.	b. A is row equivalent to the $n \times n$ identity matrix.	c. The columns of A form a linearly independent set.

g. The equation $A\mathbf{x} = \mathbf{b}$ has at least one solution for each $\mathbf{b}$ in $\mathbb{R}^n$.	*. The equation $A\mathbf{x} = \mathbf{b}$ has a unique solution for each $\mathbf{b}$ in $\mathbb{R}^n$.	*. The equation $A\mathbf{x} = \mathbf{b}$ has at most one solution for each $\mathbf{b}$ in $\mathbb{R}^n$.
i. The linear transformation $\mathbf{x} \mapsto A\mathbf{x}$ maps $\mathbb{R}^p$ onto $\mathbb{R}^n$.	*. The linear transformation $\mathbf{x} \mapsto A\mathbf{x}$ is invertible.	f. The linear transformation $\mathbf{x} \mapsto A\mathbf{x}$ is one-to-one.
	*. A is a product of elementary matrices.	d. The equation $A\mathbf{x} = \mathbf{0}$ has only the trivial solution.
	l. A^T is an invertible matrix.	*. The equation $A\mathbf{x} = \mathbf{0}$ has no free variables.

The seven statements denoted by (*) were not listed in the text as part of the IMT, mainly to avoid intimidating you with so many statements in one theorem. As part of your review, make sure you understand why these extra statements can be added to the IMT.

The text did not actually prove that for a *rectangular* matrix, statements (j) and (k) are equivalent to the other statements in their respective columns, but I listed them here anyway so you could see how they fit into the table. A matrix C such that $CA = I$ is called a **left-inverse** of A, and a matrix D such that $AD = I$ is called a **right-inverse** of A.

Checkpoint: What can you say about the statements in the first column when A has more rows than columns? (Why?) What about the statements in the third column when A has more columns than rows? (Why?)

A question such as the one in the box below is one way I test whether my students know the IMT. Test yourself. Cover up the IMT, write your answer, and then check your work. The answer is given at the end of this section.

Test Question:

Let A be an $n \times n$ matrix. Write 6 statements from the Invertible Matrix Theorem, each equivalent to the statement that A is invertible. Use the following concepts, one in each statement: (*i*) row equivalent, (*ii*) the equation $AD = I$, (*iii*) columns, (*iv*) the equation $A\mathbf{x} = \mathbf{0}$, and (*v*) linear transformation.

SOLUTIONS TO EXERCISES

1. To show that A is invertible, you have to find only **one** statement in the Invertible Matrix Theorem that is true. For instance, the text's answer used the obvious fact that the columns of A are linearly independent. The text also showed how to use $\det A$ and Theorem 4 to show that A is invertible. Until Chapter 3, this approach is available only for 2×2 matrices.

7. $\begin{bmatrix} 1 & 3 & 0 & -1 \\ 0 & 1 & -2 & -1 \\ -2 & -6 & 3 & 2 \\ 3 & 5 & 8 & -3 \end{bmatrix} \sim \begin{bmatrix} 1 & 3 & 0 & -1 \\ 0 & 1 & -2 & -1 \\ 0 & 0 & 3 & 0 \\ 0 & -4 & 8 & 0 \end{bmatrix} \sim \begin{bmatrix} 1 & 3 & 0 & -1 \\ 0 & 1 & -2 & -1 \\ 0 & 0 & 3 & 0 \\ 0 & 0 & 0 & -4 \end{bmatrix}$. The original 4×4 matrix has 4 pivot positions and so is invertible, by the IMT.

Study Tip: For theoretical purposes, the most useful characterization of an invertible matrix A is that $A\mathbf{x} = \mathbf{0}$ has only the trivial solution. For most computational work by hand, the fastest way to determine if A is invertible is to (find and) count the pivot positions.

13. Study the Invertible Matrix Theorem. The statements there are true *only for an invertible matrix*. Also, if one of the statements is true about a square matrix A, then all statements in the theorem are true; if one of the statements is false, then all are false.

a. See statements (d) and (b) of the IMT.

b. See statements (h) and (e). **c.** See statement (g).

d. See statements (d) and (c). **e.** See statement (l).

19. The full solution is in the text's answer section.

Study Tip: Learn how to recognize when a square matrix is **not** invertible. See Exercises 17 and 21. Each of the following statements was constructed by *negating* one of the statements in the IMT. If A is an $n \times n$ matrix, then each statement is true if and only if A is **not** invertible:

- The matrix A has *fewer than n* pivot positions.
- The equation $A\mathbf{x} = \mathbf{0}$ has a *nontrivial* (nonzero) solution.
- The equation $A\mathbf{x} = \mathbf{b}$ has *more than one* solution for *some* $\mathbf{b}$ in $\mathbb{R}^n$.
- The equation $A\mathbf{x} = \mathbf{b}$ has *no* solution (is inconsistent) for *some* $\mathbf{b}$ in $\mathbb{R}^n$.
- The columns of A are linearly *dependent*.
- The columns of A *do not* span $\mathbb{R}^n$.

23. By Theorem 6(b) in Section 2.2 (with A in place of B), the product AA is invertible, because both A and A are invertible.

25. Suppose that A is square and $AB = I$. Then A is invertible, by the IMT. Left-multiplying the equation $AB = I$ by A^{-1}, we have $A^{-1}AB = A^{-1}I$, $IB = A^{-1}$, and $B = A^{-1}$. By Theorem 6(a), the matrix B ($= A^{-1}$) is invertible and its inverse is $(A^{-1})^{-1} = A$.

Study Tip: Exercise 25 makes a good test question.

27. If AB is invertible, then there is a matrix W such that $(AB)W = I$ and hence $A(BW) = I$. Since A is square, A is invertible, by the IMT.

Study Tip: The solution to Exercise 27 shows a common **use** of the IMT. Just remember that statements (j) and (k) of the IMT apply only to square matrices. Whenever you use one of these statements to deduce that A is invertible, you should point out in your argument that A is square.

29. To show that T is one-to-one, suppose that $T(\mathbf{u}) = T(\mathbf{v})$ for some vectors $\mathbf{u}$ and $\mathbf{v}$ in $\mathbb{R}^n$. Then $S(T(\mathbf{u})) = S(T(\mathbf{v}))$, where S is the inverse of T. By Equation (1), $\mathbf{u} = S(T(\mathbf{u}))$ and $S(T(\mathbf{v})) = \mathbf{v}$, so $\mathbf{u} = \mathbf{v}$. Thus T is one-to-one. To show that T is onto, suppose $\mathbf{y}$ represents an arbitrary vector in $\mathbb{R}^n$ and define $\mathbf{x} = S\mathbf{y}$. Then, using Equation (2), $T(\mathbf{x}) = T(S(\mathbf{y})) = \mathbf{y}$, which shows that T maps $\mathbb{R}^n$ onto $\mathbb{R}^n$.

Second proof: By Theorem 9, the standard matrix A of T is invertible. By the IMT, the columns of A are linearly independent and span $\mathbb{R}^n$. Finally, by Theorem 12 in Section 1.8, T is one-to-one and maps $\mathbb{R}^n$ onto $\mathbb{R}^n$.

33. The standard matrix of T is $A = \begin{bmatrix} -5 & 9 \\ 4 & -7 \end{bmatrix}$, which is invertible, because

$\det A \neq 0$. By Theorem 9, the transformation T is invertible and the standard matrix of T^{-1} is A^{-1}. From the formula for a 2×2 inverse,

$$A^{-1} = \begin{bmatrix} 7 & 9 \\ 4 & 5 \end{bmatrix}. \quad \text{So } T^{-1}(x_1, x_2) = \begin{bmatrix} 7 & 9 \\ 4 & 5 \end{bmatrix}\begin{bmatrix} x_1 \\ x_2 \end{bmatrix} = (7x_1 + 9x_2, 4x_1 + 5x_2).$$

35. Given any $\mathbf{v}$ in $\mathbb{R}^n$, we may write $\mathbf{v} = T(\mathbf{x})$ for some $\mathbf{x}$, because T is an onto mapping. Then, the assumed properties of S and U show that $S(\mathbf{v}) = S(T(\mathbf{x})) = \mathbf{x}$ and $U(\mathbf{v}) = U(T(\mathbf{x})) = \mathbf{x}$. So $S(\mathbf{v})$ and $U(\mathbf{v})$ are equal for each $\mathbf{v}$. That is, S and U are the same function from $\mathbb{R}^n$ into $\mathbb{R}^n$.

38. To perform the experiment described, the following MATLAB line of instructions can be repeated easily:

```
x = rand(4,1); b = A*x; x1 = A\b; x - x1
```

Use **format long**. If you repeat this instruction line ten or more times, you will probably find that x and x1 agree to 11 digits twice as often as when they agree to 12 digits. Occasionally, they may agree to even 13 digits.

Answers to Checkpoint: If A has more rows than columns, then all statements in the first column of the table must be false, because they are equivalent and the statement about a pivot position in each row cannot be true. If A has more columns than rows, then all statements in the third column of the table must be false, because A cannot have a pivot in each of its columns.

Answers to Test Question:

(*i*) A is row equivalent to I_n. (*ii*) There exists an $n\times n$ matrix D such that $AD = I$. (*iii*) The columns of A span $\mathbb{R}^n$. (*iv*) The equation $A\mathbf{x} = \mathbf{0}$ has only the trivial solution. (*v*) The linear transformation $\mathbf{x}\mapsto A\mathbf{x}$ maps $\mathbb{R}^n$ onto $\mathbb{R}^n$.

Another answer for (*iii*) is: The columns of A are linearly independent. Similarly, (*v*) has another answer. But the following statement is unacceptable as one of the answers to the test question:

A is invertible if and only if the columns of A span $\mathbb{R}^n$.

This statement is itself a (true) theorem (assuming A is square), not a statement that is true precisely when A is invertible.

Mastering Linear Algebra: Reviewing and Reflecting

Two important steps to mastery of linear algebra are periodic review of earlier material and reflection on its relation to new material. When you reread the basic conceptual material from Chapter 1, you may be surprised to discover new insights that you missed earlier. Your broader experience now should give you a better framework within which to understand concepts such as spanning and linear independence.

Compare the review you conducted in Section 1.8 (see the *Study Guide* appendix to that section) with the three-part table at the beginning of this *Study Guide* section. (You did carry out that review, didn't you?) The left and right columns of the table should match some of your "existence" and "uniqueness" statements, respectively.

If your review in Section 1.8 was thorough, you probably anticipated some of the content of the Invertible Matrix Theorem. Existence and uniqueness threads run through the fabric of linear algebra, and they intertwine when related to square matrices (the middle column of the table). A good review procedure now is to expand the table to include references to theorems, examples, and counterexamples. This will occupy several pages. The process of constructing this table is what will help you most.

MATLAB: inv and rank

Determining whether a matrix is invertible is not always a simple matter. A fast and fairly reliable method is to use the command **inv(A)**, which computes the inverse of A. A warning is given if the matrix is singular (noninvertible) or close to singular.

Another method to test invertibility is to enter **rank(A)**. Section 4.6 discusses rank and shows that an $n \times n$ matrix A is invertible if and only if rank(A) = n. The MATLAB **rank** command requires more calculations than **inv** but is more reliable when a matrix is nearly singular.

2.4 PARTITIONED MATRICES

The ideas in this section are fairly simple. However, mark them for future reference, because you are likely to use this notation after you leave school. Partitioned matrices arise in theoretical discussions in essentially every field that makes use of matrices. Here are two examples.

1. The modern *state space* approach to control systems engineering depends on matrix calculations.[1] The problem of determining whether a system is *controllable* amounts to calculating the number of pivot positions in a *controllability matrix*

$$[B \quad AB \quad A^2B \quad \cdots \quad A^{n-1}B]$$

   where A is $n \times n$, B has n rows, and the matrices come from an equation of the form (8) in the discussion preceding Exercise 19.

2. Discussions of modern algorithms and computer software design for scientific computing naturally use the "language" of partitioned matrices. For instance, common techniques for parallel processing of large matrix calculations, such as *slicing* and *crinkling*, are described with parti-

---

[1]An understanding of control systems is important in the design of filtering circuitry, robots, process control systems, and spacecraft. Thus a control systems course is often part of the undergraduate curriculum for electrical, mechanical, chemical, and aerospace engineering. See *Control Systems Engineering*, by Norman S. Nise, Benjamin/Cummings Publishing Co., Redwood City, CA, 1992.

tioned matrices.[2] Also, the standard computer science reference on matrix calculations relies heavily on partitioned matrices.[3]

KEY IDEAS

The column-row evaluation of AB is the last of five different "views" of matrix multiplication. All five are special cases of the *block matrix* version of the *row-column* rule for matrix multiplication. Here they are:

(1) The definition of $A\mathbf{x}$ amounts to block multiplication of AB where B has only one column:

$$A\mathbf{x} = \begin{bmatrix} \mathbf{a}_1 & | & \cdots & | & \mathbf{a}_n \end{bmatrix} \begin{bmatrix} x_1 \\ \hline \vdots \\ \hline x_n \end{bmatrix} = \begin{bmatrix} x_1\mathbf{a}_1 + \cdots + x_n\mathbf{a}_n \end{bmatrix}$$

(2) Partition A as *one* row and *one* column. Then the definition of the usual product AB is a row-column block product:

$$AB = A\begin{bmatrix} \mathbf{b}_1 & | & \mathbf{b}_2 & | & \cdots & | & \mathbf{b}_p \end{bmatrix} = \begin{bmatrix} A\mathbf{b}_1 & | & A\mathbf{b}_2 & | & \cdots & | & A\mathbf{b}_p \end{bmatrix}$$

(3) Likewise, we observed in Section 2.1 that if B is partitioned as one row and one column, then

$$AB = \begin{bmatrix} \text{row}_1(A) \\ \text{row}_2(A) \\ \vdots \\ \text{row}_m(A) \end{bmatrix} B = \begin{bmatrix} \text{row}_1(A)B \\ \text{row}_2(A)B \\ \vdots \\ \text{row}_m(A)B \end{bmatrix}$$

(4) The next display can be viewed either as just the standard row-column rule in which each entry of AB is computed as the product of a row of A and a column of B, or as a multiplication of block matrices (with A having only one column of blocks and B having only one row of blocks)

---

[2]*Parallel Algorithms and Matrix Computations*, by Jagdish J. Modi, Oxford Applied Mathematics and Computing Science Series, Clarendon Press, Oxford, 1988, pp. 73-75.

[3]*Matrix Computations*, 2nd ed., by Gene H. Golub and Charles F. Van Loan, The Johns Hopkins Press, Baltimore, 1989.

$$AB = \begin{bmatrix} \text{row}_1(A) \\ \text{row}_2(A) \\ \vdots \\ \text{row}_m(A) \end{bmatrix} [\text{col}_1(B) \quad \text{col}_2(B) \quad \cdots \quad \text{col}_p(B)]$$

$$= \begin{bmatrix} \text{row}_1(A)\text{col}_1(B) & \cdots & \text{row}_1(A)\text{col}_j(B) & \cdots & \text{row}_1(A)\text{col}_p(B) \\ \vdots & & \vdots & & \vdots \\ \text{row}_i(A)\text{col}_1(B) & \cdots & \text{row}_i(A)\text{col}_j(B) & \cdots & \text{row}_i(A)\text{col}_p(B) \\ \vdots & & \vdots & & \vdots \\ \text{row}_m(A)\text{col}_1(B) & \cdots & \text{row}_m(A)\text{col}_j(B) & \cdots & \text{row}_m(A)\text{col}_p(B) \end{bmatrix}$$

(5) The final display is the column-row expansion of AB (Theorem 10 in this section). In this view, AB is expressed as a sum of *outer products* of the form $\mathbf{u}\mathbf{v}^T$, with $\mathbf{u}$ a column of A and $\mathbf{v}^T$ a row of B. But the display can also be viewed as the block version of the row-column product in which A has one row (of blocks) and B has one column (of blocks):

$$AB = [\text{col}_1(A) \quad \text{col}_2(A) \quad \cdots \quad \text{col}_n(A)] \begin{bmatrix} \text{row}_1(B) \\ \text{row}_2(B) \\ \vdots \\ \text{row}_n(b) \end{bmatrix}$$

$$= \text{col}_1(A) \cdot \text{row}_1(B) + \cdots + \text{col}_n(A) \cdot \text{row}_n(B)$$

You might say that the row-column rule computes AB as an array of inner products (view 4 above), while the column-row expansion displays AB as a sum of arrays (view 5).

SOLUTIONS TO EXERCISES

1. Apply the row-column rule as if the matrix entries were numbers, but for each product (such as EA below), always write the entry of the left block-matrix on the *left*.

$$\begin{bmatrix} I & 0 \\ E & I \end{bmatrix} \begin{bmatrix} A & B \\ C & D \end{bmatrix} = \begin{bmatrix} IA + 0C & IB + 0D \\ EA + IC & EB + ID \end{bmatrix} = \begin{bmatrix} A & B \\ EA + C & EB + D \end{bmatrix}$$

↑ This must be EA, not AE.

Checkpoint: Notice in Exercises 1 and 3 that $\begin{bmatrix} I & 0 \\ E & I \end{bmatrix}$ and $\begin{bmatrix} 0 & I \\ I & 0 \end{bmatrix}$ act as block-matrix generalizations of elementary matrices. What sort of 2×2 block matrix is the appropriate generalization of an elementary matrix that acts as a scaling operation? (Answer this carefully.)

7. Compute the left side of the equation:

$$\begin{bmatrix} X & 0 & 0 \\ Y & 0 & I \end{bmatrix} \begin{bmatrix} A & Z \\ 0 & 0 \\ B & I \end{bmatrix} = \begin{bmatrix} XA + 0 + 0B & XZ + 0 + 0 \\ YA + 0 + IB & YZ + 0 + I \end{bmatrix}$$

Set this equal to the right side of the equation:

$$\begin{bmatrix} XA & XZ \\ YA + B & YZ + I \end{bmatrix} = \begin{bmatrix} I & 0 \\ 0 & I \end{bmatrix} \quad \text{so that} \quad \begin{matrix} XA = I & XZ = 0 \\ YA + B = 0 & YZ + I = I \end{matrix}$$

Since the (1,1)-blocks are equal, $XA = I$. Since X and A are square, the IMT implies that A and X are invertible, and hence $X = A^{-1}$. From the (1,2)-entries, $XZ = 0$. Since X is invertible, Z must be 0. Therefore, the (2,2)-entries give no new information. Finally, from the (2,1)-entries, $YA + B = 0$ and $YA = -B$. Right-multiplication by A^{-1} shows that $Y = -BA^{-1}$. The order of the factors for Y is crucial.

Study Tip: Problems such as 5—10 make good exam questions. Remember to mention the IMT when appropriate, and remember that matrix multiplication is generally not commutative.

11. **a.** See the subsection *Addition and Scalar Multiplication*

b. See the paragraph before Example 3.

13. You are asked to establish an "if and only if" statement. First, suppose that A is invertible, and let $A^{-1} = \begin{bmatrix} D & E \\ F & G \end{bmatrix}$. Then

$$\begin{bmatrix} B & 0 \\ 0 & C \end{bmatrix} \begin{bmatrix} D & E \\ F & G \end{bmatrix} = \begin{bmatrix} BD & BE \\ CF & CG \end{bmatrix} = \begin{bmatrix} I & 0 \\ 0 & I \end{bmatrix}$$

Since B is square, the equation $BD = I$ implies that B is invertible, by the IMT. Similarly, $CG = I$ implies that C is invertible. Also, the equation $BE = 0$ implies that $E = B^{-1}0 = 0$. Similarly $F = 0$. Thus

$$A^{-1} = \begin{bmatrix} B & 0 \\ 0 & C \end{bmatrix}^{-1} = \begin{bmatrix} D & E \\ F & G \end{bmatrix} = \begin{bmatrix} B^{-1} & 0 \\ 0 & C^{-1} \end{bmatrix} \qquad (*)$$

This proves that A is invertible *only if* B and C are invertible. For the "*if*" part of the statement, suppose that B and C are invertible. Then (*) provides a likely candidate for A^{-1} which can be used to show that A is invertible. Compute:

$$\begin{bmatrix} B & 0 \\ 0 & C \end{bmatrix}\begin{bmatrix} B^{-1} & 0 \\ 0 & C^{-1} \end{bmatrix} = \begin{bmatrix} BB^{-1} & 0 \\ 0 & CC^{-1} \end{bmatrix} = \begin{bmatrix} I & 0 \\ 0 & I \end{bmatrix}$$

Since A is square, this calculation and the IMT imply that A is invertible. (Don't forget this final sentence. Without it, the argument is incomplete.) Instead of that sentence, you could add the equation:

$$\begin{bmatrix} B^{-1} & 0 \\ 0 & C^{-1} \end{bmatrix}\begin{bmatrix} B & 0 \\ 0 & C \end{bmatrix} = \begin{bmatrix} B^{-1}B & 0 \\ 0 & C^{-1}C \end{bmatrix} = \begin{bmatrix} I & 0 \\ 0 & I \end{bmatrix}$$

19. The matrix equation (8) in the text is equivalent to

$$(A - sI_n)\mathbf{x} + B\mathbf{u} = \mathbf{0} \quad \text{and} \quad C\mathbf{x} + \mathbf{u} = \mathbf{y}$$

Rewrite the first equation as $(A - sI_n)\mathbf{x} = -B\mathbf{u}$. When $A - sI_n$ is invertible, $\mathbf{x} = (A - sI_n)^{-1}(-B\mathbf{u}) = -(A - sI_n)^{-1}B\mathbf{u}$. Substitute this formula for $\mathbf{x}$ into the second equation above:

$$C(-(A - sI_n)^{-1}B\mathbf{u}) + \mathbf{u} = \mathbf{y}, \quad \text{so that} \quad I_m\mathbf{u} - C(A - sI_n)^{-1}B\mathbf{u} = \mathbf{y}$$

Thus $\mathbf{y} = (I_m - C(A - sI_n)^{-1}B)\mathbf{u}$. If $W(s) = I_m - C(A - sI_n)^{-1}B$, then $\mathbf{y} = W(s)\mathbf{u}$. The matrix $W(s)$ is the Schur complement of the matrix $A - sI_n$ in the system matrix in equation (8).

21. To prove a statement by induction, a good first step is to write the statement that depends on n but exclude the phrase "for all n", and label the statement for reference:

The product of two $n \times n$ *lower triangular matrices is lower triangular.* (*)

Second, verify that the statement is true for $n = 1$. In this particular case, (*) is obviously true, because *every* 1×1 matrix is lower triangular. The "induction step" is next.

Suppose that (*) is true when n is some positive integer k, and consider two $(k+1) \times (k+1)$ matrices B_1 and C_1. Partition B_1 and C_1 as

$$B_1 = \begin{bmatrix} b & \mathbf{0}^T \\ \mathbf{x} & B \end{bmatrix}, \quad C_1 = \begin{bmatrix} c & \mathbf{0}^T \\ \mathbf{y} & C \end{bmatrix}$$

where B and C are $k \times k$ matrices, $\mathbf{x}$ and $\mathbf{y}$ are in $\mathbb{R}^n$, and b and c are scalars. Since B_1 and C_1 are lower triangular, so are B and C. Now,

$$B_1C_1 = \begin{bmatrix} b & \mathbf{0}^T \\ \mathbf{x} & B \end{bmatrix}\begin{bmatrix} c & \mathbf{0}^T \\ \mathbf{y} & C \end{bmatrix} = \begin{bmatrix} bc + \mathbf{0}^T\mathbf{y} & b\mathbf{0}^T + \mathbf{0}^T C \\ \mathbf{x}c + B\mathbf{y} & \mathbf{x}\mathbf{0}^T + BC \end{bmatrix} = \begin{bmatrix} bc & \mathbf{0}^T \\ c\mathbf{x} + B\mathbf{y} & BC \end{bmatrix}$$

Assuming (*) is true for $n = k$, BC must be lower triangular. The form of B_1C_1 shows that it, too, is lower triangular. Thus, the truth of (*) for any positive integer $n = k$ implies the truth of (*) for the next positive integer $k+1$. By the principle of induction, (*) must be true for all $n \geq 1$.

Answer to Checkpoint: The block diagonal matrices $\begin{bmatrix} E & 0 \\ 0 & I \end{bmatrix}$ and $\begin{bmatrix} I & 0 \\ 0 & E \end{bmatrix}$ are obvious choices. Less obvious is the requirement that E be invertible, in order to make these block matrices invertible. (Recall that the invertibility of elementary matrices was essential for the theory in Section 2.2.)

Appendix: The Principle of Induction

Consider a statement "(*)" that depends on a positive integer n, as in Exercise 21. To prove "by induction" that (*) is true for all positive integers, you must prove two things:

(a) Statement (*) is true for $n = 1$.
(b) (The induction step) If (*) is true for any positive integer $n = k$, then (*) is also true for the next integer $n = k + 1$.

A property or axiom of the real number system, called the *principle of mathematical induction*, says that if (a) and (b) are true, then (*) is true for all integers $n \geq 1$. This is reasonable, because if (*) is true for $n = 1$, then (b) shows that (*) is true for $n = 2$. Applying (b) again with $n = 2$, we see that (*) is true for $n = 3$. Applying (b) repeatedly, we see that (*) is true for 2, 3, 4, 5, ...

MATLAB Partitioned Matrices

MATLAB uses partitioned matrix notation. For example, if A, B, C, D, E, and F are matrices of appropriate sizes, then the command

```
M = [A B C; D E F]
```

creates a larger matrix of the form $M = \begin{bmatrix} A & B & C \\ D & E & F \end{bmatrix}$. Once M is formed, there is no record of the partition that was used to create M. For instance, although B was the (1,2)-block used to form M, the number $M(1,2)$ is the same as the (1,2)-entry of A.

2.5 MATRIX FACTORIZATIONS

In a sense, Section 2.5 is the most up-to-date section in the text, because matrix factorizations lie at the heart of modern uses of matrix algebra. For instance, they are indispensable for the analysis of computational algorithms and research in parallel processing. The text focuses here on triangular factorizations, but the exercises introduce you to other important factorizations that you may encounter later.

KEY IDEAS

When a matrix A is factored as $A = LU$, the data in A are preprocessed in a way that makes the equation $A\mathbf{x} = \mathbf{b}$ easier to solve. Write $LU\mathbf{x} = \mathbf{b}$, or $L(U\mathbf{x}) = \mathbf{b}$ and let $\mathbf{y} = U\mathbf{x}$. Solve $L\mathbf{y} = \mathbf{b}$ for $\mathbf{y}$ and then solve $U\mathbf{x} = \mathbf{y}$ for $\mathbf{x}$. The two-step process is fast when L and U are triangular.

Finding L and U requires the same number of multiplications and divisions as row reducing A to an echelon form U (about $n^3/3$ operations when A is $n \times n$). After that, L and U are available for solving other equations involving A. The key to finding L is to place entries in L in such a way that the sequence of row operations reducing A to U also reduces L to the identity. In this case, LU must equal A. (See the top of page 136.)

The text discusses how to build L when no row interchanges are needed to reduce A to U. In this case, L can be unit lower triangular. An appendix below describes how to build L in permuted unit triangular form when row interchanges are needed (or desired, for numerical reasons).

SOLUTIONS TO EXERCISES

1. $L = \begin{bmatrix} 1 & 0 & 0 \\ -1 & 1 & 0 \\ 2 & -5 & 1 \end{bmatrix}$, $U = \begin{bmatrix} 3 & -7 & -2 \\ 0 & -2 & -1 \\ 0 & 0 & -1 \end{bmatrix}$, $\mathbf{b} = \begin{bmatrix} -7 \\ 5 \\ 2 \end{bmatrix}$. First, solve $L\mathbf{y} = \mathbf{b}$.

$$[L \quad \mathbf{b}] = \begin{bmatrix} 1 & 0 & 0 & -7 \\ -1 & 1 & 0 & 5 \\ 2 & -5 & 1 & 2 \end{bmatrix} \sim \begin{bmatrix} 1 & 0 & 0 & -7 \\ 0 & 1 & 0 & -2 \\ 0 & -5 & 1 & 16 \end{bmatrix}$$

The only arithmetic is in column 4

$$\sim \begin{bmatrix} 1 & 0 & 0 & -7 \\ 0 & 1 & 0 & -2 \\ 0 & 0 & 1 & 6 \end{bmatrix}, \quad \text{so } \mathbf{y} = \begin{bmatrix} -7 \\ -2 \\ 6 \end{bmatrix}$$

Next, solve $U\mathbf{x} = \mathbf{y}$, using back-substitution (with matrix notation).

$$[U \quad \mathbf{y}] = \begin{bmatrix} 3 & -7 & -2 & -7 \\ 0 & -2 & -1 & -2 \\ 0 & 0 & -1 & 6 \end{bmatrix} \sim \begin{bmatrix} 3 & -7 & -2 & -7 \\ 0 & -2 & -1 & -2 \\ 0 & 0 & 1 & -6 \end{bmatrix} \sim \begin{bmatrix} 3 & -7 & 0 & -19 \\ 0 & -2 & 0 & -8 \\ 0 & 0 & 1 & -6 \end{bmatrix}$$

$$\sim \begin{bmatrix} 3 & -7 & 0 & -19 \\ 0 & 1 & 0 & 4 \\ 0 & 0 & 1 & -6 \end{bmatrix} \sim \begin{bmatrix} 3 & 0 & 0 & 9 \\ 0 & 1 & 0 & 4 \\ 0 & 0 & 1 & -6 \end{bmatrix} \sim \begin{bmatrix} 1 & 0 & 0 & 3 \\ 0 & 1 & 0 & 4 \\ 0 & 0 & 1 & -6 \end{bmatrix}$$

So $\mathbf{x} = (3, 4, -6)$.

Checkpoint: Exercise 11 in Section 2.2 shows how to compute $A^{-1}B$ by row reduction. Describe how you could speed up this calculation if you have an LU factorization of A available (and A is invertible).

7. Place the first pivot column of $\begin{bmatrix} 2 & 5 \\ -3 & -4 \end{bmatrix}$ into L, after dividing the column by 2 (the pivot), then add 3/2 times row 1 to row 2, yielding U.

$$A = \begin{bmatrix} \textcircled{2} & 5 \\ -3 & -4 \end{bmatrix} \sim \begin{bmatrix} 2 & 5 \\ 0 & \textcircled{7/2} \end{bmatrix} = U, \qquad \begin{bmatrix} \textcircled{2} \\ -3 \end{bmatrix} \quad \begin{bmatrix} \textcircled{7/2} \end{bmatrix}$$

$$\div 2 \qquad \div 7/2$$

$$\downarrow \qquad \downarrow$$

$$\begin{bmatrix} 1 & \\ -3/2 & 1 \end{bmatrix}, \quad L = \begin{bmatrix} 1 & 0 \\ -3/2 & 1 \end{bmatrix}$$

13.

$$\begin{bmatrix} \textcircled{1} & 3 & -5 & -3 \\ -1 & -5 & 8 & 4 \\ 4 & 2 & -5 & -7 \\ -2 & -4 & 7 & 5 \end{bmatrix} \sim \begin{bmatrix} 1 & 3 & -5 & -3 \\ 0 & \textcircled{-2} & 3 & 1 \\ 0 & -10 & 15 & 5 \\ 0 & 2 & -3 & -1 \end{bmatrix} \sim \begin{bmatrix} 1 & 3 & -5 & -3 \\ 0 & -2 & 3 & 1 \\ 0 & 0 & 0 & 0 \\ 0 & 0 & 0 & 0 \end{bmatrix} = U \quad \text{No more pivots!}$$

$$\begin{bmatrix} \textcircled{1} \\ -1 \\ 4 \\ -2 \end{bmatrix} \quad \begin{bmatrix} \textcircled{-2} \\ -10 \\ 2 \end{bmatrix}$$

Use the last two columns of I_4 to make L unit lower triangular.

$$\div 1 \qquad \div -2$$

$$\downarrow \qquad \downarrow \qquad \downarrow$$

$$\begin{bmatrix} 1 & & & \\ -1 & 1 & & \\ 4 & 5 & 1 & \\ -2 & -1 & 0 & 1 \end{bmatrix}, \quad L = \begin{bmatrix} 1 & 0 & 0 & 0 \\ -1 & 1 & 0 & 0 \\ 4 & 5 & 1 & 0 \\ -2 & -1 & 0 & 1 \end{bmatrix}$$

19. A good answer will require a written paragraph or two. If you have not tried to *write* your answer, do so now, *without reading the solution below*. Explain how you would row reduce $[A \;\; I]$, knowing that A is lower triangular. Your answer to this question should contain some of the ideas shown below, although your wording might be quite different.

Let A be a lower-triangular $n \times n$ matrix with nonzero entries on the diagonal, and consider the augmented matrix $[A \quad I]$.

a. The (1,1)-entry can be scaled to 1 and the entries below it can be changed to 0 by adding multiples of row 1 to the rows below. This affects only the first column of A and the first column of I. So the (2,2)-entry in the new matrix is still nonzero and now is the only nonzero entry of row 2 in the first n columns (because A was lower triangular).

The (2,2)-entry can be scaled to 1, and the entries below it can be changed to 0 by adding multiples of row 2 to the rows below. This affects only columns 2 and $n+2$ of the augmented matrix. Now the (3,3) entry in A is the only nonzero entry of the third row in the first n columns, so it can be scaled to 1 and then used as a pivot to zero out entries below it. Continuing in this way, A is eventually reduced to I, by scaling each row with a pivot and then using only row operations that add multiples of the pivot row to rows below.

b. The row operations just described only add rows to rows below, so the I on the right in $[A \quad I]$ changes into a lower triangular matrix. By Theorem 7 in Section 2.2, that matrix is A^{-1}.

21. Suppose $A = BC$, with B invertible. Then there exist elementary matrices $E_1, \ldots, E_p$ corresponding to row operations that reduce B to I, in the sense that $E_p \cdots E_1 B = I$. Applying the same sequence of row operations to A amounts to left-multiplying A by the product $E_p \cdots E_1$. By associativity of matrix multiplication,

$$E_p \cdots E_1 A = E_p \cdots E_1 BC = IC = C$$

so the same sequence of row operations reduces A to C.

25. $A = UDV^T$. Since U and V^T are square, the equations $U^TU = I$ and $V^TV = I$ imply that U and V^T are invertible, by the IMT, and hence $U^{-1} = U^T$ and $(V^T)^{-1} = V$.. Since the diagonal entries $\sigma_1, \ldots, \sigma_n$ in D are nonzero, D is invertible, with the inverse of D being the diagonal matrix with $\sigma_1^{-1}, \ldots, \sigma_n^{-1}$ on the diagonal. Thus A is a product of invertible matrices. By Theorem 6, A is invertible and $A^{-1} = (UDV^T)^{-1} = (V^T)^{-1}D^{-1}U^{-1} = VD^{-1}U^T$.

Answer to Checkpoint: If A is an invertible $n \times n$ matrix, with an LU factorization $A = LU$, and if B is $n \times p$, then $A^{-1}B$ can be computed by first row reducing $[L \quad B]$ to a matrix $[I \quad Y]$ for some Y and then reducing $[U \quad Y]$

to $[I \quad A^{-1}B]$. One way to see that this algorithm works is to view $A^{-1}B$ as $[A^{-1}\mathbf{b}_1 \ \cdots \ A^{-1}\mathbf{b}_p]$ and use the LU algorithm to solve simultaneously the set of equations $A\mathbf{x} = \mathbf{b}_1, \ldots, A\mathbf{x} = \mathbf{b}_p$. MATLAB uses this approach to compute $A^{-1}B$ (after first finding L and U).

Appendix: Permuted LU Factorizations

Any $m \times n$ matrix A admits a factorization $A = LU$, with U in echelon form and L a *permuted unit lower triangular* matrix. That is, L is a matrix such that a permutation (rearrangement) of its rows (using row interchanges) will produce a lower triangular matrix with 1's on the diagonal.

The construction of L and U, illustrated below, depends on first using row replacements to reduce A to a *permuted echelon form* V and then using row interchanges to reduce V to an echelon form U. By watching the reduction of A to V, we can easily construct a permuted unit lower triangular matrix L with the property that the sequence of operations changing A into U also changes L into I. This property will guarantee that $A = LU$. (See the paragraph before Example 2 in the text.)

The following algorithm reduces any matrix to a permuted echelon form. In the algorithm when a row is covered, we ignore it in later calculations.

1. *Begin with the leftmost nonzero column. Choose any nonzero entry as the pivot. Designate the corresponding row as a pivot row.*
2. *Use row replacements to create zeros above and below the pivot (in all uncovered rows. Then cover that pivot row.*
3. *Repeat steps 1 and 2 on the uncovered submatrix, if any, until all nonzero entries are covered.*

This algorithm forces each pivot to be to the right of the preceding pivots; when the rows are rearranged with the pivots in stair-step fashion, all entries below each pivot will be zero. Thus, the algorithm produces a permuted echelon matrix. Whenever a pivot is selected, the column containing the pivot will be used to construct a column of L, as we shall see.

As an example, choose any entry in the first column of the following matrix as the first pivot, and use the pivot to create zeros in the rest of column 1. We choose the (3,1)-entry.

call this column **a** (first column)

$$A = \begin{bmatrix} 1 & -1 & 5 & -8 & -7 \\ -2 & -1 & -4 & 9 & 1 \\ \textcircled{4} & 8 & -4 & 0 & -8 \\ 2 & 3 & 0 & -5 & 3 \end{bmatrix} \sim \begin{bmatrix} 0 & -3 & 6 & -8 & -5 \\ 0 & 3 & -6 & 9 & -3 \\ 4 & 8 & -4 & 0 & -8 \\ 0 & -1 & 2 & -5 & 7 \end{bmatrix} \begin{matrix} \\ \\ \leftarrow \text{1st pivot row} \\ \\ \end{matrix}$$

Row 3 is the 1st pivot row. Choose the (2,2)-entry as the second pivot, and create zeros in the rest of column 2, excluding the first pivot row.

(call this column b)

$$= \begin{bmatrix} 0 & -3 & 6 & -8 & -5 \\ 0 & \textcircled{3} & -6 & 9 & -3 \\ 4 & 8 & -4 & 0 & -8 \\ 0 & -1 & 2 & -5 & 7 \end{bmatrix} \sim \begin{bmatrix} 0 & 0 & 0 & 1 & -8 \\ 0 & 3 & -6 & 9 & -3 \\ 4 & 8 & -4 & 0 & -8 \\ 0 & 0 & 0 & -2 & 6 \end{bmatrix} \begin{matrix} \\ \leftarrow \text{2nd pivot row} \\ \leftarrow \text{1st pivot row} \\ \\ \end{matrix}$$

Cover row 2, choose the (4,4)-entry as the pivot. (The row index of the pivot is relative to the original matrix.) Create zeros in the other rows (in the pivot column), excluding the first two pivot rows.

(call this column c; column d)

$$= \begin{bmatrix} 0 & 0 & 0 & 1 & -8 \\ 0 & 3 & -6 & 9 & -3 \\ 4 & 8 & -4 & 0 & -8 \\ 0 & 0 & 0 & \textcircled{-2} & 6 \end{bmatrix} \sim \begin{bmatrix} 0 & 0 & 0 & 0 & \textcircled{-5} \\ 0 & \textcircled{3} & -6 & 9 & -3 \\ \textcircled{4} & 8 & -4 & 0 & -8 \\ 0 & 0 & 0 & \textcircled{-2} & 6 \end{bmatrix} \begin{matrix} \leftarrow \text{4th pivot row} \\ \leftarrow \text{2nd pivot row} \\ \leftarrow \text{1st pivot row} \\ \leftarrow \text{3rd pivot row} \end{matrix}$$

Let V denote this permuted echelon form, and permute the rows of V to create an echelon form. The first pivot row goes to the top, the second pivot row goes next, and so on. The resulting echelon matrix U is

$$\begin{bmatrix} 4 & 8 & -4 & 0 & -8 \\ 0 & 3 & -6 & 9 & -3 \\ 0 & 0 & 0 & -2 & 6 \\ 0 & 0 & 0 & 0 & -5 \end{bmatrix} = U$$

The last step is to create L. Go back and watch the reduction of A to V. As each pivot is selected, take the pivot column, and divide the pivot into each entry in the column that is not yet in a pivot row. Place the resulting column into L. At the end, fill the holes in L with zeros.

Column: a b c d

$$\begin{bmatrix} 1 \\ -2 \\ \textcircled{4} \\ 2 \end{bmatrix} \quad \begin{bmatrix} -3 \\ \textcircled{3} \\ \\ -1 \end{bmatrix} \quad \begin{bmatrix} 1 \\ \\ \\ \textcircled{-2} \end{bmatrix} \quad \begin{bmatrix} \textcircled{-5} \\ \\ \\ \\ \end{bmatrix}$$

$$\div 4 \downarrow \quad \div 3 \downarrow \quad \div -2 \downarrow \quad \div -5 \downarrow$$

$$\begin{bmatrix} 1/4 & -1 & -1/2 & 1 \\ -1/2 & 1 & & \\ 1 & & & \\ 1/2 & -1/3 & 1 & \end{bmatrix}, \qquad L = \begin{bmatrix} 1/4 & -1 & -1/2 & 1 \\ -1/2 & 1 & 0 & 0 \\ 1 & 0 & 0 & 0 \\ 1/2 & -1/3 & 1 & 0 \end{bmatrix}$$

You can check that $LU = A$. To see why this is so, observe that L is constructed so the operations that reduce A to V also reduce L to a permuted identity matrix. Since the pivots in L are in exactly the same *rows* as in V, the sequence of row interchanges that reduces V to U also reduces the permuted identity matrix to I. Thus, the full sequence of operations that reduces A to U also reduces L to I, so that $A = LU$. (See the box on page 135 of the text.)

The next example illustrates what to do when V has one or more rows of zeros. The matrix is from the Practice Problem for Section 2.5. For the reduction of A to V, pivots were chosen to have the largest possible magnitude (the choice used for "partial pivoting"). Of course, other pivots could have been selected.

$$A = \begin{bmatrix} 2 & -4 & -2 & 3 \\ \textcircled{6} & -9 & -5 & 8 \\ 2 & -7 & -3 & 9 \\ 4 & -2 & -2 & -1 \\ -6 & 3 & 3 & 4 \end{bmatrix} \sim \begin{bmatrix} 0 & -1 & -1/3 & 1/3 \\ 6 & -9 & -5 & 8 \\ 0 & -4 & -4/3 & 19/3 \\ 0 & 4 & 4/3 & -19/3 \\ 0 & \textcircled{-6} & -2 & 12 \end{bmatrix} \sim \begin{bmatrix} 0 & 0 & 0 & -5/3 \\ 6 & -9 & -5 & 8 \\ 0 & 0 & 0 & -5/3 \\ 0 & 0 & 0 & \textcircled{5/3} \\ 0 & -6 & -2 & 12 \end{bmatrix}$$

$$\sim V = \begin{bmatrix} 0 & 0 & 0 & 0 \\ \textcircled{6} & -9 & -5 & 8 \\ 0 & 0 & 0 & 0 \\ 0 & 0 & 0 & \textcircled{5/3} \\ 0 & \textcircled{-6} & -2 & 12 \end{bmatrix} \begin{matrix} \\ \leftarrow \text{1st pivot row} \\ \\ \leftarrow \text{3rd pivot row} \\ \leftarrow \text{2nd pivot row} \end{matrix} \qquad \sim U = \begin{bmatrix} 6 & -9 & -5 & 8 \\ 0 & -6 & -2 & 12 \\ 0 & 0 & 0 & 5/3 \\ 0 & 0 & 0 & 0 \\ 0 & 0 & 0 & 0 \end{bmatrix}$$

The first three columns of L come from the three pivot columns above.

$$\begin{bmatrix} 2 \\ \textcircled{6} \\ 2 \\ 4 \\ -6 \end{bmatrix} \quad \begin{bmatrix} -1 \\ \\ -4 \\ 4 \\ \textcircled{-6} \end{bmatrix} \quad \begin{bmatrix} -5/3 \\ \\ -5/3 \\ \textcircled{5/3} \\ \\ \end{bmatrix}$$

$$\begin{matrix} \div 6 & \div -6 & \div 5/3 \\ \downarrow & \downarrow & \downarrow \end{matrix}$$

$$\begin{bmatrix} 1/3 & 1/6 & -1 & & \\ 1 & & & & \\ 1/3 & 2/3 & -1 & & \\ 2/3 & -2/3 & 1 & & \\ -1 & 1 & & & \end{bmatrix} \begin{matrix} \\ \leftarrow \text{1st pivot row} \\ \\ \leftarrow \text{3rd pivot row} \\ \leftarrow \text{2nd pivot row} \end{matrix}$$

The matrix L needs two more columns. Use columns 1 and 3 of the 5×5 identity matrix to place 1's in the "nonpivot" rows 1 and 3. Fill in the remaining holes with zeros.

$$\begin{bmatrix} 1/3 & 1/6 & -1 & 1 & 0 \\ \textcircled{1} & 0 & 0 & 0 & 0 \\ 1/3 & 2/3 & -1 & 0 & 1 \\ 2/3 & -2/3 & \textcircled{1} & 0 & 0 \\ -1 & \textcircled{1} & 0 & 0 & 0 \end{bmatrix} = L \sim \begin{bmatrix} 1 & 0 & 0 & 0 & 0 \\ -1 & 1 & 0 & 0 & 0 \\ 2/3 & -2/3 & 1 & 0 & 0 \\ 1/3 & 1/6 & -1 & 1 & 0 \\ 1/3 & 2/3 & -1 & 0 & 1 \end{bmatrix}$$

Row reduction of L using only row replacements produces a permuted identity matrix. Moving the 1's in the "pivot rows" 2, 5, and 4 into rows 1, 2, and 3 of the identity requires the same row swaps as reducing V to U. If a further row interchange on the permuted identity is required, it will involve the bottom two rows, which came from the "nonpivot" rows 1 and 3. A corresponding interchange of the bottom two rows of U has no effect on U (and the product LU is unaffected). As a result, L is reduced to I by the same operations that reduce A to V and then to U. Check that $A = LU$.

**MATLAB LU Factorization and the Backslash Operator **

Row reduction of A using the command **gauss** will produce the intermediate matrices needed for an LU factorization of A. You can try this on the matrix in Example 2, stored as Exercise 33 in the Toolbox. The matrices in (5) on page 136 in the text are produced by the commands

U = gauss(A,1)	U has 0's below the first pivot
U = gauss(U,2)	Now U has 0's below pivots 1 and 2
U = gauss(U,3)	The echelon form

You can copy the information from the screen onto your paper, and divide by the pivot entries to produce L as in the text. For most text exercises, the pivots are integers and so are displayed accurately.

To construct a permuted LU factorization, use **U = gauss(U,r,v)** , where r is the row index of the pivot and **v** is a row vector that lists the rows to be changed by replacement operations. For example, if A has 5 rows and the first pivot is in row 4, use **U = gauss(A,4,[1 2 3 5])** . If the next pivot is in row 2, use **U = gauss(U,2,[1 3 5])** . To build the permuted matrix L, use full columns from A or the partially reduced U, divided by the pivots. Then change entries to zero if they are in a row already selected as a "pivot row."

The MATLAB command **[L U] = lu(A)** produces a permuted LU factorization for any square matrix A, but it does not handle the general case.

When A is invertible, the best way to solve $A\mathbf{x} = \mathbf{b}$ with MATLAB is to use the backslash command **x = A\b** . MATLAB proceeds to compute a permuted LU factorization of A and then use L and U to compute $\mathbf{x}$. The alternative command **x = inv(A)*b** is less efficient and can be less accurate. The command **inv(A)** uses the LU factorization to compute A^{-1} in the form $U^{-1}L^{-1}$.

2.6 ITERATIVE SOLUTIONS OF LINEAR SYSTEMS

Some knowledge of iterative methods for solving linear systems is important for anyone who uses linear algebra. If your course does not have time to cover this material, you might at least glance at the discussion. Then, if you need to solve a large sparse system later in your career, you may remember (perhaps vaguely) the ideas here, and you can review this section in your text before setting out to use appropriate computer software.

KEY IDEAS

Both iterative algorithms described here are based on writing $A\mathbf{x} = \mathbf{b}$ in the form $(M - N)\mathbf{x} = \mathbf{b}$, where M is readily invertible, and then writing $M\mathbf{x} = N\mathbf{x} + \mathbf{b}$. Given an initial estimate $\mathbf{x}^{(0)}$ for the solution, the recursion

$$M\mathbf{x}^{(k+1)} = N\mathbf{x}^{(k)} + \mathbf{b} \quad (k = 0, 1, \ldots) \qquad (*)$$

defines a sequence of vectors that may converge to the desired solution.

For Jacobi's method, M is the diagonal matrix formed from the diagonal entries in A; for the Gauss-Seidel method, M is the lower triangular part of A. In both cases, of course, $N = M - A$.

SOLUTIONS TO EXERCISES

1. $A = \begin{bmatrix} 4 & 1 \\ -1 & 5 \end{bmatrix}$, $\mathbf{b} = \begin{bmatrix} 7 \\ -7 \end{bmatrix}$, $M = \begin{bmatrix} 4 & 0 \\ 0 & 5 \end{bmatrix}$, $N = \begin{bmatrix} 4 & 0 \\ 0 & 5 \end{bmatrix} - \begin{bmatrix} 4 & 1 \\ -1 & 5 \end{bmatrix} = \begin{bmatrix} 0 & -1 \\ 1 & 0 \end{bmatrix}$.

   If working by hand, the recursion (*) above can be simplified to

$$\mathbf{x}^{(k+1)} = M^{-1}(N\mathbf{x}^{(k)} + \mathbf{b}) = \begin{bmatrix} .25 & 0 \\ 0 & .20 \end{bmatrix}\left(\begin{bmatrix} 0 & -1 \\ 1 & 0 \end{bmatrix}\mathbf{x}^{(k)} + \begin{bmatrix} 7 \\ -7 \end{bmatrix}\right)$$

$$= \begin{bmatrix} 0 & -.25 \\ .2 & 0 \end{bmatrix}\mathbf{x}^{(k)} + \begin{bmatrix} 1.75 \\ -1.40 \end{bmatrix}$$

   Then, with $\mathbf{x}^{(0)} = \mathbf{0}$,

$$\mathbf{x}^{(1)} = \begin{bmatrix} 0 & -.25 \\ .2 & 0 \end{bmatrix}\mathbf{0} + \begin{bmatrix} 1.75 \\ -1.40 \end{bmatrix} = \begin{bmatrix} 1.75 \\ -1.40 \end{bmatrix}$$

$$\mathbf{x}^{(2)} = \begin{bmatrix} 0 & -.25 \\ .2 & 0 \end{bmatrix}\begin{bmatrix} 1.75 \\ -1.40 \end{bmatrix} + \begin{bmatrix} 1.75 \\ -1.40 \end{bmatrix} = \begin{bmatrix} .35 \\ .35 \end{bmatrix} + \begin{bmatrix} 1.75 \\ -1.40 \end{bmatrix} = \begin{bmatrix} 2.10 \\ -1.05 \end{bmatrix}$$

$$\mathbf{x}^{(3)} = \begin{bmatrix} 0 & -.25 \\ .2 & 0 \end{bmatrix}\begin{bmatrix} 2.10 \\ -1.05 \end{bmatrix} + \begin{bmatrix} 1.75 \\ -1.40 \end{bmatrix} = \begin{bmatrix} .26 \\ .42 \end{bmatrix} + \begin{bmatrix} 1.75 \\ -1.40 \end{bmatrix} = \begin{bmatrix} 2.01 \\ -.98 \end{bmatrix}$$

Using MATLAB,

$$\mathbf{x}^{(4)} = \begin{bmatrix} 1.9950 \\ -.9975 \end{bmatrix}, \quad \mathbf{x}^{(5)} = \begin{bmatrix} 1.9994 \\ -1.0010 \end{bmatrix}, \quad \mathbf{x}^{(6)} = \begin{bmatrix} 2.0003 \\ -1.0001 \end{bmatrix}$$

To four decimal places,

$$\mathbf{x}^{(5)} - \mathbf{x}^{(4)} = \begin{bmatrix} .0044 \\ -.0035 \end{bmatrix}, \quad \mathbf{x}^{(6)} - \mathbf{x}^{(5)} = \begin{bmatrix} .0009 \\ .0009 \end{bmatrix}$$

Thus $\mathbf{x}^{(6)}$ is the first vector in the sequence $\{\mathbf{x}^{(k)}\}$ whose entries are within .001 of the entries in its predecessor.

7. $A = \begin{bmatrix} 3 & 1 & 0 \\ -1 & -5 & 2 \\ 0 & 3 & 7 \end{bmatrix}$, $\mathbf{b} = \begin{bmatrix} 11 \\ 15 \\ 17 \end{bmatrix}$. For Gauss-Seidel, $M = \begin{bmatrix} 3 & 0 & 0 \\ -1 & -5 & 0 \\ 0 & 3 & 7 \end{bmatrix}$, and

$$N = \begin{bmatrix} 3 & 0 & 0 \\ -1 & -5 & 0 \\ 0 & 3 & 7 \end{bmatrix} - \begin{bmatrix} 3 & 1 & 0 \\ -1 & -5 & 2 \\ 0 & 3 & 7 \end{bmatrix} = \begin{bmatrix} 0 & -1 & 0 \\ 0 & 0 & -2 \\ 0 & 0 & 0 \end{bmatrix}$$

If working by hand, write the recursion (*) as $\mathbf{x}^{(k+1)} = M^{-1}N\mathbf{x}^{(k)} + M^{-1}\mathbf{b}$. This will avoid solving a 3×3 system for each iteration. The fastest way to compute $M^{-1}N$ and $M^{-1}\mathbf{b}$ is to reduce the augmented matrix $[M \;\; N \;\; \mathbf{b}]$ to a matrix of the form $[I \;\; M^{-1}N \;\; M^{-1}\mathbf{b}]$. See Exercise 11 in Section 2.2, with the matrix B replaced by $[N \;\; \mathbf{b}]$.

$$\left[\begin{array}{ccc|ccc|c} 3 & 0 & 0 & 0 & -1 & 0 & 11 \\ -1 & -5 & 0 & 0 & 0 & -2 & 15 \\ 0 & 3 & 7 & 0 & 0 & 0 & 17 \end{array}\right] \sim \begin{bmatrix} 1 & 0 & 0 & 0 & -1/3 & 0 & 11/3 \\ -1 & -5 & 0 & 0 & 0 & -2 & 15 \\ 0 & 3 & 7 & 0 & 0 & 0 & 17 \end{bmatrix}$$

$$\sim \begin{bmatrix} 1 & 0 & 0 & 0 & -1/3 & 0 & 11/3 \\ 0 & -5 & 0 & 0 & -1/3 & -2 & 56/3 \\ 0 & 3 & 7 & 0 & 0 & 0 & 17 \end{bmatrix} \sim \cdots$$

$$\sim \begin{bmatrix} 1 & 0 & 0 & 0 & -1/3 & 0 & 11/3 \\ 0 & 1 & 0 & 0 & 1/15 & 2/5 & -56/15 \\ 0 & 0 & 1 & 0 & -1/35 & -6/35 & 141/35 \end{bmatrix}$$

$$M^{-1}N = \begin{bmatrix} 0 & -1/3 & 0 \\ 0 & 1/15 & 2/5 \\ 0 & -1/35 & -6/35 \end{bmatrix} = \frac{1}{105}\begin{bmatrix} 0 & -35 & 0 \\ 0 & 7 & 42 \\ 0 & -3 & -18 \end{bmatrix} \approx \begin{bmatrix} 0 & -.33333 & 0 \\ 0 & .06667 & .40000 \\ 0 & -.02857 & -.17143 \end{bmatrix}$$

$$M^{-1}\mathbf{b} = \begin{bmatrix} 11/3 \\ -56/15 \\ 141/35 \end{bmatrix} = \frac{1}{105}\begin{bmatrix} 385 \\ -392 \\ 423 \end{bmatrix} \approx \begin{bmatrix} 3.66667 \\ -3.73333 \\ 4.02857 \end{bmatrix}$$

The first two iterations, by hand (with $\mathbf{x}^{(0)} = \mathbf{0}$), are

$$\mathbf{x}^{(1)} = M^{-1}N\mathbf{0} + M^{-1}\mathbf{b} = \begin{bmatrix} 0 \\ 0 \\ 0 \end{bmatrix} + \frac{1}{105}\begin{bmatrix} 385 \\ -392 \\ 423 \end{bmatrix} = \frac{1}{105}\begin{bmatrix} 385 \\ -392 \\ 423 \end{bmatrix}$$

$$\mathbf{x}^{(2)} = M^{-1}N(M^{-1}\mathbf{b}) + M^{-1}\mathbf{b} = (M^{-1}N + I)M^{-1}\mathbf{b}$$

Using matrix algebra to simplify the calculations

$$= \frac{1}{105}\begin{bmatrix} 105 & -35 & 0 \\ 0 & 112 & 42 \\ 0 & -3 & 87 \end{bmatrix}\left(\frac{1}{105}\right)\begin{bmatrix} 385 \\ -392 \\ 423 \end{bmatrix} = \frac{1}{105^2}\begin{bmatrix} 54145 \\ -26138 \\ 37977 \end{bmatrix} \approx \begin{bmatrix} 4.9111 \\ -2.3708 \\ 3.4446 \end{bmatrix}$$

Your answers may vary slightly if you used decimal approximations throughout your work. Next,

k	2	3	4	5	6
$\mathbf{x}^{(k)}$	$\begin{bmatrix} 4.9111 \\ -2.3708 \\ 3.4446 \end{bmatrix}$	$\begin{bmatrix} 4.4569 \\ -2.5135 \\ 3.5058 \end{bmatrix}$	$\begin{bmatrix} 4.5045 \\ -2.4986 \\ 3.4994 \end{bmatrix}$	$\begin{bmatrix} 4.49953 \\ -2.50015 \\ 3.50006 \end{bmatrix}$	$\begin{bmatrix} 4.50005 \\ -2.49998 \\ 3.49999 \end{bmatrix}$

To four decimal places,

$$\mathbf{x}^{(5)} - \mathbf{x}^{(4)} = \begin{bmatrix} -.0050 \\ -.0016 \\ .0007 \end{bmatrix}, \quad \mathbf{x}^{(6)} - \mathbf{x}^{(5)} = \begin{bmatrix} .0005 \\ .0002 \\ -.0001 \end{bmatrix}$$

Thus $\mathbf{x}^{(6)}$ is the first vector in the sequence $\{\mathbf{x}^{(k)}\}$ whose entries are within .001 of the entries in its predecessor.

13. $A = \begin{bmatrix} 4 & -2 \\ -5 & 4 \end{bmatrix}$, $\mathbf{b} = \begin{bmatrix} 10 \\ -2 \end{bmatrix}$, $M = \begin{bmatrix} 4 & 0 \\ -5 & 4 \end{bmatrix}$, $N = \begin{bmatrix} 0 & 2 \\ 0 & 0 \end{bmatrix}$

Some matrix program is essential here. To help you check your work, if necessary, the results of the first few computations are shown below. (See the text for the final answer.)

k	0	1	2	3	4	5	6	7	8
$\mathbf{x}^{(k)}$	$\begin{bmatrix} 0 \\ 0 \end{bmatrix}$	$\begin{bmatrix} 2.500 \\ 2.625 \end{bmatrix}$	$\begin{bmatrix} 3.8 \\ 4.3 \end{bmatrix}$	$\begin{bmatrix} 4.6 \\ 5.3 \end{bmatrix}$	$\begin{bmatrix} 5.1 \\ 5.9 \end{bmatrix}$	$\begin{bmatrix} 5.5 \\ 6.3 \end{bmatrix}$	$\begin{bmatrix} 5.7 \\ 6.6 \end{bmatrix}$	$\begin{bmatrix} 5.8 \\ 6.7 \end{bmatrix}$	$\begin{bmatrix} 5.9 \\ 6.8 \end{bmatrix}$

17. c. Suppose that $\mathbf{x}^{(0)} = \mathbf{0}$ and $\mathbf{x}^{(k)}$ satisfies $\mathbf{x}^{(k+1)} = C\mathbf{x}^{(k)} + \mathbf{d}$, for $k = 1,2,\ldots.$ To prove by induction that

$$\mathbf{x}^{(k)} = \mathbf{d} + C\mathbf{d} + \cdots + C^{k-1}\mathbf{d} \qquad (k = 1,2,\ldots) \qquad (*)$$

observe that (*) holds for $k = 1$, because by the recursion relation, $\mathbf{x}^{(1)} = C\mathbf{x}^{(0)} + \mathbf{d} = C\mathbf{0} + \mathbf{d} = \mathbf{d}$. Also, (*) holds for $k = 2$, because $\mathbf{x}^{(2)} = C\mathbf{x}^{(1)} + \mathbf{d} = \mathbf{d} + C\mathbf{d}$. Now suppose that (*) holds for some specific $k \geq 1$. Substitute the formula (*) for $\mathbf{x}^{(k)}$ into the recursion relation:

$$\mathbf{x}^{(k+1)} = C\mathbf{x}^{(k)} + \mathbf{d} = C(\mathbf{d} + C\mathbf{d} + \cdots + C^{k-1}\mathbf{d}) + \mathbf{d}$$

$$= C\mathbf{d} + C^2\mathbf{d} + \cdots + C^k\mathbf{d} + \mathbf{d}$$

Thus the truth of (*) for k implies the truth of (*) for $k + 1$. By the principle of induction, (*) is true for all $k \geq 1$.

MATLAB Jacobi and Gauss-Seidel Methods

The command **M = diag(diag(A))** creates a diagonal matrix from the diagonal entries of a matrix A. (The command **diag(A)** produces a column vector that lists the diagonal entries of A.) To create the lower triangular part of A for the Gauss-Seidel Method, use **M = tril(A)**.

The Jacobi and Gauss-Seidel methods use the same MATLAB commands, except for the construction of M. Enter **N = M - A** and make sure the initial data is in the vector **x**. To produce the "next" **x**, you must solve the equation $M\mathbf{x}^{(1)} = N\mathbf{x} + \mathbf{b}$ for $\mathbf{x}^{(1)}$. The correct way to do this is to use MATLAB's "backslash" operation: $\mathbf{x}^{(1)} = M\backslash(N*\mathbf{x} + \mathbf{b})$. However, instead of assigning the solution to $\mathbf{x}^{(1)}$, you should put the solution back into **x**, so you can repeat the operation *with exactly the same symbols*. That is, the command

```
x = M\(N*x + b); x'      To save space, x is displayed horizontally
```

is used over and over, creating the sequence $\mathbf{x}^{(k)}$. (The semicolon suppresses the display of the column vector **x**.) These vectors can be recorded on paper as they appear. You only type the command once. After that, use the up-arrow to recall the command, and press <Enter>.

If you wish to monitor how close the approximations are to each other, use the following sequence of commands (repeatedly):

```
t = M\(N*x + b); t'      t is a temporary storage for the new x
change = t' - x'         Compare t with the old x
x = t;                   Update x with new values; don't display them
```

2.7 THE LEONTIEF INPUT-OUTPUT MODEL

If you are in economics, you definitely will need the material in this section for later work. Although most of the discussion concerns economics, the formula for the inverse of $I - C$ is used in a variety of applications.

STUDY NOTES

The power of Leontief's model of the economy is that it compresses hundreds of equations in hundreds of variables into the simple matrix equation $(I - C)\mathbf{x} = \mathbf{d}$. You should know how to construct the consumption matrix C and know the algebra that leads from the matrix equation $\mathbf{x} = C\mathbf{x} + \mathbf{d}$ to its solution $\mathbf{x} = (I - C)^{-1}\mathbf{d}$, under the assumption that the column sums of C are less than one.

You may need to know the formula (8) for $(I - C)^{-1}$ on page 152. (Check with your instructor.) The formula is analogous to the formula for the sum of a geometric series of positive numbers:

$$1 + r + r^2 + r^3 + \cdots = (1 - r)^{-1} \quad \text{when } |r| < 1.$$

SOLUTIONS TO EXERCISES

1. Fill in C one column at a time, since each column is a unit consumption vector for one sector. Make sure that the order of the sectors is the same for the rows and columns of C. From the way the data are presented, we use the order: manufacturing, agriculture, and services. Read the sentences carefully, to get the data arranged correctly.

Purchased from:	Unit consumption vectors Manuf.	Agric.	Serv.
Manufacturing	.10	.60	.60
Agriculture	.30	.20	.00
Services	.30	.10	.10

The intermediate demands created by a production vector $\mathbf{x}$ are given by $C\mathbf{x}$. If agriculture plans to produce 100 units (and the other sectors plan to produce nothing), then the intermediate demand is

$$C\mathbf{x} = \begin{bmatrix} .10 & .60 & .60 \\ .30 & .20 & .00 \\ .30 & .10 & .10 \end{bmatrix} \begin{bmatrix} 0 \\ 100 \\ 0 \end{bmatrix} = \begin{bmatrix} 60 \\ 20 \\ 10 \end{bmatrix}$$

7. $C = \begin{bmatrix} .0 & .5 \\ .6 & .2 \end{bmatrix}$, $\mathbf{d} = \begin{bmatrix} 50 \\ 30 \end{bmatrix}$. Let $\mathbf{d}_1 = \begin{bmatrix} 1 \\ 0 \end{bmatrix}$, the demand for 1 unit of output of sector 1.

a. The production required to satisfy the demand $\mathbf{d}_1$ is the vector $\mathbf{x}_1$ such that $(I - C)\mathbf{x}_1 = \mathbf{d}_1$, namely, $\mathbf{x}_1 = (I - C)^{-1}\mathbf{d}_1$. From Exercise 5,

$$I - C = \begin{bmatrix} 1 & -.5 \\ -.6 & .8 \end{bmatrix} \quad \text{and} \quad (I - C)^{-1} = \begin{bmatrix} 1.6 & 1 \\ 1.2 & 2 \end{bmatrix}$$

so

$$\mathbf{x}_1 = \begin{bmatrix} 1.6 & 1 \\ 1.2 & 2 \end{bmatrix}\begin{bmatrix} 1 \\ 0 \end{bmatrix} = \begin{bmatrix} 1.6 \\ 1.2 \end{bmatrix}$$

b. For the final demand $\mathbf{d}_2 = \begin{bmatrix} 51 \\ 30 \end{bmatrix}$, the corresponding production $\mathbf{x}_2$ is given by

$$\mathbf{x}_2 = (I - C)^{-1}\mathbf{d}_2 = \begin{bmatrix} 1.6 & 1 \\ 1.2 & 2 \end{bmatrix}\begin{bmatrix} 51 \\ 30 \end{bmatrix} = \begin{bmatrix} 111.6 \\ 121.2 \end{bmatrix}$$

c. From Exercise 5, the production $\mathbf{x}$ corresponding to the demand $\mathbf{d}$ is given by $\mathbf{x} = \begin{bmatrix} 110 \\ 120 \end{bmatrix}$. Observe from (a) and (b) that $\mathbf{x}_2 = \mathbf{x} + \mathbf{x}_1$. Also, as pointed out in the text, $\mathbf{d}_2 = \mathbf{d} + \mathbf{d}_1$. The sum of the production vectors $\mathbf{x}$ and $\mathbf{x}_1$ gives the production needed to satisfy the sum of the demands $\mathbf{d}$ and $\mathbf{d}_1$. This is expressing the *linearity* between final demand and production. This relation is true in general, because

$$\begin{aligned} \mathbf{x}_2 &= (I - C)^{-1}\mathbf{d}_2 = (I - C)^{-1}(\mathbf{d} + \mathbf{d}_1) \\ &= (I - C)^{-1}\mathbf{d} + (I - C)^{-1}\mathbf{d}_1 \\ &= \mathbf{x} + \mathbf{x}_1 \end{aligned}$$

11. Following the hint in the text, you should obtain $\mathbf{p}^T\mathbf{x} = \mathbf{p}^TC\mathbf{x} + \mathbf{v}^T\mathbf{x}$ (from the price equation). Then, from the production equation, $\mathbf{p}^T\mathbf{x} = \mathbf{p}^T(C\mathbf{x} + \mathbf{d}) = \mathbf{p}^TC\mathbf{x} + \mathbf{p}^T\mathbf{d}$. Equate the two expressions for $\mathbf{p}^T\mathbf{x}$ to yield $\mathbf{p}^T\mathbf{d} = \mathbf{v}^T\mathbf{x}$.

Another solution: Take transposes in the price equation,

$$\mathbf{p}^T = (C^T\mathbf{p})^T + \mathbf{v}^T = \mathbf{p}^TC + \mathbf{v}^T, \quad \text{so} \quad \mathbf{v}^T = \mathbf{p}^T - \mathbf{p}^TC$$

and right-multiply by $\mathbf{x}$ to obtain

$$\mathbf{v}^T\mathbf{x} = \mathbf{p}^T\mathbf{x} - \mathbf{p}^TC\mathbf{x} = \mathbf{p}^T(I - C)\mathbf{x} = \mathbf{p}^T\mathbf{d} \qquad \text{From the production equation}$$

Warning: In Exercise 9, don't multiply the consumption matrix C by 10, to get rid of the decimals. That changes the equation $C\mathbf{x} = \mathbf{x} + \mathbf{d}$ into $10C\mathbf{x} = \mathbf{x} + \mathbf{d}$, whose solution is different. However, you *may* multiply the *augmented matrix* $[(I - C) \quad \mathbf{0}]$ by 10, because the solution of an equation is not affected when both sides are multiplied by a nonzero number.

13. The data for this exercise are in the Toolbox. To solve the equation $(I - C)x = \mathbf{d}$, row reduce the augmented matrix $[(I - C) \quad \mathbf{d}]$ rather than compute $(I - C)^{-1}$. (Another reasonable solution method is suggested in Exercise 15.) The numerical solution is given in the text.

2.8 APPLICATIONS TO COMPUTER GRAPHICS

According to my students over the past few years, this section is one of the most interesting application sections in the text, because it shows how matrix calculations, performed millions of times per second, can create the illusion of 3D-motion on a computer screen or in a movie theater. Of course, one short section cannot begin to indicate the vast scope of computer graphics. I encourage you to look at the book by Foley, et al., referenced in your text. Chapters 5, 6, and 11 are filled with matrices! The rest of the 1100 pages in the book contains lots of interesting mathematics, detailed discussions of computer algorithms, and scores of spectacular (in some cases, almost unbelievable) color plates.

KEY IDEAS

A graphical object, represented by a number of key points, can be stored in a data matrix D, one column for each key point. When a linear transformation, determined by a matrix A, acts on the object, the transformed object is determined by the data matrix AD, because A acts separately on each column of D.

Homogeneous coordinates are needed to make translation act as a linear transformation and to provide perspective projections that give the illusion of three dimensions. The matrix for these transformations has the form

$$\begin{bmatrix} A & \mathbf{p} \\ \mathbf{q}^T & r \end{bmatrix}$$

For 2D-graphics, A is a 2×2 matrix, $\mathbf{q}^T = [0 \quad 0]$, and $\mathbf{p}$ is a translation vector. For 3D-graphics, A is 3×3, $\mathbf{q}$ is associated with a perspective transformation (if any), and **p** is a translation vector. For simplicity, the only perspective transformations considered have their center of projection at $(0,0,d)$, and $\mathbf{q}^T = [0 \quad 0 \quad -1/d]$. In fact, $d = 10$ in Example 8 and Exercises 19 and 20.

SOLUTIONS TO EXERCISES

1. From Example 5, the matrix $\begin{bmatrix} 1 & .25 & 0 \\ 0 & 1 & 0 \\ 0 & 0 & 1 \end{bmatrix}$ has the same effect on homogeneous coordinates for $\mathbb{R}^2$ that the matrix $\begin{bmatrix} 1 & .25 \\ 0 & 1 \end{bmatrix}$ of Example 2 has on ordinary vectors in $\mathbb{R}^2$. Partitioned matrix notation explains why this is true. Let A be a 2×2 matrix. The following diagram shows that the action of $\begin{bmatrix} A & \mathbf{0} \\ \mathbf{0}^T & 1 \end{bmatrix}$ on $\begin{bmatrix} \mathbf{x} \\ 1 \end{bmatrix}$ corresponds to the action of A on $\mathbf{x}$.

$$\mathbf{x} \longmapsto A\mathbf{x} \qquad \text{Coordinates in } \mathbb{R}^2$$

$$\updownarrow \qquad\qquad\qquad \updownarrow$$

$$\begin{bmatrix} \mathbf{x} \\ 1 \end{bmatrix} \longmapsto \begin{bmatrix} A & \mathbf{0} \\ \mathbf{0}^T & 1 \end{bmatrix}\begin{bmatrix} \mathbf{x} \\ 1 \end{bmatrix} = \begin{bmatrix} A\mathbf{x} + \mathbf{0}\cdot 1 \\ \mathbf{0}^T\mathbf{x} + 1\cdot 1 \end{bmatrix} = \begin{bmatrix} A\mathbf{x} \\ 1 \end{bmatrix} \qquad \text{Homogeneous coordinates}$$

7. A 60% rotation about the origin in $\mathbb{R}^2$ is given by $\begin{bmatrix} \cos 60^\circ & -\sin 60^\circ \\ \sin 60^\circ & \cos 60^\circ \end{bmatrix} = \begin{bmatrix} 1/2 & -\sqrt{3}/2 \\ \sqrt{3}/2 & 1/2 \end{bmatrix}$, so the 3×3 matrix for rotation about $\begin{bmatrix} 6 \\ 8 \end{bmatrix}$ is

$$\underbrace{\begin{bmatrix} 1 & 0 & 6 \\ 0 & 1 & 8 \\ 0 & 0 & 1 \end{bmatrix}}_{\substack{\text{Finally,} \\ \text{translate} \\ \text{back}}} \underbrace{\begin{bmatrix} 1/2 & -\sqrt{3}/2 & 0 \\ \sqrt{3}/2 & 1/2 & 0 \\ 0 & 0 & 1 \end{bmatrix}}_{\substack{\text{Then, rotate} \\ \text{about the} \\ \text{origin}}} \underbrace{\begin{bmatrix} 1 & 0 & -6 \\ 0 & 1 & -8 \\ 0 & 0 & 1 \end{bmatrix}}_{\substack{\text{First,} \\ \text{translate} \\ \text{by -p}}}$$

$$= \begin{bmatrix} 1 & 0 & 6 \\ 0 & 1 & 8 \\ 0 & 0 & 1 \end{bmatrix} \begin{bmatrix} 1/2 & -\sqrt{3}/2 & -3 + 4\sqrt{3} \\ \sqrt{3}/2 & 1/2 & -4 - 3\sqrt{3} \\ 0 & 0 & 1 \end{bmatrix} = \begin{bmatrix} 1/2 & -\sqrt{3}/2 & 3 + 4\sqrt{3} \\ \sqrt{3}/2 & 1/2 & 4 - 3\sqrt{3} \\ 0 & 0 & 1 \end{bmatrix}$$

13. The answer is given in the text. Notice that the order of the transformations is important. If the translation is done first (that is, if the matrix for the translation is on the right), then

$$\begin{bmatrix} A & \mathbf{0} \\ \mathbf{0}^T & 1 \end{bmatrix}\begin{bmatrix} I & \mathbf{p} \\ \mathbf{0}^T & 1 \end{bmatrix} = \begin{bmatrix} AI + \mathbf{0}\mathbf{0}^T & A\mathbf{p} + \mathbf{0}\cdot 1 \\ \mathbf{0}^T I + 1\mathbf{0}^T & \mathbf{0}^T\mathbf{p} + 1\cdot 1 \end{bmatrix} = \begin{bmatrix} A & A\mathbf{p} \\ \mathbf{0}^T & 1 \end{bmatrix} \neq \begin{bmatrix} A & \mathbf{p} \\ \mathbf{0}^T & 1 \end{bmatrix}$$

Here, $\mathbf{0}^T$ is a zero row vector, and so the outer product $\mathbf{0}\mathbf{0}^T$ is a zero matrix.

19. The matrix P for the perspective transformation with center of projection at (0,0,10) and the data matrix D using homogeneous coordinates are shown below. The data matrix for the image of the triangle is PD:

$$PD = \begin{bmatrix} 1 & 0 & 0 & 0 \\ 0 & 1 & 0 & 0 \\ 0 & 0 & 0 & 0 \\ 0 & 0 & -.1 & 1 \end{bmatrix}\begin{bmatrix} 4.2 & 6 & 2 \\ 1.2 & 4 & 2 \\ 4 & 2 & 6 \\ 1 & 1 & 1 \end{bmatrix} = \begin{bmatrix} 4.2 & 6 & 2 \\ 1.2 & 4 & 2 \\ 0 & 0 & 0 \\ .6 & .8 & .4 \end{bmatrix}$$

The $\mathbb{R}^3$ coordinates of the image points come from the top three entries in each column, divided by the corresponding entries in the fourth row.

$$\begin{bmatrix} 4.2/.6 & 6/.8 & 2/.4 \\ 1.2/.6 & 4/.8 & 2/.4 \\ 0 & 0 & 0 \end{bmatrix} = \begin{bmatrix} 7 & 7.5 & 5 \\ 2 & 5 & 5 \\ 0 & 0 & 0 \end{bmatrix}$$

2.9 SUBSPACES OF R^N

This section is a condensed version of the ideas from Chapter 4 needed for Chapters 5—7. You need to study this section only if you plan to skip most or all of Chapter 4. If you reviewed carefully after Sections 1.8 and 2.3, you should be well prepared for the new material here.

STUDY NOTES

There are seven fundamental concepts in this section: subspace, column space, null space, basis, coordinate vector, dimension, and rank. (The term *isomorphism* is also used briefly, to explain why a p-dimensional subspace "acts like" $\mathbb{R}^p$.) Actually, there is really not so much to learn, because you have already been working with most of these concepts for several weeks, without the terminology. But you do need to know the precise definitions of these seven terms, and you need to understand much more than just mechanical computations. See "Mastering Linear Algebra Concepts" at the end of this *Study Guide* section for help.

With many terms to learn, you have to be especially careful not to combine ideas in ways that are undefined, even though they may sound correct to you. For example, after you finish your work on this section you should be able to recognize that the following phrases (which have appeared on student papers) are meaningless: "the basis of a matrix," "the dimension of a basis," and "the rank of a basis."

You should pay particular attention to the notion of a *subspace*. The best mental image for a subspace is a plane in $\mathbb{R}^3$ through the origin. The distinguishing feature of such a plane is that the sum of any two vectors in the plane is another vector in the same plane (by the parallelogram rule), and any scalar multiple of any vector in the plane is also in the plane. The main examples of subspaces are column spaces (defined explicitly) and null spaces (defined implicitly).

SOLUTIONS TO EXERCISES

1. The vector $\mathbf{u} = (1,2)$ is in the set H, but $(-1)\cdot\mathbf{u}$ is not in H.

7. a. $\{\mathbf{v}_1, \mathbf{v}_2, \mathbf{v}_3\}$ is a set that contains three vectors.

b. Col A, which is the same as Span$\{\mathbf{v}_1, \mathbf{v}_2, \mathbf{v}_3\}$, contains infinitely many vectors: all possible linear combinations of $\mathbf{v}_1$, $\mathbf{v}_2$, and $\mathbf{v}_3$.

c. $\mathbf{p}$ is a linear combination of the columns of A if and only if the equation $A\mathbf{x} = \mathbf{p}$ has a solution. The following row reduction shows that $\mathbf{p}$ is indeed in Col A:

$$[A \quad \mathbf{p}] = \begin{bmatrix} 2 & -3 & -4 & 6 \\ -8 & 8 & 6 & -10 \\ 6 & -7 & -7 & 11 \end{bmatrix} \sim \begin{bmatrix} 2 & -3 & -4 & 6 \\ 0 & -4 & -10 & 14 \\ 0 & 2 & 5 & -7 \end{bmatrix} \sim \begin{bmatrix} 2 & -3 & -4 & 6 \\ 0 & -4 & -10 & 14 \\ 0 & 0 & 0 & 0 \end{bmatrix}$$

13. For Col A, select any column of A, or any linear combination of the columns of A. To find vectors in Nul A, solve the equation $A\mathbf{x} = \mathbf{0}$:

$$\begin{bmatrix} 3 & 2 & 1 & -5 & 0 \\ -9 & -4 & 1 & 7 & 0 \\ 9 & 2 & -5 & 1 & 0 \end{bmatrix} \sim \begin{bmatrix} 3 & 2 & 1 & -5 & 0 \\ 0 & 2 & 4 & -8 & 0 \\ 0 & -4 & -8 & 16 & 0 \end{bmatrix} \sim \begin{bmatrix} 3 & 2 & 1 & -5 & 0 \\ 0 & 2 & 4 & -8 & 0 \\ 0 & 0 & 0 & 0 & 0 \end{bmatrix} \sim$$

$$\begin{bmatrix} 3 & 2 & 1 & -5 & 0 \\ 0 & 1 & 2 & -4 & 0 \\ 0 & 0 & 0 & 0 & 0 \end{bmatrix} \sim \begin{bmatrix} 1 & 0 & -1 & 1 & 0 \\ 0 & 1 & 2 & -4 & 0 \\ 0 & 0 & 0 & 0 & 0 \end{bmatrix}, \qquad \begin{aligned} x_1 \quad - x_3 + x_4 &= 0 \\ x_2 + 2x_3 - 4x_4 &= 0 \\ 0 &= 0 \end{aligned}$$

The general solution is $x_1 = x_3 - x_4$, $x_2 = -2x_3 + 4x_4$, with x_3, x_4 free. Choose any values of x_3 and x_4 (not both zero) to produce a specific nonzero vector. For instance, if $x_3 = 1$ and $x_4 = 0$, the vector $(1, -2, 1, 0)$ is in Nul A.

19. No. Be careful *not* to say that the vectors are a basis for $\mathbb{R}^2$. They are not *in* $\mathbb{R}^2$, because they each have three (not two) entries. The vectors *do* form a basis for a two-dimensional subspace of $\mathbb{R}^3$, but that is not the same as $\mathbb{R}^2$.

Warning: $\mathbb{R}^2$ is *not* a subspace of $\mathbb{R}^3$. The notation $\mathbb{R}^2$ refers explicitly to lists of numbers with exactly two entries. $\mathbb{R}^3$ is the set of all lists (of numbers) with three entries.

Warning for Exercises 21-24: A common error is to think that the nonpivot columns of an $m \times n$ matrix A form a basis for Nul A. This is not true in general, even when $m = n$.

25. Make a matrix A from the four vectors. Then Col A is the space spanned by the four vectors. Look for the pivot columns of A:

$$\begin{bmatrix} 1 & -2 & 2 & -3 \\ -4 & 8 & -5 & 4 \\ 2 & -4 & 4 & -1 \\ -5 & 10 & -4 & 4 \end{bmatrix} \sim \begin{bmatrix} 1 & -2 & 2 & -3 \\ 0 & 0 & 3 & -8 \\ 0 & 0 & 0 & 5 \\ 0 & 0 & 6 & -11 \end{bmatrix} \sim \begin{bmatrix} 1 & -2 & 2 & -3 \\ 0 & 0 & 3 & -8 \\ 0 & 0 & 0 & 5 \\ 0 & 0 & 0 & 0 \end{bmatrix}$$

The pivot columns arc 1, 3, and 4. So a basis for the subspace generated by the four vectors consists of

$$\begin{bmatrix} 1 \\ -4 \\ 2 \\ -5 \end{bmatrix}, \begin{bmatrix} 2 \\ -5 \\ 4 \\ -4 \end{bmatrix}, \begin{bmatrix} -3 \\ 4 \\ -1 \\ 4 \end{bmatrix}$$

Warning: A standard mistake in Exercise 25 is to confuse Col A with Nul A. This easily happens when the definitions of these spaces are not known precisly.

31. The entries in $[\mathbf{x}]_{\mathcal{B}}$ are the numbers x_1, x_2 such that $x_1\mathbf{b}_1 + x_2\mathbf{b}_2 = \mathbf{x}$. To find x_1 and x_2, row reduce the augmented matrix $[\mathbf{b}_1 \;\; \mathbf{b}_2 \;\; \mathbf{x}]$:

$$\begin{bmatrix} 1 & -3 & -2 \\ 5 & -7 & 0 \\ -3 & 5 & 1 \end{bmatrix} \sim \begin{bmatrix} 1 & -3 & -2 \\ 0 & 8 & 10 \\ 0 & -4 & -5 \end{bmatrix} \sim \begin{bmatrix} 1 & -3 & -2 \\ 0 & 8 & 10 \\ 0 & 0 & 0 \end{bmatrix} \sim \begin{bmatrix} 1 & 0 & 7/4 \\ 0 & 1 & 5/4 \\ 0 & 0 & 0 \end{bmatrix}$$

So $x_1 = 7/4$, $x_2 = 5/4$ and the coordinate vector of $\mathbf{x}$ is $[\mathbf{x}]_{\mathcal{B}} = \begin{bmatrix} 7/4 \\ 5/4 \end{bmatrix}$.

37. **a.** See the definition at the beginning of the section.

b. See the paragraph before Example 4.

c. See Theorem 12.

d. See the Warning after Theorem 13.

e. See Example 5.

43. Suppose $W = \text{Span}\{\mathbf{b}_1, \ldots, \mathbf{b}_p\}$ and let $\{\mathbf{a}_1, \ldots, \mathbf{a}_q\}$ be any set in W with more than p vectors. Let $B = [\mathbf{b}_1 \ \cdots \ \mathbf{b}_p]$ and $A = [\mathbf{a}_1 \ \cdots \ \mathbf{a}_q]$. To show that $\{\mathbf{a}_1, \ldots, \mathbf{a}_q\}$ is linearly dependent, we will show that the equation $A\mathbf{x} = \mathbf{0}$ has a nontrivial solution.

a. By hypothesis, the columns of B span W, so for each vector $\mathbf{a}_j$ in W, there exists a vector $\mathbf{c}_j$ of weights such that $\mathbf{a}_j = B\mathbf{c}_j$. Note that $\mathbf{c}_j$ is in $\mathbb{R}^p$ because B has p columns, and there are q vectors $\mathbf{c}_j$, one for each of the $\mathbf{a}_j$.

b. Let $C = [\mathbf{c}_1 \ \cdots \ \mathbf{c}_q]$. Since C has q columns and p rows, and $q > p$, the equation $C\mathbf{x} = \mathbf{0}$ must have a nontrivial solution; call it $\mathbf{u}$.

c. By definition of matrix multiplication,

$$A = [\mathbf{a}_1 \ \cdots \ \mathbf{a}_q] = [B\mathbf{c}_1 \ \cdots \ B\mathbf{c}_q] = BC$$

Hence, $A\mathbf{u} = (BC)\mathbf{u} = B(C\mathbf{u}) = B\mathbf{0} = \mathbf{0}$. Since $\mathbf{u}$ is nonzero, the equation $A\mathbf{u} = \mathbf{0}$ shows that the columns of A are linearly dependent.

Mastering Linear Algebra Concepts: Seven Basic Ideas

For each of the seven key concepts in this section, assemble notes that will help you form a complete image of the idea. For each concept, include as many of the following categories as possible:

▸ definitions	Always write the definitions
▸ equivalent descriptions	For example, Theorem 12
▸ geometric interpretations	For example, Fig. 1
▸ special cases	The zero subspace, the standard basis
▸ examples and counterexamples	Fig. 2
▸ algorithms and computations	Check the examples and exercises
▸ connections with other concepts	For instance, Theorems 13 and 14

Finally, strengthen your understanding of column space, null space, basis, dimension, and rank, by using each of these terms in a sentence that could be added to the Invertible Matrix Theorem. (For instance, assuming A is $n \times n$, write what must be true about Col A if and only if A is invertible.) Write as many new statements for the IMT as possible. When you are finished, check the table that appears on page 4-21 in this *Study Guide*.

MATLAB ref and rank

The command **ref(A)** produces the reduced echelon form of A. From that you can write a basis for Col A or write the homogeneous equations that describe Nul A. (Don't forget that A is a coefficient matrix, not an augmented matrix.) MATLAB has another command, **rref**, that works basically the same as **ref** but is often much slower because it checks for rational entries in the matrix.

You can use **ref(A)** to check the rank of A, but roundoff error or an extremely small pivot entry can produce an incorrect echelon form. A more reliable command is **rank(A)**.

CHAPTER 2 SUPPLEMENTARY EXERCISES

1. Justifications for the answers to the True/False questions are given below. Other justifications are possible.

a. True. If A and B are $m\times n$, then B^T has as many rows as A has columns, so AB^T is defined; also A^TB is defined because A^T has m columns and B has m rows.

b. False. B must have 2 columns. A has as many columns as B has rows.

c. True. The ith row of A has the form $(0,\dots,d_i,\dots,0)$. So the ith row of AB is $(0,\dots,d_i,\dots,0)B$, which is d_i times the ith row of B.

d. False. Take the zero matrix for B. Or, construct a matrix B such that the equation $B\mathbf{x} = \mathbf{0}$ has nontrivial solutions, and construct C and D so that $C \neq D$ and the columns of $C - D$ satisfy the equation $B\mathbf{x} = 0$. Then $B(C - D) = 0$ and $BC = BD$.

e. False. Counterexample: $A = \begin{bmatrix} 1 & 0 \\ 0 & 0 \end{bmatrix}$ and $C = \begin{bmatrix} 0 & 0 \\ 0 & 1 \end{bmatrix}$.

f. False. $(A + B)(A - B) = A^2 - AB + BA - B^2$. This equals $A^2 - B^2$ if and only if A commutes with B.

g. True. An $n\times n$ replacement matrix has $n+1$ nonzero entries. The $n\times n$ scale and swap matrices have n nonzero entries.

h. True. The transpose of an elementary matrix is an elementary matrix of the same type.

i. True. An $n \times n$ elementary matrix is obtained by a row operation on I_n.

j. False. Elementary matrices are invertible, so a product of such matrices is invertible. But not every square matrix is invertible.

k. True. If A is 3×3 with three pivot positions, then A is row equivalent to I_3.

l. False. A must be square in order to conclude from the equation $AB = I$ that A is invertible.

m. False. AB is invertible, but $(AB)^{-1} = B^{-1}A^{-1}$, and this product is not always equal to $A^{-1}B^{-1}$.

n. True. Given $AB = BA$, left-multiply by A^{-1} to get $B = A^{-1}BA$, and then right-multiply by A^{-1} to have $BA^{-1} = A^{-1}B$.

o. False. The correct equation is $(rA)^{-1} = r^{-1}A^{-1}$, because $(rA)(r^{-1}A^{-1}) = (rr^{-1})(AA^{-1}) = 1 \cdot I = I$.

p. True. If the equation $A\mathbf{x} = \begin{bmatrix} 1 \\ 0 \\ 0 \end{bmatrix}$ has a unique solution, then there are no free variables in this equation, which means that A must have three pivot positions (since A is 3×3). By the IMT, A is invertible.

8. A satisfies $A\begin{bmatrix} 1 & 2 \\ 3 & 7 \end{bmatrix} = \begin{bmatrix} 1 & 3 \\ 1 & 1 \end{bmatrix}$. Right-multiply by $\begin{bmatrix} 1 & 2 \\ 3 & 7 \end{bmatrix}^{-1}$:

$$A = \begin{bmatrix} 1 & 3 \\ 1 & 1 \end{bmatrix} \cdot \begin{bmatrix} 7 & -2 \\ -3 & 1 \end{bmatrix} = \begin{bmatrix} -2 & 1 \\ 4 & -1 \end{bmatrix}$$

11. The solution is in the text. However, note that in practice, interpolating polynomials are usually found from special formulas, because the equation $V\mathbf{c} = \mathbf{y}$ is ill-conditioned for large values of n, in the sense that minor errors in the entries of V and $\mathbf{y}$ can greatly affect the solution.

18. If J denotes the $n \times n$ matrix of ones, then

$$A_n = J - I_n, \text{ and } A_n^{-1} = \frac{1}{n-1} \cdot J - I_n$$

CHAPTER 2 GLOSSARY CHECKLIST

Check your knowledge by attempting to write definitions of the terms below. Then compare your work with the definitions given in the text's Glossary. Ask your instructor which definitions, if any, might appear on a test.

associative law of multiplication: ....

basis (for a subspace H of $\mathbb{R}^n$, §2.9): A set $\mathcal{B} = \{\mathbf{v}_1, \ldots, \mathbf{v}_p\}$ in $\mathbb{R}^n$ such that: ....

block matrix: *See* partitioned matrix.

block matrix multiplication: The ... multiplication of ... as if ....

column space (of an $m \times n$ matrix A, §2.9): The set Col A of ....

column sum: The sum of ....

commuting matrices: Two matrices A and B such that ....

composition of linear transformations: A mapping produced by applying ....

conformable for block multiplication: Two partitioned matrices A and B such that ....

consumption matrix: A matrix in the ... model whose columns are ....

coordinate vector of x relative to a basis $\mathcal{B} = \{\mathbf{b}_1, \ldots, \mathbf{b}_p\}$, (§2.9): The vector $[\mathbf{x}]_{\mathcal{B}}$ whose entries $c_1, \ldots, c_p$ satisfy ....

diagonal entries (in a matrix): Entries having ....

diagonal matrix: A square matrix whose ... entries are ....

dimension (of a vector space V, §2.9): The number ....

determinant of $A = \begin{bmatrix} a & b \\ c & d \end{bmatrix}$: The number ..., denoted by ....

distributive laws: (left) ... (right) ....

elementary matrix: An invertible matrix that results by ....

final demand vector (or **bill of final demands**): The vector **d** in the ... model that lists .... The vector **d** can represent ....

flexibility matrix: A matrix whose jth column gives ... of an elastic beam at specified points when ... is applied at ....

Gauss-Seidel algorithm: An iterative method that produces ... that in certain cases converges to a ...; based on the decomposition $A = M - N$, with M ....

Householder reflection: A transformation $\mathbf{x} \mapsto Q\mathbf{x}$, where $Q = \ldots.$

identity matrix: The $n \times n$ matrix I or I_n with ....

inner product: A matrix product ... where **u** and **v** are ....

input-output matrix: *See* consumption matrix.

input-output model: *See* Leontief input-output model.

intermediate demands: Demands for goods or services that ....

inverse (of an $n \times n$ matrix A): An $n \times n$ matrix A^{-1} such that ....

invertible linear transformation: A linear transformation $T: \mathbb{R}^n \to \mathbb{R}^n$ such that there exists ....

invertible matrix: A square matrix that ....

Jacobi's method: An iterative method that produces ... that in certain cases converges to a ...; based on the decomposition $A = M - N$, with M ....

ladder network: An electrical network assembled by connecting ....

left inverse (of A): Any rectangular matrix C such that ....

Leontief input-output model (or **Leontief production equation**): The equation ..., where ....

lower triangular matrix: A matrix with ....

lower triangular part (of A): A ... matrix whose entries on ....

LU factorization: The representation of a matrix A in the form $A = LU$, where L is ... and U is ....

main diagonal (of a matrix): The location of the ....

null space (of an $m \times n$ matrix A, §2.9): The set Nul A of all ....

outer product: A matrix product ... where **u** and **v** are ....

partitioned matrix: A matrix whose entries are .... Sometimes called ....

permuted lower triangular matrix: A matrix such that ....

permuted LU factorization: The representation of a matrix A in the form $A = LU$ where L is ... and U is ....

production vector: The vector in the ... model that lists ....

rank (of a matrix A, §2.9): ....

right inverse (of A): Any rectangular matrix C such that ....

row-column rule: The rule for computing a product AB in which ....

Schur complement: A certain matrix formed from the blocks of a 2×2 partitioned matrix $A = [A_{ij}]$. If A_{11} is invertible, its Schur complement is given by .... If A_{22} is invertible, its Schur complement is given by ....

stiffness matrix: The inverse of a ... matrix. The *j*th column of a stiffness matrix gives ... at specified points on an elastic beam in order to produce ....

subspace (of $\mathbb{R}^n$, §2.9): A subset H of $\mathbb{R}^n$ with the properties: ....

transfer matrix: A matrix A associated with an electrical circuit having input and output terminals, such that ....

transpose (of A): An $n \times m$ matrix A^T whose ... are the corresponding ... of ....

unit consumption vector: A column vector in the ... model that lists ....

unit lower triangular matrix: A ... matrix with ....

upper triangular matrix: A matrix U with ....

Vandermonde matrix: An $n \times n$ matrix V or its transpose, of the form ....

3 DETERMINANTS

3.1 INTRODUCTION TO DETERMINANTS

This section is relatively short and easy. Some exercises provide computational practice and others allow you to discover properties of determinants to be studied in the next section. You will enjoy Section 3.2 more if you finish your work on this section first.

KEY IDEAS

The second paragraph of the section sets the stage for what follows. Read it quickly, without worrying about the details of the row operations. The main idea is that the determinant of A is a number that appears along the diagonal of an echelon form of A, and this number is nonzero if and only if the matrix is invertible. Later (on pages 189 and 199) you will understand why this idea is important. For now, this 3×3 case is only used to motivate the definition of $\det A$.

Determinants are defined here via a cofactor expansion along the first row. Since the cofactors involve determinants of smaller matrices, the definition is said to be *recursive.* For each $n \geq 2$, the determinant of an $n\times n$ matrix is based on the definition of the determinant of an $(n-1)\times(n-1)$ matrix. There are other equivalent definitions of the determinant, but we shall not digress to discuss them.

Study Tip: Watch how parentheses are used in Example 2 to avoid a common mistake. The cofactor expansion puts a minus sign in front of a_{32} because $(-1)^{3+2} = -1$. Since a_{32} happens to be negative, the correct term in the expansion is $-(-2)\det A_{32}$, *not* $-2\det A_{32}$.

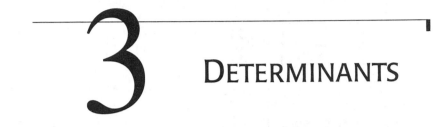

SOLUTIONS TO EXERCISES

1. By definition, det A is computed via a cofactor expansion along the first row:

$$\begin{vmatrix} 3 & 0 & 4 \\ 2 & 3 & 2 \\ 0 & 5 & -1 \end{vmatrix} = 3\begin{vmatrix} 3 & 2 \\ 5 & -1 \end{vmatrix} - 0\begin{vmatrix} 2 & 2 \\ 0 & -1 \end{vmatrix} + 4\begin{vmatrix} 2 & 3 \\ 0 & 5 \end{vmatrix}$$

$$= 3(-3 - 10) - 0 + 4(10 - 0) = -39 + 40 = 1$$

For comparison, a cofactor expansion down the second column yields

$$\begin{vmatrix} 3 & 0 & 4 \\ 2 & 3 & 2 \\ 0 & 5 & -1 \end{vmatrix} = (-1)^{1+2}\cdot 0\begin{vmatrix} 2 & 2 \\ 0 & -1 \end{vmatrix} + (-1)^{2+2}\cdot 3\begin{vmatrix} 3 & 4 \\ 0 & -1 \end{vmatrix} + (-1)^{3+2}\cdot 5\begin{vmatrix} 3 & 4 \\ 2 & 2 \end{vmatrix}$$

$$= 0 + 3(-3 - 0) - 5(6 - 8) = -9 + 10 = 1$$

Study Tip: To save time, omit the zero terms in a cofactor expansion, but be careful to use the proper plus or minus signs with the nonzero terms.

7. By definition,

$$\begin{vmatrix} 4 & 3 & 0 \\ 6 & 5 & 2 \\ 9 & 7 & 3 \end{vmatrix} = 4\begin{vmatrix} 5 & 2 \\ 7 & 3 \end{vmatrix} - 3\begin{vmatrix} 6 & 2 \\ 9 & 3 \end{vmatrix} = 4(15 - 14) -3(18 - 18) = 4$$

Using the second column of A instead,

$$\begin{vmatrix} 4 & 3 & 0 \\ 6 & 5 & 2 \\ 9 & 7 & 3 \end{vmatrix} = -3\begin{vmatrix} 6 & 2 \\ 9 & 3 \end{vmatrix} + 5\begin{vmatrix} 4 & 0 \\ 9 & 3 \end{vmatrix} - 7\begin{vmatrix} 4 & 0 \\ 6 & 2 \end{vmatrix}$$

$$= -3(18 - 18) + 5(12 - 0) - 7(8 - 0) = 0 + 60 - 56 = 4$$

13. Row 2 or column 2 are the best choices because they contain the most zeros. We'll use row 2. Since the only nonzero entry in that row is 2, the determinant is $(-1)^{2+3}\cdot 2\cdot A_{23}$.

$$\det A = (-1)^{2+3}\cdot 2\cdot\begin{vmatrix} 4 & 0 & -7 & 3 & -5 \\ 0 & 0 & 2 & 0 & 0 \\ 7 & 3 & -6 & 4 & -8 \\ 5 & 0 & 5 & 2 & -3 \\ 0 & 0 & 9 & -1 & 2 \end{vmatrix} = (-2)\cdot\begin{vmatrix} 4 & 0 & 3 & -5 \\ 7 & 3 & 4 & -8 \\ 5 & 0 & 2 & -3 \\ 0 & 0 & -1 & 2 \end{vmatrix}$$

The best choice for this 4×4 determinant is to expand down the second column. Notice that the cofactor associated with the 3 in the (2,2)-position is the (2,2)-cofactor of the 4×4 matrix. The original location of the "3" in the 5×5 matrix is irrelevant.

$$\det A = (-2)\cdot(-1)^{2+2}(3)\cdot\begin{vmatrix} 4 & 3 & -5 \\ 5 & 2 & -3 \\ 0 & -1 & 2 \end{vmatrix}$$

Finally, use column 1 (although row 3 would work as well).

$$\det A = (-2)\cdot(-1)^{2+2}(3)\cdot\left(4\begin{vmatrix} 2 & -3 \\ -1 & 2 \end{vmatrix} - 5\begin{vmatrix} 3 & -5 \\ -1 & 2 \end{vmatrix}\right)$$

$$= -6\Big[4(4-3) - 5(6-5)\Big] = -\,6(4-5) = 6$$

Checkpoint: Try to complete the following statement: "If the kth column of the $n \times n$ identity matrix is replaced by a column vector $\mathbf{x}$ whose entries are $x_1, \ldots, x_n$, then the determinant of the resulting matrix is _____." To discover the answer, compute the determinants of the following matrices:

a. $\begin{bmatrix} 1 & 3 & 0 & 0 \\ 0 & 4 & 0 & 0 \\ 0 & 5 & 1 & 0 \\ 0 & 6 & 0 & 1 \end{bmatrix}$ b. $\begin{bmatrix} 1 & 0 & 3 & 0 \\ 0 & 1 & 4 & 0 \\ 0 & 0 & 5 & 0 \\ 0 & 0 & 6 & 1 \end{bmatrix}$ c. $\begin{bmatrix} 1 & 0 & 3 & 0 & 0 \\ 0 & 1 & 4 & 0 & 0 \\ 0 & 0 & 5 & 0 & 0 \\ 0 & 0 & 6 & 1 & 0 \\ 0 & 0 & 7 & 0 & 1 \end{bmatrix}$

19. $\det\begin{bmatrix} a & b \\ c & d \end{bmatrix} = ad - bc$, and $\det\begin{bmatrix} c & d \\ a & b \end{bmatrix} = cb - da = -\det\begin{bmatrix} a & b \\ c & d \end{bmatrix}$.

Interchanging two rows reverses the sign of the determinant, at least for the 2×2 case. Perhaps this is true for larger matrices.

25. The matrix is triangular, so use Theorem 2.

$$\det\begin{bmatrix} 1 & 0 & 0 \\ 0 & 1 & 0 \\ 0 & k & 1 \end{bmatrix} = 1\cdot1\cdot1 = 1 \quad \text{Product of the diagonal entries}$$

31. A 3×3 row replacement matrix has one of the following forms:

$$\begin{bmatrix} 1 & 0 & 0 \\ k & 1 & 0 \\ 0 & 0 & 1 \end{bmatrix}, \begin{bmatrix} 1 & 0 & 0 \\ 0 & 1 & 0 \\ k & 0 & 1 \end{bmatrix}, \begin{bmatrix} 1 & 0 & 0 \\ 0 & 1 & 0 \\ 0 & k & 1 \end{bmatrix},$$

$$\begin{bmatrix}1 & k & 0\\0 & 1 & 0\\0 & 0 & 1\end{bmatrix}, \begin{bmatrix}1 & 0 & k\\0 & 1 & 0\\0 & 0 & 1\end{bmatrix}, \begin{bmatrix}1 & 0 & 0\\0 & 1 & k\\0 & 0 & 1\end{bmatrix}$$

In each case the matrix is triangular with 1's on the diagonal, so its determinant equals 1. The determinant of a row replacement matrix is 1, at least for the 3×3 case. Perhaps this is true for larger matrices.

37. $\det A = \det\begin{bmatrix}3 & 1\\4 & 2\end{bmatrix} = 3(2) - 1(4) = 6 - 4 = 2.$ Since $5A = \begin{bmatrix}5\cdot3 & 5\cdot1\\5\cdot4 & 5\cdot2\end{bmatrix}$,

$$\det 5A = (5\cdot3)(5\cdot2) - (5\cdot1)(5\cdot4) = 150 - 100 = 50$$

So $\det 5A \neq 5\cdot\det A$. Can you see what the true relation between $\det 5A$ and $\det A$ really is, at least for this example? What about $\det 5A$ for any 2×2 matrix? Try to guess (and perhaps verify) a formula for $\det rA$, where r is any scalar and A is any $n\times n$ matrix.

39. **a.** See the paragraph preceding the definition of $\det A$.

b. See the definition of cofactor, preceding Theorem 1.

41. $\det[\mathbf{u}\ \ \mathbf{v}] = \det\begin{bmatrix}3 & 1\\0 & 2\end{bmatrix} = 6$, $\det[\mathbf{u}\ \ \mathbf{x}] = \det\begin{bmatrix}3 & x\\0 & 2\end{bmatrix} = 6$, and the areas of the parallelograms determined by $[\mathbf{u}\ \ \mathbf{v}]$ and $[\mathbf{u}\ \ \mathbf{x}]$ both equal 6. To see why the areas are equal, consider the parallelograms determined by $\mathbf{u} = (3,0)$ and $\mathbf{v} = (1,2)$ and by $\mathbf{u}$ and $\mathbf{x} = (x,2)$:

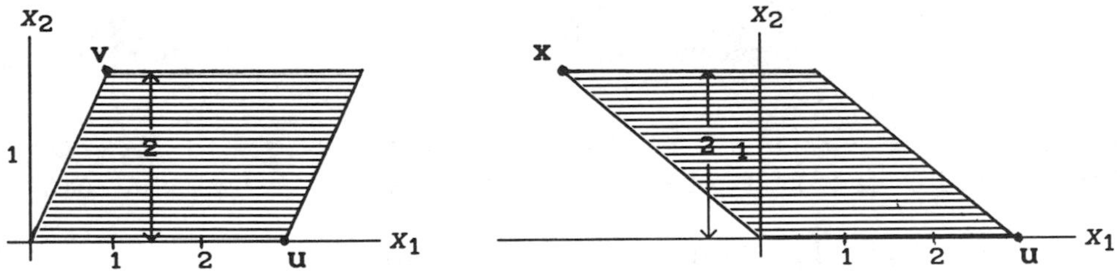

The parallelogram on the left is determined by $\mathbf{u}$ and $\mathbf{v}$ (and the vertices $\mathbf{u}+\mathbf{v}$ and $\mathbf{0}$). Its base is 3 and its altitude is 2, so the area is (base)(altitude) = 6. The parallelogram on the right, determined by $\mathbf{u}$ and $\mathbf{x} = (x,2)$, has the same base. Also, the altitude is 2 for any value of x, so the area again equals $3\cdot2 = 6$.

Answer to Checkpoint: a. 4 b. 5 c. 5 "If the kth column of the $n\times n$ identity matrix is replaced by a column vector $\mathbf{x}$ whose entries are $x_1, \ldots, x_n$, then the determinant of the resulting matrix is x_k." Can you explain why this is true? You'll learn the answer when you begin Section 3.3.

3.2 PROPERTIES OF DETERMINANTS

This section presents the main properties of determinants, gives an efficient method of computation, and proves that a matrix is invertible if and only if its determinant is nonzero.

KEY IDEAS

It is not surprising that row operations relate nicely to determinants. After all, we found the definition of a 3×3 determinant by row reducing a 3×3 matrix. Theorem 3 can be rephrased informally as follows:

a. *Adding a multiple of one row (or column) of* A *to another does not change the determinant.*

b. *Interchanging two rows (or columns) of* A *reverses the sign of the determinant.*

c. *A constant may be factored out of one row (or column) of the determinant of* A.

The other properties to learn are stated in Theorems 4, 5, and 6, together with the boxed formula for det A on page 188. Theorem 4 is sometimes stated as: *A square matrix is nonsingular if and only if its determinant is nonzero.* Theorems 4 and 6 will be used extensively in Chapter 5.

Your instructor may or may not want you to know the (multi-) linearity property on page 191. This property is important in more advanced courses but is not used later in the text. *Warning*: in general, det$(A + B)$ is *unequal* to det A + det B.

SOLUTIONS TO EXERCISES

1. Rows 1 and 2 are interchanged, so the determinant changes sign.

7.
$$\begin{vmatrix} 1 & 3 & 0 & 2 \\ -2 & -5 & 7 & 4 \\ 3 & 5 & 2 & 1 \\ 1 & -1 & 2 & -3 \end{vmatrix} = \begin{vmatrix} 1 & 3 & 0 & 2 \\ 0 & 1 & 7 & 8 \\ 0 & -4 & 2 & -5 \\ 0 & -4 & 2 & -5 \end{vmatrix} = \begin{vmatrix} 1 & 3 & 0 & 2 \\ 0 & 1 & 7 & 8 \\ 0 & 0 & 30 & 27 \\ 0 & 0 & 30 & 27 \end{vmatrix} = \begin{vmatrix} 1 & 3 & 0 & 2 \\ 0 & 1 & 7 & 8 \\ 0 & 0 & 30 & 27 \\ 0 & 0 & 0 & 0 \end{vmatrix} = 0$$

Note, the second array already shows that the determinant is zero, because two rows are equal, as in Example 3.

Study Tip: In general, computation of a 3×3 determinant by row reduction takes 10 multiplications (and divisions), but cofactor expansion only takes 9 multiplications. At $n = 4$, the advantage switches to row reduction, which requires 23 multiplications, cofactor expansion 40 (9 for each 3×3

determinant, plus four multiplications of a_{1j} times det A_{1j}). Often, the best strategy is to combine the two techniques, as in Exercises 11—14.

13. Use row or column operations whenever convenient to create a row or column that has only one nonzero entry. (I recommend using only row operations, because you already have experience with them.) Then use a cofactor expansion to reduce the size of the matrix.

$$\begin{vmatrix} 2 & 5 & 4 & 1 \\ 4 & 7 & 6 & 2 \\ 6 & -2 & -4 & 0 \\ -6 & 7 & 7 & 0 \end{vmatrix} = \begin{vmatrix} 2 & 5 & 4 & 1 \\ 0 & -3 & -2 & 0 \\ 6 & -2 & -4 & 0 \\ -6 & 7 & 7 & 0 \end{vmatrix} \quad \text{Zero created in column 4}$$

$$= -\begin{vmatrix} 0 & -3 & -2 \\ 6 & -2 & -4 \\ -6 & 7 & 7 \end{vmatrix} \quad \text{Result of cofactor expansion down column 4}$$

$$= -\begin{vmatrix} 0 & -3 & -2 \\ 6 & -2 & -4 \\ 0 & 5 & 3 \end{vmatrix} \quad \text{Zero created in column 1}$$

$$= -(-6)\begin{vmatrix} -3 & -2 \\ 5 & 3 \end{vmatrix} \quad \text{Result of cofactor expansion down column 1}$$

$$= 6\cdot(-9 + 10) = 6$$

19. $$\begin{vmatrix} a & b & c \\ 2d+a & 2e+b & 2f+c \\ g & h & i \end{vmatrix} = \begin{vmatrix} a & b & c \\ 2d & 2e & 2f \\ g & h & i \end{vmatrix} \quad \text{(-1)·row 1 added to row 2}$$

$$= 2\begin{vmatrix} a & b & c \\ d & e & f \\ g & h & i \end{vmatrix} \quad \text{2 factored out of row 2}$$

$$= 2\cdot 7 = 14$$

25. By Theorem 4 and the IMT, the set $\{\mathbf{v}_1, \mathbf{v}_2, \mathbf{v}_3\}$ is linearly independent if and only if det $[\mathbf{v}_1\ \ \mathbf{v}_2\ \ \mathbf{v}_3] \neq 0$. Rather than use row operations on $[\mathbf{v}_1\ \ \mathbf{v}_2\ \ \mathbf{v}_3]$, you might choose to expand the determinant by cofactors of the third column:

$$\begin{vmatrix} 7 & -8 & 7 \\ -4 & 5 & 0 \\ -6 & 7 & -5 \end{vmatrix} = 7\begin{vmatrix} -4 & 5 \\ -6 & 7 \end{vmatrix} + (-5)\begin{vmatrix} 7 & -8 \\ -4 & 5 \end{vmatrix} = 7(-28 + 30) - 5(35 - 32)$$

$$= 7(2) - 5(3) = -1$$

The determinant is nonzero, so the vectors are linearly independent.

Study Tip: For 3×3 matrices, some students tend to prefer the special trick suggested for Exercises 15—18 in Section 3.1, even though in general there are 12 multiplications instead of the 9 multiplications needed for cofactor expansion. Note, however, that numbers in the special method can sometimes be large. For comparison, here are those computations for the matrix studied above in Exercise 25:

$$\begin{vmatrix} 7 & -8 & 7 \\ -4 & 5 & 0 \\ -6 & 7 & -5 \end{vmatrix} \begin{matrix} 7 & -8 \\ -4 & 5 \\ -6 & 7 \end{matrix}$$

$$\det [\mathbf{v}_1 \quad \mathbf{v}_2 \quad \mathbf{v}_3] = 7(5)(-5) + (-8)(0)(-6) + 7(-4)(7) \\ -(-6)(5)(7) - 7(0)(7) - (-5)(-4)(-8)$$

$$= -175 + 0 + (-196) - (-210) - 0 - (-160) = -1$$

27. **a.** See Theorem 3.

b. See the paragraph following Example 2.

c. See the remark following Theorem 4.

d. See the warning after Example 5.

31. Since the determinant is multiplicative (Theorem 6),

$(\det A)(\det A^{-1}) = \det(AA^{-1}) = \det I = 1$. So $\det A^{-1} = 1/\det A$.

Study Tip: The result of Exercise 31 might come in handy on a test.

33. By Theorem 6 (twice), $\det AB = (\det A)(\det B) = (\det B)(\det A) = \det BA$.

35. By Theorem 5, $\det U^T = \det U$. So, by Theorem 6,

$$\det U^TU = (\det U^T)(\det U) = (\det U)^2$$

If $U^TU = I$, then $(\det U)^2 = \det I = 1$, which implies that $\det U = \pm 1$.

37. The solution is in the text. (The determinant of a triangular matrix is the product of the entries on the main diagonal.)

Study Tip: Exercises 15—26, 39 and 40 make good test questions because they check your knowledge of determinant properties without requiring much computation. Exercise 39(b) is the one most likely to be answered incorrectly. What would be the answer to 39(b) if A were 4×4?

43. Compute det A by a cofactor expansion down column 3:

$$\det A = (u_1 + v_1)\cdot\det A_{13} - (u_2 + v_2)\cdot\det A_{23} + (u_3 + v_3)\cdot\det A_{33}$$

$$= u_1\cdot\det A_{13} - u_2\cdot\det A_{23} + u_3\cdot\det A_{33} + v_1\cdot\det A_{13} - v_2\cdot\det A_{23} + v_3\cdot\det A_{33}$$

$$= u_1\cdot\det B_{13} - u_2\cdot\det B_{23} + u_3\cdot\det B_{33} + v_1\cdot\det C_{13} - v_2\cdot\det C_{23} + v_3\cdot\det C_{33}$$

$$= \det B + \det C$$

45. Suppose A is $m \times n$ with more columns than rows. Then A^TA is $n \times n$ and must be singular. If A is generated with random entries, then A^TA will be nonsingular (invertible) practically all the time. Try to explain why these statements should be true. (Use the IMT.)

MATLAB Computing Determinants

To compute det A, set **U = A** and then repeatedly use the commands **U = gauss(U,r)** and **U = swap(U,r,s)** as needed to reduce A to an echelon form U. Then, except for a + or - sign (depending on how many times you swap rows), the determinant of A is given by the command

prod(diag(U))

The command **diag(U)** extracts the diagonal entries of U and places them in a column vector, and **prod** computes the product of those entries. You can also use **det(A)** to check your work, but the longer sequence of commands helps you think about the *process* of computing det A.

3.3 CRAMER'S RULE, VOLUME, AND LINEAR TRANSFORMATIONS

This section will be a valuable reference for students who plan to take a course in multivariable calculus. Mathematics and statistics majors probably will encounter the material here several times. Also, economics students and engineers (particularly electrical engineers) are likely to need Cramer's rule and some of the supplementary exercises in later courses.

KEY IDEAS

The main results of the section are stated in Theorems 7, 8, 9, and 10. The proof of Theorem 7 is simple and yet involves three important ideas: the definition of a matrix product, the multiplicative property of the determinant, and the evaluation of a determinant by cofactors. Check with your instructor about whether you should be able to reproduce the proof of Theorem 7.

A heuristic proof of Theorem 9 for 2×2 matrices is given in an appendix at the end of this section. Theorem 10 provides a key idea in calculus and physics needed for the study of double and triple integrals. The matrix used there is called a *Jacobian.*

STUDY NOTE

In Exercise 25, you are asked to use Theorem 9 to explain why the determinant of a 3×3 matrix A is zero if and only if A is not invertible. (A similar explanation holds for the 2×2 case.) The answer is in the text, so be sure to work on this before looking at the answer section. *Work on Exercise 25, even if it is not assigned.*

Remember, learning *does* take place when you think hard about an exercise, even when you are unsuccessful, if you try to look at the problem from different angles, browse back through the text, and perhaps look at earlier exercises. *Write* your solution, don't just talk to yourself about what you would write if you had to.

SOLUTIONS TO EXERCISES

1. The system is equivalent to $A\mathbf{x} = \mathbf{b}$, where $A = \begin{bmatrix} 5 & 7 \\ 2 & 4 \end{bmatrix}$ and $\mathbf{b} = \begin{bmatrix} 3 \\ 1 \end{bmatrix}$.

   Write

$$A_1(\mathbf{b}) = \begin{bmatrix} 3 & 7 \\ 1 & 4 \end{bmatrix}, \quad A_2(\mathbf{b}) = \begin{bmatrix} 5 & 3 \\ 2 & 1 \end{bmatrix}$$
$$\quad\;\uparrow \qquad\qquad\qquad\qquad \uparrow$$
$$\quad\;\mathbf{b} \qquad\qquad\qquad\qquad \mathbf{b}$$

   and compute

$$\det A = 20 - 14 = 6, \quad \det A_1(\mathbf{b}) = 12 - 7 = 5, \quad \det A_2(\mathbf{b}) = 5 - 6 = -1$$

$$x_1 = \frac{\det A_1(\mathbf{b})}{\det A} = \frac{5}{6}, \quad x_2 = \frac{\det A_2(\mathbf{b})}{\det A} = \frac{-1}{6} = -\frac{1}{6}$$

7. The system is equivalent to $A\mathbf{x} = \mathbf{b}$, where $A = \begin{bmatrix} 6s & 4 \\ 9 & 2s \end{bmatrix}$ and $\mathbf{b} = \begin{bmatrix} 5 \\ -2 \end{bmatrix}$. Write

$$A_1(\mathbf{b}) = \begin{bmatrix} 5 & 4 \\ -2 & 2s \end{bmatrix}, \quad A_2(\mathbf{b}) = \begin{bmatrix} 6s & 5 \\ 9 & -2 \end{bmatrix}$$

and compute

$$\det A = 12s^2 - 36 = 12(s^2 - 3) = 12(s - \sqrt{3})(s + \sqrt{3})$$

$$\det A_1(\mathbf{b}) = 10s + 8, \quad \det A_2(\mathbf{b}) = -12s - 45$$

The system has a unique solution when $\det A \neq 0$, that is, when $s \neq \pm\sqrt{3}$. For such a system, the solution is $\mathbf{x} = (x_1, x_2)$, where

$$x_1 = \frac{\det A_1(\mathbf{b})}{\det A} = \frac{10s + 8}{12(s^2 - 3)} = \frac{5s + 4}{6(s^2 - 3)}$$

$$x_2 = \frac{\det A_2(\mathbf{b})}{\det A} = \frac{-12s - 45}{12(s^2 - 3)} = \frac{-4s - 15}{4(s^2 - 3)}$$

13. First, find the cofactors of $A = \begin{bmatrix} 3 & 5 & 4 \\ 1 & 0 & 1 \\ 2 & 1 & 1 \end{bmatrix}$.

$$C_{11} = +\begin{vmatrix} 0 & 1 \\ 1 & 1 \end{vmatrix} = -1, \quad C_{12} = -\begin{vmatrix} 1 & 1 \\ 2 & 1 \end{vmatrix} = 1, \quad C_{13} = +\begin{vmatrix} 1 & 0 \\ 2 & 1 \end{vmatrix} = 1$$

$$C_{21} = -\begin{vmatrix} 5 & 4 \\ 1 & 1 \end{vmatrix} = -1, \quad C_{22} = +\begin{vmatrix} 3 & 4 \\ 2 & 1 \end{vmatrix} = -5, \quad C_{23} = -\begin{vmatrix} 3 & 5 \\ 2 & 1 \end{vmatrix} = 7$$

$$C_{31} = +\begin{vmatrix} 5 & 4 \\ 0 & 1 \end{vmatrix} = 5, \quad C_{32} = -\begin{vmatrix} 3 & 4 \\ 1 & 1 \end{vmatrix} = 1, \quad C_{33} = +\begin{vmatrix} 3 & 5 \\ 1 & 0 \end{vmatrix} = -5$$

Then, arrange the *transpose* of the array of cofactors into the adjugate of A.

$$\operatorname{adj} A = \begin{bmatrix} -1 & -1 & 5 \\ 1 & -5 & 1 \\ 1 & 7 & -5 \end{bmatrix}$$

Were you to compute $\det A$ now, you could write A^{-1}, but you would still need to check whether your calculations are correct. To build in this check, compute

$$A\cdot\text{adj}\,A = \begin{bmatrix} 3 & 5 & 4 \\ 1 & 0 & 1 \\ 2 & 1 & 1 \end{bmatrix}\begin{bmatrix} -1 & -1 & 5 \\ 1 & -5 & 1 \\ 1 & 7 & -5 \end{bmatrix} = \begin{bmatrix} 6 & 0 & 0 \\ 0 & 6 & 0 \\ 0 & 0 & 6 \end{bmatrix}$$

If any off-diagonal entries in the product are nonzero, or if the diagonal entries are not all the same, then some errors have been made, and you can recheck your cofactor calculations. (One possible mistake is to forget the ± signs in front of the 2×2 determinants. Another error is to *not* transpose the array of cofactors.) In this case, the calculations above *are* correct and det A must be 6. So

$$A^{-1} = \frac{1}{\det A}\text{adj}\,A = \frac{1}{6}\begin{bmatrix} -1 & -1 & 5 \\ 1 & -5 & 1 \\ 1 & 7 & -5 \end{bmatrix}$$

19. The parallelogram with vertices (0,0), (5,2), (6,4), (11,6) is shown below. If no vertex were zero, we would have to translate the parallelogram to the origin by subtracting one vertex from all four vertices. Since one vertex already is zero, use the two vertices adjacent to the origin to construct the columns of A, and compute $|\det A|$.

$$A = \begin{bmatrix} 5 & 6 \\ 2 & 4 \end{bmatrix}, \quad \begin{Bmatrix}\text{area of the} \\ \text{parallelogram}\end{Bmatrix} = |\det A| = |20 - 12| = 8$$

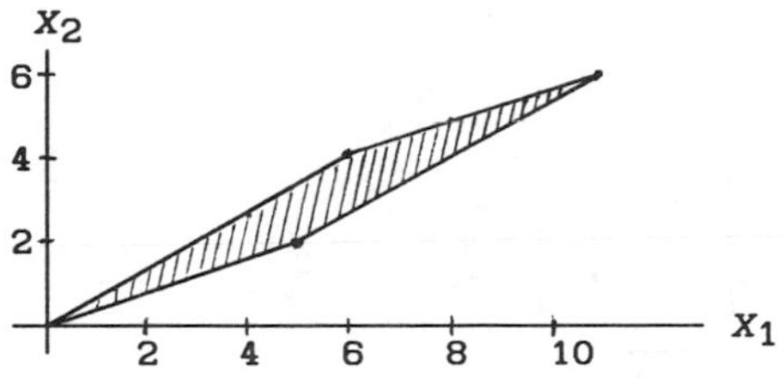

25. The answer is in the text. I hope you took the advice at the beginning of this Study Guide section and worked the problem (or at least tried hard to work the problem) before checking the answer section. If you were successful, you should be proud of yourself; you are mastering the material—not only determinants but also linear dependence!

31. Let $\mathbf{x} = \begin{bmatrix} x_1 \\ x_2 \\ x_3 \end{bmatrix}$, $\mathbf{u} = \begin{bmatrix} u_1 \\ u_2 \\ u_3 \end{bmatrix}$, and $A = \begin{bmatrix} a & 0 & 0 \\ 0 & b & 0 \\ 0 & 0 & c \end{bmatrix}$. Also, let S be the unit ball in $\mathbb{R}^3$, whose bounding surface consists of all vectors $\mathbf{u}$ such that $u_1^2 + u_2^2 + u_3^2 = 1$, and let S' be the image of S under the mapping $\mathbf{u} \mapsto A\mathbf{u}$.

a. If **x** is in S', then $\mathbf{x} = A\mathbf{u}$ for some **u** in S, and $\mathbf{u} = A^{-1}\mathbf{x} = \begin{bmatrix} x_1/a \\ x_2/b \\ x_3/c \end{bmatrix}$.

The condition on u_1, u_2, u_3 shows that $\left[\dfrac{x_1}{a}\right]^2 + \left[\dfrac{x_2}{b}\right]^2 + \left[\dfrac{x_3}{c}\right]^2 = 1$.

b. Since the volume of the unit ball bounded by S is $4\pi/3$ and the determinant of A is abc, Theorem 10 shows that the volume of the region bounded by S' is $4\pi abc/3$.

Appendix: A Geometric Proof of a Determinant Property

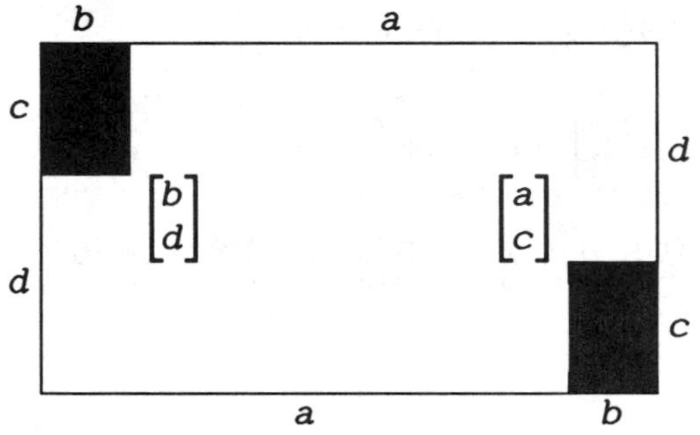

$$\det \begin{bmatrix} a & b \\ c & d \end{bmatrix} = ad - bc$$

$$(a + b)(c + d) \quad = \quad ac + ad + bc + bd$$

$$-\ 2 \left(\left[\begin{matrix} b \\ \end{matrix}\ c\right]\right) - 2\left(\left[\begin{matrix} d \\ b \end{matrix}\right]\right) = \quad -\ 2bc - bd$$

$$-\ 2\left(\left[\begin{matrix} a \\ \end{matrix}\ c\right]\right) = -\ ac$$

(Area of the Parallelogram)   = $ad - bc$

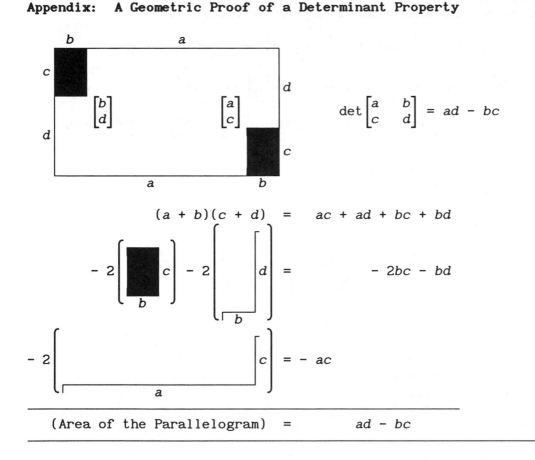

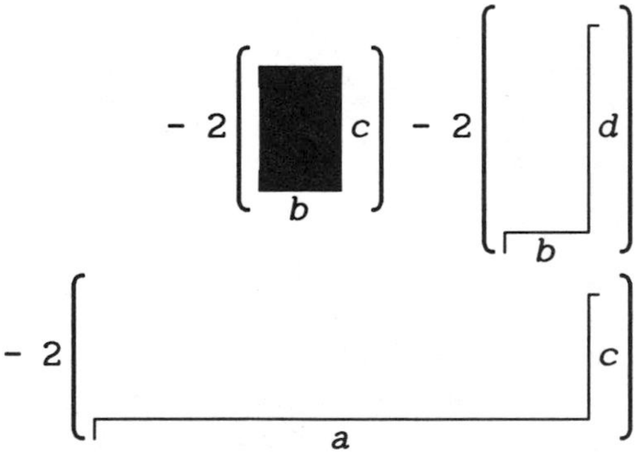

CHAPTER 3 SUPPLEMENTARY EXERCISES

1. Justifications for the answers to the True/False questions are given below. Other justifications are possible.

a. True. If $\det A = 0$, then the columns of A are linearly dependent. Since there are only two columns, one column must be a multiple of the other.

b. True. (See Exercise 30 in Section 3.2.) If the rows of A are linearly dependent, then the reduced echelon form of A will have a row of zeros, in which case $\det A = 0$, by Theorem 4.

c. False. If A is 3×3, then $\det 5A = 5^3 \det A$, by Theorem 3(c).

d. False. Consider $A = \begin{bmatrix} 2 & 0 \\ 0 & 1 \end{bmatrix}$, $B = \begin{bmatrix} 1 & 0 \\ 0 & 3 \end{bmatrix}$, and $A + B = \begin{bmatrix} 3 & 0 \\ 0 & 4 \end{bmatrix}$.

e. False. If A is $n\times n$ and $\det A = 2$, then $\det A^3 = 2^3$, by Theorem 6.

f. False, by Theorem 3(b).

g. True, by Theorem 3(c).

h. True, by using Theorem 3(a) several times.

i. False, by Theorem 5.

j. False. The statement is false for $n \times n$ invertible matrices when n is an even integer, by Theorem 3(c).

k. True, because $\det A^TA = (\det A^T)(\det A) = (\det A)^2 \geq 0$, by Theorems 6 and 5.

l. False. Cramer's rule applies only to an $n \times n$ system with an invertible coefficient matrix.

m. False. The area of the triangle mentioned is 5.

n. True. By Theorem 6, $(\det A)^3 = \det A^3 = \det 0 = 0$. Hence $\det A = 0$.

o. False. The correct formula is given in Exercise 31 of Section 3.2.

p. True. By Theorem 6, $(\det A)(\det A^{-1}) = \det(AA^{-1}) = \det I = 1$.

CHAPTER 3 GLOSSARY CHECKLIST

Check your knowledge by attempting to write definitions of the terms below. Then compare your work with the definitions given in the text's Glossary. Ask your instructor which definitions, if any, might appear on a test.

adjugate (or **classical adjoint**): The matrix adj A formed from a square matrix A by replacing the (i, j)-entry of A by ....

cofactor: A number C_{ij} = ..., called the *(i, j)-cofactor of A*, where A_{ij} is the submatrix formed by deleting ....

cofactor expansion: A formula for det A using cofactors associated with one row or one column, such as for row 1: ....

Cramer's Rule: A formula for each entry in ....

interchange (matrix): An elementary matrix obtained by interchanging ....

row replacement (matrix): An elementary matrix obtained from the identity matrix by ....

scale by r (matrix): An elementary matrix obtained by multiplying ....

4 VECTOR SPACES

4.1 VECTOR SPACES AND SUBSPACES

The main focus of the chapter is on $\mathbb{R}^n$ and its subspaces. However, Section 4.1 builds a framework within which the theory for $\mathbb{R}^n$ rests. Most of the exercises in the chapter concern subspaces of $\mathbb{R}^n$, but some are designed to help you learn gradually about other important vector spaces.

KEY IDEAS

A *vector space* is any collection of objects that behave as vectors do in $\mathbb{R}^n$. (The precise meaning of "behave" is described by the axioms on page 211.) A *vector* is simply any object that belongs to a vector space. Lists of numbers, arrows, and polynomials are all examples of vectors, in different vector spaces.

The most important vector spaces in this text are subspaces of $\mathbb{R}^n$. Visualize subspaces as lines or planes *through the origin*, the origin by itself, or the entire space $\mathbb{R}^n$. To show that a set is a vector space, use Theorem 1. To show that a set is *not* a subspace, show that one of the properties in the subspace definition is violated. (See Exercises 1-4.)

STUDY NOTES

Parts of this chapter are somewhat more theoretical than the earlier chapters, but that is necessary in order to give a solid foundation for the rest of the course. Learn the key definitions and theorems as they appear in the text (rather than waiting until just before an exam). You need this knowledge to get through the conceptual exercises and to be prepared for later sections.

In Example 5, the concept of a function as a single "vector" in a vector space is difficult to absorb on a first reading, and you should not expect to master it in a few days.

The word *subspace* should usually be accompanied by a phrase such as "of V", or "of $\mathbb{R}^n$". Without such a phrase, the nature of the elements in the subspace is unknown. For instance, the statement "H is a subspace" does not tell us whether the elements in H are pairs of numbers, or polynomials, or something else. But the phrase "H is a subspace of $\mathbb{P}_3$" includes the information that each vector in H is a polynomial of degree three or less.

Set Notation: The notation introduced in Example 8 is sometimes used in this chapter as an efficient way to describe a set. The symbols and phrases inside the set brackets describe the set. The part to the left of the colon displays the basic nature of the elements in the set (such as vectors in $\mathbb{R}^3$) while the part to the right adds any qualifying conditions that must be met in order for an element to belong to the set (such as all vector entries must be positive). For instance, the set in Example 8 can also be written as

$$H = \left\{ \begin{bmatrix} s \\ t \\ u \end{bmatrix} : s \text{ and } t \text{ are real, and } u = 0 \right\}$$

As another example, the set in Exercise 11 can be written as

$$W = \left\{ \begin{bmatrix} a \\ b \\ c \end{bmatrix} : a = 5b + 2c, \text{ where } b \text{ and } c \text{ are arbitrary} \right\}$$

Here, and elsewhere, reference to the scalars b and c as "arbitrary" means that the scalars can be any real numbers.

Study Tip: Review Sections 1.3 to 1.6 before you reach Section 4.3.

SOLUTIONS TO EXERCISES

1. **a.** V is a subset of $\mathbb{R}^2$. The defining property of V is that the entries of every vector in V are nonnegative. So, if **u** and **v** are in V, their entries are nonnegative. Since a sum of nonnegative numbers is nonnegative, the vector **u** + **v** satisfies the condition that defines V. That is, **u** + **v** is in V.

   b. The text's solution gives a specific **u** and c. One specific "counterexample" suffices to show that V is not a vector space. However, you could also simply say, "If any nonzero vector **v** in V is multiplied by a negative scalar c, then c**v** is not in V because at least one of its entries is negative."

7. Most examples in the text involve integers, to make calculations simple, and one can easily overlook the fact that a subspace must be closed under multiplication by *all* real numbers.

Checkpoint: Let H be the set of all points (x, y, z) in $\mathbb{R}^3$ that satisfy the condition $x^2 + y^2 - z^2 = 0$. Is H a subspace of $\mathbb{R}^3$?

13. **a.** $\mathbf{w}$ is certainly not one of the three vectors in $\{\mathbf{v}_1, \mathbf{v}_2, \mathbf{v}_3\}$.

b. Span$\{\mathbf{v}_1, \mathbf{v}_2, \mathbf{v}_3\}$ contains infinitely many vectors.

c. $\mathbf{w}$ is in the subspace (of $\mathbb{R}^3$) spanned by $\mathbf{v}_1, \mathbf{v}_2, \mathbf{v}_3$ if and only if the equation $x_1\mathbf{v}_1 + x_2\mathbf{v}_2 + x_3\mathbf{v}_3 = \mathbf{w}$ is consistent (has a solution). Row reduce the augmented matrix:

$$\begin{bmatrix} 1 & 2 & 4 & 3 \\ 0 & 1 & 2 & 1 \\ -1 & 3 & 6 & 2 \end{bmatrix} \sim \begin{bmatrix} 1 & 2 & 4 & 3 \\ 0 & 1 & 2 & 1 \\ 0 & 5 & 10 & 5 \end{bmatrix} \sim \begin{bmatrix} 1 & 2 & 4 & 3 \\ 0 & 1 & 2 & 1 \\ 0 & 0 & 0 & 0 \end{bmatrix}$$

There is no pivot in the augmented column, so the vector equation is consistent, and $\mathbf{w}$ is in Span$\{\mathbf{v}_1, \mathbf{v}_2, \mathbf{v}_3\}$.

19. Let H be the set of all functions described in (5). Then H is a subset of the vector space V of all real-valued functions, and H consists of all linear combinations of the functions $\cos \omega t$ and $\sin \omega t$. By Theorem 1, H is a subspace of V, and hence is a vector space.

23. **a.** See Example 5. This is an important problem, so write your answer and justification before you read a note I have added at the end of this section.

b. See the definition of a vector.

c. See Exercises 1, 2, or 3.

d. See the paragraph before Example 6.

e. See Example 3.

25. Axiom 4 (plus Axiom 2) shows that $\mathbf{0} + \mathbf{w} = \mathbf{w}$. Exercises 25—30 show how facts that we take for granted in $\mathbb{R}^n$ depend only on a few basic properties of $\mathbb{R}^n$, properties that are now axioms for a general vector space.

31. Let H be a subspace of V that contains the vectors $\mathbf{u}$ and $\mathbf{v}$. Since H is closed under multiplication by scalars, H must contain all scalar multiples of $\mathbf{u}$ and $\mathbf{v}$. Since H is also closed under vector addition, it contains all sums of scalar multiples of $\mathbf{u}$ and $\mathbf{v}$. That is, H contains all vectors in Span$\{\mathbf{u}, \mathbf{v}\}$.

32. (Comments, not the solution) If H and K are subspaces of a vector space V, then $H \cap K$ is the largest subspace of V that is contained in both H and K. A related subspace, not mentioned in the text, is the smallest subspace of V that contains both H and K. This subspace is denoted by $H + K$, and it consists of all vectors $\mathbf{v}$ in V that can be written in the form $\mathbf{v} = \mathbf{h} + \mathbf{k}$, where $\mathbf{h}$ is some vector in H and $\mathbf{k}$ is some vector in K. In set notation,

$$H + K = \{\mathbf{h} + \mathbf{k} : \mathbf{h} \text{ is in } H \text{ and } \mathbf{k} \text{ is in } K\}$$

If Exercise 32 is assigned as homework, there is a chance that somewhere along the way you might encounter the subspace $H + K$. (Guess where!)

23. Note for part (a): I included this True/False question to point out a mistake that many students make in their writing. The statement in (a) is false because the equation $\mathbf{f}(t) = 0$ is not accompanied by the phrase "for all t" or "for all t in $\mathbb{R}$." The zero vector in a space of real-valued functions is not the *number* 0 but is instead a *function* $\mathbf{f}(t)$ that has the value zero for all t in its domain.

An equation such as $\mathbf{f}(t) = 0$ must be accompanied by a context or specific phrase that explains what the equation "$\mathbf{f}(t) = 0$" represents. For instance, t might represent a specific known number that satisfies $\mathbf{f}(t) = 0$, or t might represent an unknown number or numbers to be found by solving $\mathbf{f}(t) = 0$.

Answer to Checkpoint: H is not a subspace. Counterexample: $(1,0,1)$ and $(0,1,1)$ are in H and yet their sum, $(1,1,2)$, is not in H (because $1^2 + 1^2 - 2^2 \neq 0$). Many counterexamples are possible. Only one is needed.

4.2 NULL SPACES, COLUMN SPACES, AND LINEAR TRANSFORMATIONS

Many problems in linear algebra involve a subspace in one way or another. This section provides an opportunity to become comfortable with the concept. The foundation for this section was laid in Sections 1.3 and 1.5. Have you reviewed those sections yet?

KEY IDEAS

Theorems 2 and 3 describe the main types of subspaces. (The proof of Theorem 2 makes a good exam question, because it tests both the definition of Nul A and the definition of a subspace.)

Theorem 3 actually has two conclusions: Col A is a subspace of $\mathbb{R}^m$.

The first phrase tells us that linear combinations of vectors in Col A remain in Col A. The phrase "of $\mathbb{R}^m$" reminds us that each vector has m entries (because A has m rows). A similar remark applies to the statement from Theorem 2 that Nul A is a subspace of $\mathbb{R}^n$.

The box after Example 4 shows that the statement "Col $A = \mathbb{R}^m$" can be added to the list of equivalent statements in Theorem 4 of Section 1.4.

STUDY NOTES

In Example 3, the statement "Nul A = Span{**u**, **v**, **w**}" would be an *explicit* description of Nul A (provided you specify what **u**, **v**, and **w** are).

In some applications, it is important to know that for a given $m \times n$ matrix A, every equation $A\mathbf{x} = \mathbf{b}$ has a solution (assuming **b** is in $\mathbb{R}^m$). Yet it may require some effort to determine whether this is the case. If not every equation $A\mathbf{x} = \mathbf{b}$ has a solution, then not every **b** belongs to Col A, and hence Col A is a proper subspace of $\mathbb{R}^m$. One of the goals of the next few sections is to obtain a method for determining when Col $A = \mathbb{R}^m$.

Checkpoint 1: How many pivot positions does an $m \times n$ matrix A have if Col $A = \mathbb{R}^m$?

Study Tip: Theorems 1, 2, and 3 are the main tools for showing that a set is a vector space (that is, a subspace of some known vector space). Review these theorems now, before starting the exercises.

SOLUTIONS TO EXERCISES

1. Now is the time to learn the *definition* of Nul A. A vector **x** is in Nul A precisely when the product $A\mathbf{x}$ is defined and $A\mathbf{x} = \mathbf{0}$. Given **x**, simply compute $A\mathbf{x}$ to determine whether $A\mathbf{x}$ is zero.

$$A\mathbf{x} = \begin{bmatrix} 3 & -5 & -3 \\ 6 & -2 & 0 \\ -8 & 4 & 1 \end{bmatrix} \begin{bmatrix} 1 \\ 3 \\ -4 \end{bmatrix} = \begin{bmatrix} 0 \\ 0 \\ 0 \end{bmatrix}, \quad \text{so} \quad \begin{bmatrix} 1 \\ 3 \\ -4 \end{bmatrix} \text{ is in Nul } A.$$

Warning: In Exercises 3—6, writing an equation $\mathbf{x} = c\mathbf{u} + d\mathbf{v}$ is not the same as listing, say, the vectors **u** and **v** that span the null space. The appropriate answer for these exercises is a list of a small finite number of vectors that span Nul A, not a description of *all* the vectors in Nul A.

Study Tip: Try Practice Problem 1 before you work on Exercises 7-14. If you can't find *two* ways to work the practice problem, reread the first paragraph of Section 4.2 (but don't look at it until you have attempted the practice problem).

7. The set W is a subset of $\mathbb{R}^3$. If W were a vector space (under the standard operations in $\mathbb{R}^3$), it would be a subspace of $\mathbb{R}^3$. But W fails *every* property of a subspace, so it is not a vector space. For instance, the vector (0,0,0) does not satisfy the condition $a + b + c = 2$, and so the zero vector is not in W.

13. A typical element of W can be written as follows:

$$\begin{bmatrix} c-6d \\ d \\ c \end{bmatrix} = \begin{bmatrix} c \\ 0 \\ c \end{bmatrix} + \begin{bmatrix} -6d \\ d \\ 0 \end{bmatrix} = c\underset{\substack{\uparrow \\ \mathbf{u}}}{\begin{bmatrix} 1 \\ 0 \\ 1 \end{bmatrix}} + d\underset{\substack{\uparrow \\ \mathbf{v}}}{\begin{bmatrix} -6 \\ 1 \\ 0 \end{bmatrix}} = \begin{bmatrix} 1 & -6 \\ 0 & 1 \\ 1 & 0 \end{bmatrix}\begin{bmatrix} c \\ d \end{bmatrix}$$

Since c and d are any real numbers, this calculation shows that W is a subspace of $\mathbb{R}^3$ (and hence a vector space) by Theorem 3. Alternatively, this calculation shows that W is the same as Span{$\mathbf{u}, \mathbf{v}$}, so W is a subspace of $\mathbb{R}^3$, by Theorem 1.

Checkpoint 2: Why is W a subspace "of $\mathbb{R}^3$"?

19. The matrix A is 2×5, so vectors in Nul A must have 5 entries and vectors in Col A must have 2 entries. Thus Nul A is a subspace of $\mathbb{R}^5$ and Col A is a subspace of $\mathbb{R}^2$.

Study Tip: Exercises 17—20 may seem simple, but they will help you in Section 4.5.

25. **a.** Check the definition before Example 1.

 b. See Theorem 2.

 c. See the remark just before Example 4.

 d. See the table that contrasts Nul A and Col A.

 e. See Fig. 2.

 f. See the remark after Theorem 3.

31. The solution in the text shows that T is a linear transformation. If $T(\mathbf{p})$ is the zero vector, then $\mathbf{p}(0) = 0$ and $\mathbf{p}(1) = 0$, by definition of T. One such polynomial is $\mathbf{p}(t) = t(t - 1)$. Any other *quadratic* polynomial that vanishes at 0 and 1 must be a multiple of $\mathbf{p}$, so $\mathbf{p}$ spans the kernel of T.

For the range of T, observe that the image of the constant 1 function is $\begin{bmatrix}1\\1\end{bmatrix}$, and the image of the polynomial t is $\begin{bmatrix}0\\1\end{bmatrix}$. Denote these two images by $\mathbf{u}$ and $\mathbf{v}$, respectively. Since the range of T is a subspace of $\mathbb{R}^2$ that contains $\mathbf{u}$ and $\mathbf{v}$, the range must contain all linear combinations of $\mathbf{u}$ and $\mathbf{v}$. By inspection, $\mathbf{u}$ and $\mathbf{v}$ are linearly independent, so they span $\mathbb{R}^2$. Thus the range of T must contain all of $\mathbb{R}^2$.

37. The vector $\mathbf{w}$ is in Col A because row reduction of the augmented matrix $[A \;\; \mathbf{w}]$ shows that the equation $A\mathbf{x} = \mathbf{w}$ is consistent. The vector $\mathbf{w}$ is not in Nul A because $A\mathbf{x}$ is not the zero vector. (The first printing of the text has an error: the (1,1)-entry in A should be 7, not -7.)

Answers to Checkpoints:

1. An $m \times n$ matrix A must have m pivot positions in order for Col A to be all of $\mathbb{R}^m$.

2. W is a subspace "of $\mathbb{R}^3$" because each vector in W has three entries.

Mastering Linear Algebra Concepts: Subspace

You will need a clear mental image of a subspace throughout the rest of the text. Organize what you have learned in Sections 4.1 and 4.2 (together with Sections 1.3—1.5) on a review sheet that covers the following categories:

▶ definition	Page 214
▶ equivalent description	Paragraph before Example 6 in Sec. 4.1
▶ geometric interpretations	Figs. 6, 7, 9 in Sec. 4.1
▶ subspaces defined explicitly	Theorems 1, 3; Exercises 15-18 in Sec. 4.1
subspaces defined implicitly	Theorem 2; Exercise 20(b), 22 in Sec. 4.1
▶ special cases	Theorems 1, 2; Fig. 2 in Sec. 4.2
▶ examples and counterexamples	Examples and exercises in Sec. 4.1, 4.2
▶ algorithms and computations	Exercises 7-14 in Sec. 4.2
▶ connections with other concepts	Fig. 2 and Example 8 in Sec. 4.2

The references in the right column above are not exhaustive. You can find more facts and examples in the exercises for Sections 4.1 and 4.2.

4.3 LINEARLY INDEPENDENT SETS; BASES

The definition of linear independence carries over from $\mathbb{R}^n$ to any vector space. And the geometric interpretations in Chapter 1 of linearly independent and dependent sets should help you visualize these concepts here.

KEY IDEAS

In general, you cannot use an ordinary matrix equation $A\mathbf{x} = \mathbf{0}$ to study linear dependence of $\{\mathbf{v}_1, \ldots, \mathbf{v}_p\}$. You have to work with the vector equation $c_1\mathbf{v}_1 + \cdots + c_p\mathbf{v}_p = \mathbf{0}$, or with Theorem 4, unless the vectors happen to be n-tuples of numbers.

A set $\{\mathbf{v}\}$ with $\mathbf{v} \neq \mathbf{0}$ is linearly independent because the vector equation $x_1\mathbf{v} = \mathbf{0}$ has only the trivial solution. (See Exercise 30 in Section 4.1.) The set $\{\mathbf{0}\}$ is linearly dependent, because the equation $x_1\mathbf{0} = \mathbf{0}$ has many nontrivial solutions.

Theorem 6 is important for later work, but its proof is rather subtle. Study Examples 8 and 9 carefully, as well as the proof of the theorem.

A basis for a vector space V is a set in V that is large enough to span V and small enough to be linearly independent. See also the subsection *Two Views of a Basis*. The plural of *basis* is *bases*.

Warning: If a set in V does not span V, the set may or may not be linearly dependent.

The table below adds three new statements to the Invertible Matrix Theorem, at the bottom of the list from Section 2.3 in this *Study Guide*.

STATEMENTS FROM THE INVERTIBLE MATRIX THEOREM

Equivalent statements, for any $n \times p$ matrix A.	Equivalent only for an $n \times n$ square matrix A.	Equivalent statements, for any $n \times p$ matrix A.
k. There is a matrix D such that $AD = I$.	a. A is an invertible matrix.	j. There is a matrix C such that $CA = I$.
*. A has a pivot position in every row.	c. A has n pivot positions.	*. A has a pivot position in every column
h. The columns of A span $\mathbb{R}^n$.	b. A is row equivalent to the $n \times n$ identity matrix.	e. The columns of A form a linearly independent set.

g. The equation $A\mathbf{x} = \mathbf{b}$ has at least one solution for each $\mathbf{b}$ in $\mathbb{R}^n$.	*. The equation $A\mathbf{x} = \mathbf{b}$ has a unique solution for each $\mathbf{b}$ in $\mathbb{R}^n$.	*. The equation $A\mathbf{x} = \mathbf{b}$ has at most one solution for each $\mathbf{b}$ in $\mathbb{R}^n$.
i. The linear transformation $\mathbf{x} \mapsto A\mathbf{x}$ maps $\mathbb{R}^p$ onto $\mathbb{R}^n$.	*. The linear transformation $\mathbf{x} \mapsto A\mathbf{x}$ is invertible.	f. The linear transformation $\mathbf{x} \mapsto A\mathbf{x}$ is one-to-one.
	*. A is a product of elementary matrices.	d. The equation $A\mathbf{x} = \mathbf{0}$ has only the trivial solution.
	l. A^T is an invertible matrix.	*. The equation $A\mathbf{x} = \mathbf{0}$ has no free variables.
*. Col $A = \mathbb{R}^n$	*. The columns of A form a basis of $\mathbb{R}^n$.	*. Nul $A = \{\mathbf{0}\}$

SOLUTIONS TO EXERCISES

1. The complete solution is in the text. For a general set of n vectors in $\mathbb{R}^n$, row operations on a matrix will usually be needed to determine if the matrix has n pivot positions.

7. Again, the solution is in the text. Any set in $\mathbb{R}^n$ with fewer than n vectors cannot span $\mathbb{R}^n$ and therefore cannot be a basis for $\mathbb{R}^n$. Such a set may or may not be linearly independent. What similar statement can you make about a set in $\mathbb{R}^n$ with more than n vectors? See Exercise 8 for ideas.

Study Tip: Theorem 4 in Section 1.4 may help you decide whether a set of vectors spans $\mathbb{R}^n$.

13. The matrix B is in echelon form and displays the pivot columns. A basis for Col A consists of columns 1 and 2 of A: $\begin{bmatrix}-2\\2\\-3\end{bmatrix}, \begin{bmatrix}4\\-6\\8\end{bmatrix}$. This is not the only correct choice, but it is the "standard" choice. A *wrong* choice would be columns 1 and 2 of B. See the Warning after Theorem 6.

For the null space, solve $A\mathbf{x} = \mathbf{0}$:

$$[A \quad \mathbf{0}] \sim [B \quad \mathbf{0}] = \begin{bmatrix}1 & 0 & 6 & 5 & 0\\0 & 2 & 5 & 3 & 0\\0 & 0 & 0 & 0 & 0\end{bmatrix} \sim \begin{bmatrix}1 & 0 & 6 & 5 & 0\\0 & 1 & 5/2 & 3/2 & 0\\0 & 0 & 0 & 0 & 0\end{bmatrix}$$

Then $x_1 = -6x_3 - 5x_4$, $x_2 = -(5/2)x_3 - (3/2)x_4$, with x_3 and x_4 free. The general solution is

$$\mathbf{x} = \begin{bmatrix} x_1 \\ x_2 \\ x_3 \\ x_4 \end{bmatrix} = \begin{bmatrix} -6x_3 - 5x_4 \\ -(5/2)x_3 - (3/2)x_4 \\ x_3 \\ x_4 \end{bmatrix} = x_3\begin{bmatrix} -6 \\ -5/2 \\ 1 \\ 0 \end{bmatrix} + x_4\begin{bmatrix} -5 \\ -3/2 \\ 0 \\ 1 \end{bmatrix}$$

This equation describes *all* vectors in Nul A, not just a basis for Nul A. For a basis, "standard" choice is (-6, -5/2, 1, 0) and (-5, -3/2, 0, 1). Another choice is (-12, -5, 2, 0) and (-10, -3, 0, 2), which avoids fractions.

Warning: You really need to know the definition of Nul A and the definition of Col A, not just the procedures for finding bases for these spaces. The definitions will help you avoid the fairly common mistake of attempting to use the null space procedure to find a basis for a column space, or vice-versa.

19. We can solve the equation $4\mathbf{v}_1 + 5\mathbf{v}_2 - 3\mathbf{v}_3 = \mathbf{0}$ for any one of the three vectors in terms of the others. By the Spanning Set Theorem, the set spanned by all three vectors is the same as the set spanned by any two of the vectors—any one of the three vectors can be discarded. If we discard $\mathbf{v}_3$, then $\mathbf{v}_1$ and $\mathbf{v}_2$ span H and are obviously linearly independent. Hence $\{\mathbf{v}_1, \mathbf{v}_2\}$ is a basis for H. The same reasoning applies to $\{\mathbf{v}_1, \mathbf{v}_3\}$ and $\{\mathbf{v}_2, \mathbf{v}_3\}$.

21. **a.** See the paragraph preceding Theorem 4.

b. See the definition of a basis. **c.** See Example 3.

d. See the subsection *Two Views of a Basis.*

e. See the box before Example 9.

23. Let $A = [\mathbf{v}_1\ \ \mathbf{v}_2\ \ \mathbf{v}_3\ \ \mathbf{v}_4]$. Since A is square and its columns span $\mathbb{R}^4$, the columns of A must be linearly independent, by the Invertible Matrix Theorem. So $\{\mathbf{v}_1, \mathbf{v}_2, \mathbf{v}_3, \mathbf{v}_4\}$ is a basis for $\mathbb{R}^4$.

Checkpoint: Suppose $\{\mathbf{v}_1, \ldots, \mathbf{v}_n\}$ is a basis for $\mathbb{R}^n$ and A is an invertible $n \times n$ matrix. Explain why $\{A\mathbf{v}_1, \ldots, A\mathbf{v}_n\}$ is a basis for $\mathbb{R}^n$.

25. The displayed equation shows only that Span$\{\mathbf{v}_1, \mathbf{v}_2, \mathbf{v}_3\}$ *contains* H. In fact, the vectors $\mathbf{v}_1, \mathbf{v}_2, \mathbf{v}_3$ are not all in H, so Span$\{\mathbf{v}_1, \mathbf{v}_2, \mathbf{v}_3\}$ cannot be H. Therefore $\{\mathbf{v}_1, \mathbf{v}_2, \mathbf{v}_3\}$ cannot be a basis for H. (It is easy to check that $\{\mathbf{v}_1, \mathbf{v}_2, \mathbf{v}_3\}$ is a basis for $\mathbb{R}^3$.)

31. (This generalizes Exercise 31 in Section 1.7.) Suppose $\{\mathbf{v}_1, \ldots, \mathbf{v}_p\}$ is linearly dependent. Then there exist $c_1, \ldots, c_p$, not all zero, such that

$$c_1\mathbf{v}_1 + \cdots + c_p\mathbf{v}_p = \mathbf{0}$$

Then, since T is linear,

$$T(c_1\mathbf{v}_1 + \cdots + c_p\mathbf{v}_p) = T(\mathbf{0}) = \mathbf{0}$$ See the boxed statement on page 71.

and

$$c_1T(\mathbf{v}_1) + \cdots + c_pT(\mathbf{v}_p) = \mathbf{0}$$

Since not all the c_i are zero, $\{T(\mathbf{v}_1), \ldots, T(\mathbf{v}_p)\}$ is linearly dependent.

Study Tip: The solution of Exercise 31 illustrates how to use linear dependence: If $\{\mathbf{v}_1, \ldots, \mathbf{v}_p\}$ is known to be linearly dependent, then you can write $c_1\mathbf{v}_1 + \cdots + c_p\mathbf{v}_p = \mathbf{0}$, assume that not all the c_k are zero, and use this equation in some way.

Answer to Checkpoint: Let $B = [\mathbf{v}_1 \ \cdots \ \mathbf{v}_n]$. Then B is invertible because $\{\mathbf{v}_1, \ldots, \mathbf{v}_n\}$ is a basis for $\mathbb{R}^n$. If A is an invertible $n \times n$ matrix, then so is AB; hence the columns of AB, namely $A\mathbf{v}_1, \ldots, A\mathbf{v}_n$, form a basis for $\mathbb{R}^n$.

Mastering Linear Algebra Concepts: Basis

To the review sheet(s) you have on linear independence, add Examples 1, 2, and 6 from this section. The definition and geometric interpretations are unchanged. Add a note about not using the matrix equation $A\mathbf{x} = \mathbf{0}$ in the general case.

Start a separate review sheet for "basis", even though it involves two other concepts (span and linear independence) already being reviewed.

▶ definition	Page 232
▶ geometric interpretation	Fig. 1
▶ special cases	Standard bases for $\mathbb{R}^n$ and $\mathbb{P}_n$
▶ examples and counterexamples	Example 10, Exercise 25
▶ algorithms and computations	Examples 7 and 9; Example 3 in Sec. 4.2
▶ connections with other concepts	Invertible Matrix Theorem
	Unique Representation Theorem (in Sec. 4.4)

MATLAB ref and cos

The command **ref(A)** produces the reduced echelon form of A. From that you can write a basis for Col A or write the homogeneous equations that describe Nul A. (Don't forget that A is a coefficient matrix, not an augmented matrix.) MATLAB has another command, **rref**, that works basically the same as **ref** but often is much slower. See page 2-39.

In some cases, roundoff error or an extremely small pivot entry can cause **ref** to produce an incorrect echelon form. The more reliable singular value decomposition (see Section 7.4) can produce bases for Col A and Nul A, but **ref** is satisfactory for our purposes.

For Exercise 34: If **t** is a vector and k is a positive integer, then **cos(t).^k** is a vector the same size as **t**, formed entrywise from **t** using the function $\cos^k(\ )$.

4.4 COORDINATE SYSTEMS

This section contains a variety of geometric and algebraic explanations of the idea of a coordinate system for a vector space.

KEY IDEAS

The coordinate mapping from a vector space V (with a basis of n elements) onto $\mathbb{R}^n$ is a rule for giving "$\mathbb{R}^n$-names" to vectors in V in such a way that the vector space structure of V is still visible in $\mathbb{R}^n$. Every vector space calculation in V is precisely mirrored by the same calculation in $\mathbb{R}^n$.

An important special case is when V is itself $\mathbb{R}^n$, and each vector $\mathbf{x}$ and its coordinate vector $[\mathbf{x}]_{\mathcal{B}}$ are related by a matrix equation $\mathbf{x} = P_{\mathcal{B}}[\mathbf{x}]_{\mathcal{B}}$.

Everything in the section depends on the Unique Representation Theorem. The proof of that theorem could appear on an exam because it shows precisely why the two properties of a basis $\mathcal{B}$ are important, and it illustrates how linear independence can be used in an argument:

> Any vector in V has "coordinates" because $\mathcal{B}$ spans V, and the coordinates are uniquely determined because $\mathcal{B}$ is linearly independent.

If you are asked to prove the theorem, make sure your proof shows exactly where each property of a basis is needed in the proof. Also, be careful *not* to use a matrix in the proof. The vectors $\mathbf{v}_1, \ldots, \mathbf{v}_n$ cannot be arranged as the columns of an ordinary matrix when the vectors are in some abstract vector space.

Checkpoint: Let $\mathcal{B} = \{\mathbf{b}_1, \ldots, \mathbf{b}_n\}$ be a basis for $\mathbb{R}^n$. Apply the Invertible Matrix Theorem to the matrix $A = [\mathbf{b}_1 \ \cdots \ \mathbf{b}_n]$ and deduce the Unique Representation Theorem for the case when $V = \mathbb{R}^n$.

STUDY NOTES

Be careful to distinguish between $\mathbf{x}$ and $[\mathbf{x}]_{\mathcal{B}}$. They are *not equal* in general, even if $\mathbf{x}$ itself is in $\mathbb{R}^n$ (unless $\mathcal{B}$ is the standard basis for $\mathbb{R}^n$).

Theorem 8 and Exercises 25 and 26 show that the coordinate mapping translates vector space statements or calculations in V into equivalent (and familiar) calculations in $\mathbb{R}^n$. The table below lists some examples of typical linear algebra statements.

CORRESPONDING STATEMENTS IN ISOMORPHIC VECTOR SPACES

Linear Algebra in V	Matrix Algebra in $\mathbb{R}^n$
a. $\mathbf{u}$, $\mathbf{v}$, and $\mathbf{w}$ are in V	$[\mathbf{u}]_{\mathcal{B}}$, $[\mathbf{v}]_{\mathcal{B}}$, and $[\mathbf{w}]_{\mathcal{B}}$ are in $\mathbb{R}^n$
b. $\mathbf{w}$ is in Span$\{\mathbf{u}, \mathbf{v}\}$, or	$[\mathbf{w}]_{\mathcal{B}}$ is in Span$\{[\mathbf{u}]_{\mathcal{B}}, [\mathbf{v}]_{\mathcal{B}}\}$, or
$\mathbf{w}$ is in the subspace of V spanned by $\mathbf{u}$ and $\mathbf{v}$	$[\mathbf{w}]_{\mathcal{B}}$ is in the subspace of $\mathbb{R}^n$ spanned by $[\mathbf{u}]_{\mathcal{B}}$ and $[\mathbf{v}]_{\mathcal{B}}$
c. $\mathbf{w} = c\mathbf{u} + d\mathbf{v}$	$[\mathbf{w}]_{\mathcal{B}} = c[\mathbf{u}]_{\mathcal{B}} + d[\mathbf{v}]_{\mathcal{B}}$
d. $\{\mathbf{v}_1, \ldots, \mathbf{v}_p\}$ is lin. indep.	$\{[\mathbf{v}_1]_{\mathcal{B}}, \ldots, [\mathbf{v}_p]_{\mathcal{B}}\}$ is lin. indep.
e. $\{\mathbf{v}_1, \ldots, \mathbf{v}_p\}$ spans V	$\{[\mathbf{v}_1]_{\mathcal{B}}, \ldots, [\mathbf{v}_p]_{\mathcal{B}}\}$ spans $\mathbb{R}^n$
f. $\{\mathbf{v}_1, \ldots, \mathbf{v}_n\}$ is a basis for V	$\{[\mathbf{v}_1]_{\mathcal{B}}, \ldots, [\mathbf{v}_n]_{\mathcal{B}}\}$ is a basis for $\mathbb{R}^n$

SOLUTIONS TO EXERCISES

1. Since $[\mathbf{x}]_{\mathcal{B}} = \begin{bmatrix} 5 \\ 3 \end{bmatrix}$, we have $\mathbf{x} = 5\mathbf{b}_1 + 3\mathbf{b}_2 = 5\begin{bmatrix} 3 \\ -5 \end{bmatrix} + 3\begin{bmatrix} -4 \\ 6 \end{bmatrix} = \begin{bmatrix} 3 \\ -7 \end{bmatrix}$.

7. The $\mathcal{B}$-coordinates of $\mathbf{x}$ are scalars c_1, c_2, c_3 that satisfy $c_1\mathbf{b}_1 + c_2\mathbf{b}_2 + c_3\mathbf{b}_3 = \mathbf{x}$. To solve this vector equation, row reduce the augmented matrix:

$$\begin{bmatrix} 1 & -3 & 2 & 8 \\ -1 & 4 & -2 & -9 \\ -3 & 9 & 4 & 6 \end{bmatrix} \sim \begin{bmatrix} 1 & -3 & 2 & 8 \\ 0 & 1 & 0 & -1 \\ 0 & 0 & 10 & 30 \end{bmatrix} \sim \begin{bmatrix} 1 & -3 & 0 & 2 \\ 0 & 1 & 0 & -1 \\ 0 & 0 & 1 & 3 \end{bmatrix} \sim \begin{bmatrix} 1 & 0 & 0 & -1 \\ 0 & 1 & 0 & -1 \\ 0 & 0 & 1 & 3 \end{bmatrix}$$

$$\uparrow \ \mathbf{b}_1 \quad \uparrow \ \mathbf{b}_2 \quad \uparrow \ \mathbf{b}_3 \quad \uparrow \ \mathbf{x}$$

So $[\mathbf{x}]_{\mathcal{B}} = \begin{bmatrix} -1 \\ -1 \\ 3 \end{bmatrix}$.

13. Using the method of Practice Problem 2, look for c_1, c_2, c_3 such that

$$c_1(1 + t^2) + c_2(t + t^2) + c_3(1 + 2t + t^2) = \mathbf{p}(t) = 1 + 4t + 7t^2 \quad (*)$$

Multiply out terms on the left, equate coefficients of like powers of t, and obtain the system

$$\begin{aligned} c_1 + c_3 &= 1 \\ c_2 + 2c_3 &= 4 \\ c_1 + c_2 + c_3 &= 7 \end{aligned}$$

Row reduction of the augmented matrix produces

$$\begin{bmatrix} 1 & 0 & 1 & 1 \\ 0 & 1 & 2 & 4 \\ 1 & 1 & 1 & 7 \end{bmatrix} \sim \begin{bmatrix} 1 & 0 & 1 & 1 \\ 0 & 1 & 2 & 4 \\ 0 & 0 & -2 & 2 \end{bmatrix} \sim \begin{bmatrix} 1 & 0 & 0 & 2 \\ 0 & 1 & 0 & 6 \\ 0 & 0 & 1 & -1 \end{bmatrix}, \text{ and } [\mathbf{p}]_{\mathcal{B}} = \begin{bmatrix} 2 \\ 6 \\ -1 \end{bmatrix}$$

A slightly faster solution uses Theorem 8 and the fact that a calculation in $\mathbb{P}_2$ can be done instead with coordinate vectors of all the given polynomials relative to the standard basis $\{1, t, t^2\}$. Equation (*) in such coordinate vectors becomes

$$c_1\begin{bmatrix} 1 \\ 0 \\ 1 \end{bmatrix} + c_2\begin{bmatrix} 0 \\ 1 \\ 1 \end{bmatrix} + c_3\begin{bmatrix} 1 \\ 2 \\ 1 \end{bmatrix} = \begin{bmatrix} 1 \\ 4 \\ 7 \end{bmatrix}$$

This vector equation is, of course, equivalent to the system of equations above, and it is solved by the same row reduction process.

15. **a.** See the definition of a $\mathcal{B}$-coordinate vector.

b. See equation (4).

c. See Example 5.

19. The set S spans V because every $\mathbf{x}$ in V has a representation as a (unique) linear combination of elements of S. To show linear independence, suppose that $S = \{\mathbf{v}_1, \ldots, \mathbf{v}_n\}$ and $c_1\mathbf{v}_1 + \cdots + c_n\mathbf{v}_n = \mathbf{0}$ for some scalars $c_1, \ldots, c_n$. The case when $c_1 = \cdots = c_n = 0$ is one possibility. By hypothesis, this is the *only* possible representation of the zero vector as a linear combination of the elements of S. So S is linearly independent and hence is a basis for V.

21. We want A to satisfy $A\mathbf{x} = [\mathbf{x}]_{\mathcal{B}}$ for each $\mathbf{x}$, and we know that the change-of-coordinates matrix satisfies $P_{\mathcal{B}}[\mathbf{x}]_{\mathcal{B}} = \mathbf{x}$ for each $\mathbf{x}$. Comparing these two equations, we see that

$$A = P_{\mathcal{B}}^{-1} = [\mathbf{b}_1 \quad \mathbf{b}_2]^{-1} = \begin{bmatrix} 1 & -2 \\ -4 & 9 \end{bmatrix}^{-1} = \begin{bmatrix} 9 & 2 \\ 4 & 1 \end{bmatrix}$$

23. Suppose that $[\mathbf{u}]_{\mathcal{B}} = [\mathbf{w}]_{\mathcal{B}}$ for some $\mathbf{u}$ and $\mathbf{w}$ in V, and denote the entries in this coordinate vector by $c_1, \ldots, c_n$. By definition of the coordinate vectors,

$$\mathbf{u} = c_1\mathbf{b}_1 + \cdots + c_n\mathbf{b}_n \quad \text{and} \quad \mathbf{w} = c_1\mathbf{b}_1 + \cdots + c_n\mathbf{b}_n$$

which shows that $\mathbf{u} = \mathbf{w}$. Since $\mathbf{u}$ and $\mathbf{w}$ were arbitrary elements of V, this shows that the coordinate mapping is one-to-one.

25. Since the coordinate mapping is one-to-one, the following equations have the same solutions, $c_1, \ldots, c_p$:

$$c_1\mathbf{u}_1 + \cdots + c_p\mathbf{u}_p = \mathbf{0} \qquad \text{(the zero vector in } V) \tag{1}$$

$$[c_1\mathbf{u}_1 + \cdots + c_p\mathbf{u}_p]_{\mathcal{B}} = [\mathbf{0}]_{\mathcal{B}} \qquad \text{(the zero vector in } \mathbb{R}^n) \tag{2}$$

Since the coordinate mapping is linear, (2) is equivalent to

$$c_1[\mathbf{u}_1]_{\mathcal{B}} + \cdots + c_p[\mathbf{u}_p]_{\mathcal{B}} = \begin{bmatrix} 0 \\ \vdots \\ 0 \end{bmatrix} \tag{3}$$

Hence $c_1, \ldots, c_p$ satisfy (1) if and only if they satisfy (3). So (1) has only the trivial solution if and only if (3) has only the trivial solution. It follows that $\{\mathbf{u}_1, \ldots, \mathbf{u}_p\}$ is linearly independent if and only if $\{[\mathbf{u}_1]_{\mathcal{B}}, \ldots, [\mathbf{u}_p]_{\mathcal{B}}\}$ is linearly independent. (This fact is also an immediate consequence of Exercises 31 and 32 in Section 4.3.)

Warning: The standard mistake in Exercises 27-32 is to write the coordinate vectors as the *rows* of a matrix and then to turn around and check the linear independence or dependence of the *columns*. Since we mainly work with column vectors, it is wise to write the coordinate vectors first (as columns) and afterwards write a matrix that can be row reduced to check the linear independence of its columns.

27. The coordinate vectors of the polynomials are $\begin{bmatrix}1\\0\\0\\1\end{bmatrix}, \begin{bmatrix}3\\1\\-2\\0\end{bmatrix}, \begin{bmatrix}0\\-1\\3\\-1\end{bmatrix}$ (relative to the standard basis). One can verify that these vectors are linearly independent in $\mathbb{R}^4$. This will show that the corresponding polynomials in $\mathbb{P}_3$ are linearly independent. This same type of analysis is needed for Exercises 28-32. The details are omitted.

Study Tip: Exercises 27-32 are easily changed into a question whether the given polynomials form a basis for $\mathbb{P}_3$. What else would you have to write or calculate in this case? Remember that although you may look at vectors in $\mathbb{R}^4$, your answer must include a discussion about the polynomials themselves. That is, you must explain why the given polynomials form, or do not form, a basis for the space of polynomials.

Answer to Checkpoint: The columns of the matrix $A = [\mathbf{b}_1 \ \cdots \ \mathbf{b}_n]$ form a basis for $\mathbb{R}^n$, so A is invertible, by the Invertible Matrix Theorem. By Theorem 5 in Section 2.2, for each $\mathbf{x}$ in $\mathbb{R}^n$ there exists a unique vector $\mathbf{c} = (c_1, \ldots, c_n)$ such that $\mathbf{x} = A\mathbf{c}$, that is, $\mathbf{x} = c_1\mathbf{b}_1 + \cdots + c_n\mathbf{b}_n$.

**MATLAB The Backslash Operator **

If an equation $A\mathbf{x} = \mathbf{b}$ has a unique solution, MATLAB will automatically produce $\mathbf{x}$ if you use the command

x = A\b

In this section, the equation probably will have the form $P\mathbf{u} = \mathbf{x}$, with $\mathbf{u}$ the $\mathcal{B}$-coordinate vector of $\mathbf{x}$, and the command will be **u = P\x**.

The "backslash" command works in two different ways. When A is square, the command **A\b** causes MATLAB to create an LU factorization of A (see Section 2.5); if A is invertible, the factorization is used to produce the unique solution to $A\mathbf{x} = \mathbf{b}$; and if A is not invertible, MATLAB gives the error message "*matrix is singular*" (even if the system $A\mathbf{x} = \mathbf{b}$ has a solution). When A is not square, **A\b** creates a least-squares solution (see Section 6.5).

4.5 THE DIMENSION OF A VECTOR SPACE

This short section provides a convenient way to compare the "sizes" of various subspaces of a vector space without having first to choose specific bases for the subspaces and then to count the number of basis elements.

KEY IDEAS

Theorem 10 shows that the dimension of a finite-dimensional vector space does not depend on the particular basis for the space. Example 4 shows how to visualize subspaces of various dimensions.

Theorems 9 and 12 are important for later theory and applications. You might remember Theorem 9 more easily in this form:

> In an n-dimensional vector space, any set of more than n vectors must be linearly dependent.

You already know this, of course, for $\mathbb{R}^n$. And if you read about isomorphisms in Section 4.4, then you can see why the result is true in general, because any n-dimensional vector space is isomorphic to $\mathbb{R}^n$. The Basis Theorem (Theorem 12) may be restated as follows:

> If $\dim V = p \geq 1$ and if S is a subset of V that contains exactly p elements, then S is linearly independent if and only if S spans V.

Warning: Suppose H is a subspace of a finite-dimensional space V. Theorem 11 shows that any basis of H can be extended to a basis of V. But it is *not* true that any basis of V can be cut down to a basis for H. That is, if S is a basis for V, it is not likely that a subset of S is a basis for H. For instance, consider the standard basis for $\mathbb{R}^3$ and any plane H that contains the origin but none of the coordinate axes.

SOLUTIONS TO EXERCISES

1. Since $\begin{bmatrix} s-2t \\ s+t \\ 3t \end{bmatrix} = s\begin{bmatrix} 1 \\ 1 \\ 0 \end{bmatrix} + t\begin{bmatrix} -2 \\ 1 \\ 3 \end{bmatrix}$ for all s, t, the set $\left\{\begin{bmatrix} 1 \\ 1 \\ 0 \end{bmatrix}, \begin{bmatrix} -2 \\ 1 \\ 3 \end{bmatrix}\right\}$ certainly spans the subspace, call it H. Also, the set is obviously linearly independent (because the vectors are not multiples), so the set is a basis for H. Hence, $\dim H = 2$.

3. The given subspace, call it H, is the set of all linear combinations of the vectors

$$\mathbf{v}_1 = \begin{bmatrix} 0 \\ 1 \\ 0 \\ 1 \end{bmatrix}, \quad \mathbf{v}_2 = \begin{bmatrix} 0 \\ -1 \\ 1 \\ 2 \end{bmatrix}, \quad \mathbf{v}_3 = \begin{bmatrix} 2 \\ 0 \\ -3 \\ 0 \end{bmatrix}$$

First determine if $\{\mathbf{v}_1, \mathbf{v}_2, \mathbf{v}_3\}$ is linearly independent. One way to do this is to row reduce the augmented matrix $[\mathbf{v}_1 \ \ \mathbf{v}_2 \ \ \mathbf{v}_3 \ \ \mathbf{0}]$. A faster way is to use Theorem 4 in Section 4.3. Clearly, $\mathbf{v}_1 \neq \mathbf{0}$, and $\mathbf{v}_2$ is not a multiple of $\mathbf{v}_1$, and $\mathbf{v}_3$ is not a linear combination of the vectors $\mathbf{v}_1, \mathbf{v}_2$ that precede it, because the first entry in $\mathbf{v}_3$ is not zero. Hence $\{\mathbf{v}_1, \mathbf{v}_2, \mathbf{v}_3\}$ is linearly independent and thus is a basis for the space H it spans. Thus $\dim H = 3$.

7. Standard calculations show that the set of solutions of the homogeneous system consists of only the trivial solution. So the subspace is $\{\mathbf{0}\}$, and it has no basis. (The vector $\mathbf{0}$ spans the space, but $\{\mathbf{0}\}$ is a linearly dependent set.) By definition, the dimension is zero. [*Note*: Persons who want every subspace to have a basis often define the empty set to be a basis for $\{\mathbf{0}\}$. The number of vectors in this basis is zero, so the dimension of $\{\mathbf{0}\}$ is still zero.]

13. A has three pivot columns, so $\dim \operatorname{Col} A = 3$. There are two columns without pivot positions, so the equation $A\mathbf{x} = \mathbf{0}$ has two free variables, and $\dim \operatorname{Nul} A = 2$.

19. **a**. See the box before Example 5. **b**. Read Example 4 carefully.

    c. See Example 1. **d**. See Theorem 10.

    e. See Practice Problem 2. (You should be working the practice problems before you start the exercises.)

21. Form the matrix whose columns are the coordinate vectors of the Hermite polynomials, relative to the standard basis $\{1, t, t^2, t^3\}$:

$$A = \begin{bmatrix} 1 & 0 & -2 & 0 \\ 0 & 2 & 0 & -12 \\ 0 & 0 & 4 & 0 \\ 0 & 0 & 0 & 8 \end{bmatrix}$$

The matrix has four pivots and hence is invertible. So its columns, the coordinate vectors, are linearly independent. Hence the Hermite polynomials themselves are linearly independent in $\mathbb{P}_3$. Since there are *four* Hermite polynomials, and $\dim \mathbb{P}_3 = 4$, we conclude from The Basis Theorem that the Hermite polynomials form a *basis* for $\mathbb{P}_3$.

Note: You could, of course, say that the columns of the matrix A span $\mathbb{R}^4$. But you cannot stop with that assertion, because you need the polynomials to span $\mathbb{P}_3$. You have to go on and point out that because of the isomorphism between $\mathbb{P}_3$ and $\mathbb{R}^4$, a set of vectors spans $\mathbb{P}_3$ if and only if the set of coordinate vectors (the columns of A) spans $\mathbb{R}^4$. So the solution is shorter if you appeal to the Basis Theorem.

25. Note that $n \geq 1$, because S cannot have fewer than 0 vectors. If $\dim V = n \geq 1$, then $V \neq \{0\}$. If S spans V, then a subset S' of S is a basis for V, by the Spanning Set Theorem. But if S has fewer than n vectors, then S' also has fewer than n vectors. This is impossible, by Theorem 10, because $\dim V = n$. So S cannot span V.

27. If $\mathbb{P}$ were finite-dimensional, then Theorem 11 would imply that $n + 1 = \dim \mathbb{P}_n \leq \dim \mathbb{P}$ for each n, because each $\mathbb{P}_n$ is a subspace of $\mathbb{P}$. This is impossible, so $\mathbb{P}$ must be infinite-dimensional.

29. **a.** True. Apply the Spanning Set Theorem to the set $\{\mathbf{v}_1,\ldots,\mathbf{v}_p\}$ and produce a basis for V. This basis will have no more than p elements in it, so $\dim V$ must be no more than p.

b. True. By Theorem 11, $\{\mathbf{v}_1,\ldots,\mathbf{v}_p\}$ can be expanded to a basis for V. The basis will have at least p elements in it, so $\dim V$ must be at least p.

c. True. Take any basis (of p vectors) for V and adjoin the zero vector. Spanning sets can be arbitrarily large. The dimension of V being p only keeps spanning sets from having *fewer* than p elements.

31. Since H is a nonzero subspace of a finite-dimensional space, H is finite-dimensional and has a basis, say, $\mathbf{v}_1,\ldots,\mathbf{v}_p$. Any vector in $T(H)$ has the form $T(\mathbf{y})$ for some $\mathbf{y}$ in H. Since $\{\mathbf{v}_1,\ldots,\mathbf{v}_p\}$ spans H, there exist scalars $c_1,\ldots,c_p$ such that $\mathbf{y} = c_1\mathbf{v}_1 + \cdots + c_p\mathbf{v}_p$. Since T is linear, $T(\mathbf{y}) = c_1T(\mathbf{v}_1) + \cdots + c_pT(\mathbf{v}_p)$. This shows that $\{T(\mathbf{v}_1),\ldots, T(\mathbf{v}_p)\}$ spans $T(H)$. By Exercise 29(a), $\dim T(H) \leq p = \dim H$.

Second proof: Let $k = \dim T(H)$. If $k = 0$, then $k < \dim H$. Otherwise, $T(H)$ has a basis, which can be written in the form $T(\mathbf{v}_1),\ldots, T(\mathbf{v}_k)$ for some vectors $\mathbf{v}_1,\ldots,\mathbf{v}_k$ in H. Since $\{T(\mathbf{v}_1),\ldots,T(\mathbf{v}_k)\}$ is linearly independent, so is $\{\mathbf{v}_1,\ldots,\mathbf{v}_k\}$, by Exercise 31 in Section 4.3. Since $\mathbf{v}_1,\ldots,\mathbf{v}_k$ are in H, the dimension of H must be at least k.

Hint for Exercise 32*:* Use an exercise in Section 4.3.

Study Tip: The next section is quite important. Do your best to get caught up now. Otherwise, you may have difficulty relating the various concepts and facts about matrices that will be reviewed in Section 4.6.

4.6 RANK

This section gives you a chance to put together most of the ideas of the chapter in the same way that Section 2.3 collected the main ideas of the sections that preceded it.

KEY IDEAS

The Rank Theorem is the main result. By definition, rank A = dim Col A. But because rank A is also the dimension of Row A, the displayed equation in the theorem leads to the equation: dim Row A + dim Nul A = n.

Equivalent Descriptions of Rank

The rank of an $m \times n$ matrix A may be described in several ways:

▶the dimension of the column space of A, (our definition)
▶the number of pivot positions in A, (from Theorem 6)
▶the maximum number of linearly independent columns in A,
▶the dimension of the row space of A, (from the Rank Theorem)
▶the maximum number of linearly independent rows in A,
▶the number of nonzero rows in an echelon form of A,
▶the maximum number of columns in an invertible submatrix of A.
(Supplementary Exercise 12 at the end of the chapter)

Pay attention to how Theorem 13 differs from the results in Section 4.3 about Col A: If you are interested in *rows* of A, use the nonzero rows of an echelon form B as a basis for Row A; if you are interested in the *columns* of A, only use B to obtain *information* about A (namely, to identify the pivot columns), and use the pivot columns of A as a basis for Col A. For Nul A, it is important to use the *reduced* echelon form of A.

When a matrix A is changed into a matrix B by one or more elementary row operations, the row space, null space, and column space of A may or may not be the same as the corresponding subspaces for B. The following table summarizes what can happen in this situation.

Effects of Elementary Row Operations

▶Row operations do not affect the linear dependence relations among the columns. (That is, the columns of A have exactly the same linear dependence relations as the columns of any matrix row-equivalent to A.)
▶Row operations usually change the column space.
▶Row operations never change the row space.
▶Row operations never change the null space.

The four subspaces shown in Figure 1 in the text are called the *fundamental subspaces* associated with A. (See Exercises 27-29.) The main difficulty here is to avoid confusion between Row A, Nul A, and Col A. The fourth subspace will appear again in Sections 6.1 and 7.4.

The table below adds three more statements to the bottom of the list for the Invertible Matrix Theorem, Section 4.3 in this *Study Guide*.

STATEMENTS FROM THE INVERTIBLE MATRIX THEOREM

Equivalent statements, for any $n \times p$ matrix A.	Equivalent only for an $n \times n$ square matrix A.	Equivalent statements, for any $n \times p$ matrix A.
k. There is a matrix D such that $AD = I$.	a. A is an invertible matrix.	j. There is a matrix C such that $CA = I$.
*. A has a pivot position in every row.	c. A has n pivot positions.	*. A has a pivot position in every column
h. The columns of A span $\mathbb{R}^n$.	b. A is row equivalent to the $n \times n$ identity matrix.	e. The columns of A form a linearly independent set.
g. The equation $A\mathbf{x} = \mathbf{b}$ has at least one solution for each $\mathbf{b}$ in $\mathbb{R}^n$.	*. The equation $A\mathbf{x} = \mathbf{b}$ has a unique solution for each $\mathbf{b}$ in $\mathbb{R}^n$.	*. The equation $A\mathbf{x} = \mathbf{b}$ has at most one solution for each $\mathbf{b}$ in $\mathbb{R}^n$.
i. The linear transformation $\mathbf{x} \mapsto A\mathbf{x}$ maps onto $\mathbb{R}^n$.	*. The linear transformation $\mathbf{x} \mapsto A\mathbf{x}$ is invertible.	f. The linear transformation $\mathbf{x} \mapsto A\mathbf{x}$ is one-to-one.
	*. A is a product of elementary matrices.	d. The equation $A\mathbf{x} = \mathbf{0}$ has only the trivial solution.
	l. A^T is an invertible matrix.	*. The equation $A\mathbf{x} = \mathbf{0}$ has no free variables.
n. Col $A = \mathbb{R}^n$	m. The columns of A form a basis of $\mathbb{R}^n$.	q. Nul $A = \{\mathbf{0}\}$
o. dim Col $A = n$	p. rank $A = n$.	r. dim Nul $A = 0$

SOLUTIONS TO EXERCISES

1. $A = \begin{bmatrix} 1 & -4 & 9 & -7 \\ -1 & 2 & -4 & 1 \\ 5 & -6 & 10 & 7 \end{bmatrix} \sim B = \begin{bmatrix} 1 & 0 & -1 & 5 \\ 0 & -2 & 5 & -6 \\ 0 & 0 & 0 & 0 \end{bmatrix}$

Look at B, and conclude that A has two pivot columns and the equation $A\mathbf{x} = \mathbf{0}$ has two free variables. So rank $A = 2$ and dim Nul $A = 2$. In fact, the first two columns of A are pivot columns, so

$$\text{Basis for Col } A: \left\{ \begin{bmatrix} 1 \\ -1 \\ 5 \end{bmatrix}, \begin{bmatrix} -4 \\ 2 \\ -6 \end{bmatrix} \right\}$$

For the row space, use the rows in the echelon form B. That is,

Basis for Row A: $\{(1,0,-1,5),\ (0,-2,5,-6)\}$

For the null space, use the *reduced* echelon form of A to solve $A\mathbf{x} = \mathbf{0}$:

$$A \sim B \sim \begin{bmatrix} 1 & 0 & -1 & 5 \\ 0 & 1 & -5/2 & 3 \\ 0 & 0 & 0 & 0 \end{bmatrix}; \quad \begin{aligned} x_1 \quad\ - \quad\quad\ x_3 + 5x_4 &= 0 \\ x_2 - (5/2)x_3 + 3x_4 &= 0 \\ 0 &= 0 \end{aligned}$$

Thus $x_1 = x_3 - 5x_4$, $x_2 = (5/2)x_3 - 3x_4$, with x_3, x_4 free. The general solution of $A\mathbf{x} = \mathbf{0}$ is

$$\begin{bmatrix} x_1 \\ x_2 \\ x_3 \\ x_4 \end{bmatrix} = \begin{bmatrix} x_3 - 5x_4 \\ (5/2)x_3 - 3x_4 \\ x_3 \\ x_4 \end{bmatrix} = x_3 \underset{\substack{\uparrow \\ \mathbf{u}}}{\begin{bmatrix} 1 \\ 5/2 \\ 1 \\ 0 \end{bmatrix}} + x_4 \underset{\substack{\uparrow \\ \mathbf{v}}}{\begin{bmatrix} -5 \\ -3 \\ 0 \\ 1 \end{bmatrix}}$$

Thus $\{\mathbf{u}, \mathbf{v}\}$ is a basis for Nul A.

Study Tip: Because rank $A = 2$ in Exercise 1, *any* two linearly independent columns of A form a basis for Col A, and any two linearly independent rows of A form a basis for Row A. When the rank of a matrix exceeds 2, selecting bases in this way is not so easy. (That is why you examine an echelon form of A.) On an exam, you should always choose the pivot columns of A as the basis for Col A and the nonzero rows of an echelon form of A as the basis for Row A. This will show that you can handle matrices with any rank.

7. Yes, Col $A = \mathbb{R}^4$, because Col A is a 4-dimensional subspace of $\mathbb{R}^4$ and hence coincides with $\mathbb{R}^4$. No, Nul A cannot be $\mathbb{R}^3$, because the vectors in Nul A have 7 entries. Nul A is a 3-dimensional subspace of $\mathbb{R}^7$, by the Rank Theorem.

13. If A is either a 7×5 matrix or a 5×7 matrix, then A has at most 5 pivot positions. So 5 is the largest possible value for rank A.

17. **a.** See the paragraph before Example 1.

b. See the warning after Example 2.

c. See the Rank Theorem. **d.** See the Rank Theorem.

e. See the Numerical Note before the Practice Problem.

19. Visualize the system as $A\mathbf{x} = \mathbf{0}$ where A is a 5×6 matrix. The information in the problem implies that the solution space is one-dimensional. By the Rank Theorem, rank $A = 6 - 1 = 5$. So dim Col A = 5. But Col A is a subspace of $\mathbb{R}^5$. Hence Col $A = \mathbb{R}^5$. Thus $A\mathbf{x} = \mathbf{b}$ has a solution for all $\mathbf{b}$.

Study Tip: Exercises 19—25 make good exam questions.

21. Visualize the system as $A\mathbf{x} = \mathbf{b}$, where A is 9×10 matrix. You are told that the system has a solution for all $\mathbf{b}$, so A must have a pivot position in each row. (That is, rank $A = 9$.) Since A has 10 columns, the Rank Theorem implies that dim Nul $A = 1$. So it is not possible to find two linearly independent vectors in Nul A.

23. The set of interest is the null space of a 12×8 matrix A. The description of this set implies that dim Nul $A = 2$. By the Rank Theorem, rank $A = 8 - 2 = 6$. So the equation $A\mathbf{x} = \mathbf{0}$ is equivalent to $B\mathbf{x} = \mathbf{0}$, where B is an echelon form of A with 6 nonzero rows. The answer to the question is that six homogeneous equations are sufficient.

25. Let A be the 10×12 coefficient matrix. By hypothesis, there are three free variables in the system $A\mathbf{x} = \mathbf{b}$, so dim Nul $A = 3$. By the Rank Theorem, dim Col $A = 12 - 3 = 9$. Since Col A is a subspace of $\mathbb{R}^{10}$ (because A has 10 rows), Col A cannot be all of $\mathbb{R}^{10}$, so some nonhomogeneous equations $A\mathbf{x} = \mathbf{b}$ will *not* have solutions.

31. The solution is in the text.

33. Let $A = [\mathbf{u}\ \ \mathbf{u}_2\ \ \mathbf{u}_3]$. If $\mathbf{u} \neq \mathbf{0}$, then $\mathbf{u}$ must be a basis for Col A, since Col A is one-dimensional. Hence there exist scalars r and s such that $\mathbf{u}_2 = r\mathbf{u}$ and $\mathbf{u}_3 = s\mathbf{u}$, so that

$$A = [\mathbf{u}\ \ r\mathbf{u}\ \ s\mathbf{u}] = \mathbf{u}[1\ \ r\ \ s] = \mathbf{u}\mathbf{v}^T, \text{ where } \mathbf{v} = \begin{bmatrix} 1 \\ r \\ s \end{bmatrix}$$

If the first column of A is zero and the second column, call it $\mathbf{u}$, is nonzero, then $A = [\mathbf{0}\ \ \mathbf{u}\ \ r\mathbf{u}]$ for some r. In this case, take $\mathbf{v} = (0, 1, r)$. If $A = [\mathbf{0}\ \ \mathbf{0}\ \ \mathbf{u}]$, take $\mathbf{v} = (0, 0, 1)$.

35. a. Let C and N be the matrices you construct whose columns are bases for Col A and Nul A, respectively. For the specific 5×7 matrix A in this problem, rank $A = 4$ and dim Nul $A = 3$. So C should be 5×4 (because Col A is a four-dimensional subspace of $\mathbb{R}^5$), and N should be 7×3 (because Nul A is a three-dimensional subspace of $\mathbb{R}^7$. Also, if the *rows* of R form a basis for Row A, then R should be 4×7, because dim Row A = rank $A = 4$ and Row A is a subspace of $\mathbb{R}^7$. (Make sure you understand these statements.)

b. The matrix $S = [R^T \quad N]$ is 7×7, because R^T is 7×4 and N is 7×3. If M is a matrix whose columns form a basis for Nul A^T, then M should be 5×1, because Nul A^T is a one-dimensional subspace of $\mathbb{R}^5$, by Exercise 28(b). Since C is 5×4, the matrix $T = [C \quad \text{M}]$ should be 5×5.

In general, the matrix S is $n \times n$ because the dimensions of Row A (spanned by the columns of R^T) and Nul A add up to n, the number of columns of A, and both Row A and Nul A are subspaces of $\mathbb{R}^n$. The matrix T is $m \times m$ because the dimensions of the column space and Nul A^T add up to m, the number of rows of A, and both Col A and Nul A^T are subspaces of $\mathbb{R}^m$.

37. Giving an answer here would spoil the problem for you.

Mastering Linear Algebra Concepts: Eight Basic Ideas

Sometime between now and when you finish the chapter, you should do a major review of the eight key concepts introduced in this chapter: vector space, subspace, column space, null space, basis, coordinate vector, dimension, and rank. (The row space of A is not really a separate concept; it is just the column space of A^T.) Study your old review sheets, and prepare new summary sheets for coordinate vector, dimension, and rank. Use as many of the standard categories (special cases, examples, algorithms, etc.) as possible. The tables in this section will be helpful. Also, add cross-references about dimension and rank to other sheets (subspace, column space, etc.) and update your summary sheet for the Invertible Matrix Theorem.

MATLAB ref and rank

In this course, you can use either **ref(A)** or **rank(A)** to check the rank of A. In practical work, you should use the more reliable command **rank(A)**, based on the singular value decomposition (Section 7.4).

4.7 CHANGE OF BASIS

This section will help you better understand coordinate systems. A review of Section 4.4 now is strongly recommended.

KEY IDEAS

Figure 1 and the accompanying discussion will help you visualize the main idea of the section. Imagine superimposing the $\mathcal{C}$-graph paper (Figure 1-b) on the $\mathcal{B}$-graph paper (Figure 1-a). Can you see where $\mathbf{b}_1$ will lie on the $\mathcal{C}$-coordinate system? Four units in the $\mathbf{c}_1$-direction and one unit in the $\mathbf{c}_2$-direction. That is the geometric interpretation of the equation $[\mathbf{b}_1]_{\mathcal{C}} = \begin{bmatrix} 4 \\ 1 \end{bmatrix}$ in Example 1. Similarly, since $[\mathbf{b}_2]_{\mathcal{C}} = \begin{bmatrix} -6 \\ 1 \end{bmatrix}$, $\mathbf{b}_2$ lies six units in the negative $\mathbf{c}_1$-direction and one unit in the $\mathbf{c}_2$-direction.

In general, the locations of $\mathbf{b}_1$ and $\mathbf{b}_2$ on the $\mathcal{C}$-graph paper are precisely what you must find in order to build the columns of the change-of-coordinates matrix:

$$\underset{\mathcal{C}\leftarrow\mathcal{B}}{P} = [[\mathbf{b}_1]_{\mathcal{C}} \quad [\mathbf{b}_2]_{\mathcal{C}}]$$

The notation for this matrix should help you remember the basic equation for changing $\mathcal{B}$-coordinates into $\mathcal{C}$-coordinates:

$$[\mathbf{x}]_{\mathcal{C}} = \underset{\mathcal{C}\leftarrow\mathcal{B}}{P}[\mathbf{x}]_{\mathcal{B}}$$

The calculations are simple when $\mathcal{B}$ and $\mathcal{C}$ are bases for $\mathbb{R}^n$. The box after Example 2 illustrates the algorithm for computing the change-of-coordinates matrix. In general,

$$[\mathbf{c}_1 \ \cdots \ \mathbf{c}_n \ \vdots \ \mathbf{b}_1 \ \cdots \ \mathbf{b}_n] \sim [\ I \ \vdots \ \underset{\mathcal{C}\leftarrow\mathcal{B}}{P}\ ]$$

Equivalently, using the notation of Section 4.4,

$$[P_{\mathcal{C}} \quad P_{\mathcal{B}}] \sim [\ \mathrm{I} \quad \underset{\mathcal{C}\leftarrow\mathcal{B}}{P}]$$

where $P_{\mathcal{B}}$ is the matrix $[\mathbf{b}_1 \ \cdots \ \mathbf{b}_n]$ that changes $\mathcal{B}$-coordinates to *standard coordinates*, and $P_{\mathcal{C}}$ is similarly defined. If you refer back to Exercise 11 of Section 2.2, you will see that $\underset{\mathcal{C}\leftarrow\mathcal{B}}{P}$ is the same as $(P_{\mathcal{C}})^{-1}P_{\mathcal{B}}$. Since $(P_{\mathcal{C}})^{-1}$ changes standard coordinates to $\mathcal{C}$-coordinates, we can obtain $[\mathbf{x}]_{\mathcal{C}}$ from $[\mathbf{x}]_{\mathcal{B}}$ as follows:

$$[\mathbf{x}]_{\mathcal{B}} \longrightarrow P_{\mathcal{B}}[\mathbf{x}]_{\mathcal{B}} \longrightarrow (P_{\mathcal{C}})^{-1}P_{\mathcal{B}}[\mathbf{x}]_{\mathcal{B}} = \underset{\mathcal{C}\leftarrow\mathcal{B}}{P}[\mathbf{x}]_{\mathcal{B}} = [\mathbf{x}]_{\mathcal{C}}$$

B-coordinates | standard coordinates | C-coordinates

This diagram provides another way of viewing the change of coordinates.

SOLUTIONS TO EXERCISES

1. a. From $\mathbf{b}_1 = 6\mathbf{c}_1 - 2\mathbf{c}_2$ and $\mathbf{b}_2 = 9\mathbf{c}_1 - 4\mathbf{c}_2$, write

$$[\mathbf{b}_1]_{\mathcal{C}} = \begin{bmatrix} 6 \\ -2 \end{bmatrix}, \quad \mathbf{b}_2 = \begin{bmatrix} 9 \\ -4 \end{bmatrix}, \text{ and } \underset{\mathcal{C}\leftarrow\mathcal{B}}{P} = \begin{bmatrix} 6 & 9 \\ -2 & -4 \end{bmatrix}$$

b. Since $\mathbf{x} = -3\mathbf{b}_1 + 2\mathbf{b}_2$,

$$[\mathbf{x}]_{\mathcal{B}} = \begin{bmatrix} -3 \\ 2 \end{bmatrix} \text{ and } [\mathbf{x}]_{\mathcal{C}} = \begin{bmatrix} 6 & 9 \\ -2 & -4 \end{bmatrix}\begin{bmatrix} -3 \\ 2 \end{bmatrix} = \begin{bmatrix} 0 \\ -2 \end{bmatrix}$$

7. Unlike Exercise 1, you do not have direct information from which you can write $[\mathbf{b}_1]_{\mathcal{C}}$ and $[\mathbf{b}_2]_{\mathcal{C}}$. Rather than compute these two coordinate vectors separately, use the algorithm from Example 2:

$$[\mathbf{c}_1 \quad \mathbf{c}_2 \quad \mathbf{b}_1 \quad \mathbf{b}_2] = \begin{bmatrix} 1 & -2 & 7 & -3 \\ -5 & 2 & 5 & -1 \end{bmatrix} \sim \begin{bmatrix} 1 & -2 & 7 & -3 \\ 0 & -8 & 40 & -16 \end{bmatrix}$$

$$\sim \begin{bmatrix} 1 & -2 & 7 & -3 \\ 0 & 1 & -5 & 2 \end{bmatrix} \sim \begin{bmatrix} 1 & 0 & -3 & 1 \\ 0 & 1 & -5 & 2 \end{bmatrix}$$

Thus $\underset{\mathcal{C}\leftarrow\mathcal{B}}{P} = \begin{bmatrix} -3 & 1 \\ -5 & 2 \end{bmatrix}$. The change-of-coordinates matrix from $\mathcal{C}$ to $\mathcal{B}$ is

$$\underset{\mathcal{B}\leftarrow\mathcal{C}}{P} = (\underset{\mathcal{C}\leftarrow\mathcal{B}}{P})^{-1} = \begin{bmatrix} -3 & 1 \\ -5 & 2 \end{bmatrix}^{-1} = \frac{1}{-1}\begin{bmatrix} 2 & -1 \\ 5 & -3 \end{bmatrix} = \begin{bmatrix} -2 & 1 \\ -5 & 3 \end{bmatrix}$$

11. **a.** See Theorem 15.

b. See the first paragraph in the subsection on *Change of Basis in* $\mathbb{R}^n$.

13. Let $\mathbf{b}_1$ represent the polynomial $1 - 2t + t^2$, let $\mathbf{b}_2$ be $3 - 5t + 4t^2$, let $\mathbf{b}_3$ be $2t + 3t^2$, and let $\mathcal{C}$ be the standard basis $\{1, t, t^2\}$ for $\mathbb{P}_2$. The $\mathcal{C}$-coordinate vectors of the vectors $\mathbf{b}_1, \mathbf{b}_2, \mathbf{b}_3$ are

$$[\mathbf{b}_1]_{\mathcal{C}} = \begin{bmatrix} 1 \\ -2 \\ 1 \end{bmatrix}, \quad [\mathbf{b}_2]_{\mathcal{C}} = \begin{bmatrix} 3 \\ -5 \\ 4 \end{bmatrix}, \quad [\mathbf{b}_3]_{\mathcal{C}} = \begin{bmatrix} 0 \\ 2 \\ 3 \end{bmatrix}$$

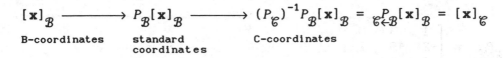

So

$$\underset{\mathcal{C}\leftarrow\mathcal{B}}{P} = \begin{bmatrix} 1 & 3 & 0 \\ -2 & -5 & 2 \\ 1 & 4 & 3 \end{bmatrix}$$

The coordinate vector $[-1 + 2t]_{\mathcal{B}}$ satisfies

$$\underset{\mathcal{C}\leftarrow\mathcal{B}}{P}[-1 + 2t]_{\mathcal{B}} = [-1 + 2t]_{\mathcal{C}} = \begin{bmatrix} -1 \\ 2 \\ 0 \end{bmatrix}$$

This equation can be solved by row reduction:

$$\begin{bmatrix} 1 & 3 & 0 & -1 \\ -2 & -5 & 2 & 2 \\ 1 & 4 & 3 & 0 \end{bmatrix} \sim \cdots \sim \begin{bmatrix} 1 & 0 & 0 & 5 \\ 0 & 1 & 0 & -2 \\ 0 & 0 & 1 & 1 \end{bmatrix}; \quad [-1 + 2t]_{\mathcal{B}} = \begin{bmatrix} 5 \\ -2 \\ 1 \end{bmatrix}$$

19. [M] a. Let $\mathcal{C}$ be the basis $\{\mathbf{v}_1, \mathbf{v}_2, \mathbf{v}_3\}$. Then the columns of P are $[\mathbf{u}_1]_{\mathcal{C}}$, $[\mathbf{u}_2]_{\mathcal{C}}$, and $[\mathbf{u}_3]_{\mathcal{C}}$. By definition of $\mathcal{C}$-coordinate vectors,

$$\mathbf{u}_j = [\mathbf{v}_1 \quad \mathbf{v}_2 \quad \mathbf{v}_3][\mathbf{u}_j]_{\mathcal{C}} = V[\mathbf{u}_j]_{\mathcal{C}}$$

where V is the matrix $V = [\mathbf{v}_1 \quad \mathbf{v}_2 \quad \mathbf{v}_3]$. Then, by the definition of matrix multiplication,

$$[\mathbf{u}_1 \quad \mathbf{u}_2 \quad \mathbf{u}_3] = V[\,[\mathbf{u}_1]_{\mathcal{C}} \quad [\mathbf{u}_2]_{\mathcal{C}} \quad [\mathbf{u}_3]_{\mathcal{C}}\,] = VP$$

You know V and P, so you can compute $\mathbf{u}_1$, $\mathbf{u}_2$, and $\mathbf{u}_3$.

$$VP = \begin{bmatrix} -2 & -8 & -7 \\ 2 & 5 & 2 \\ 3 & 2 & 6 \end{bmatrix}\begin{bmatrix} 1 & 2 & -1 \\ -3 & -5 & 0 \\ 4 & 6 & 1 \end{bmatrix} = \begin{bmatrix} -6 & -6 & -5 \\ -5 & -9 & 0 \\ 21 & 32 & 3 \end{bmatrix} = [\mathbf{u}_1 \quad \mathbf{u}_2 \quad \mathbf{u}_3]$$

b. By analogy to part (a), $V = WP$, where $W = [\mathbf{w}_1 \quad \mathbf{w}_2 \quad \mathbf{w}_3]$. You know V and P, so compute W from $W = VP^{-1}$.

MATLAB The Toolbox has data for Exercises 7—10 and 17—19. The command **ref(M)** row reduces a matrix such as $[\mathbf{c}_1 \quad \mathbf{c}_2 \quad \mathbf{b}_1 \quad \mathbf{b}_2]$ to a useful form.

4.8 APPLICATIONS TO DIFFERENCE EQUATIONS

Difference equations are the discrete analogues of differential equations. Both are important in science and engineering. The discrete and continuous theories are remarkably parallel, and linear algebra is applied in similar ways, although the calculations are somewhat easier for difference equations. A variety of examples and exercises here illustrate some difference equations you may encounter later in your work.

KEY IDEAS

Each signal in $\mathbb{S}$ is an infinite list of numbers. Linear independence of a set of signals can often be demonstrated by looking at short segments of the signals, that is, by showing that a Casorati matrix is invertible. A Casorati matrix cannot be used in general to demonstrate linear *dependence* of a set. However, see the appendix at the end of this *Study Guide* section.

The main focus of the section is on difference equations. Given a homogeneous difference equation, you should be able to:

- determine whether a specified signal is a solution of the equation;
- find solutions of the equation, using the auxiliary equation;
- give the general solution (when the auxiliary equation has no multiple roots and no complex roots).

Theorem 18 is the key result that enables you to write the general solution. Just finding some specific solutions is not enough; you must show that they *span* the set of all solutions. But Theorem 18 and the Basis Theorem in Section 4.5 together show that for an nth order equation, you only need to find *n linearly independent* solutions. (See Example 5.) The same principle apples to an nth order differential equation (discussed later in Section 5.7.) This principle is one of the most powerful applications of linear algebra in the text.

The subsection on nonhomogenous equations is optional. If this is covered, you should be able to work Exercises 25—28. The general principle is illustrated in Figure 4, page 278:

$$\left\{\begin{matrix}\text{General solution of}\\ \text{nonhomogeneous eqn.}\end{matrix}\right\} = \left\{\begin{matrix}\text{Particular solution of}\\ \text{nonhomogeneous eqn.}\end{matrix}\right\} + \left\{\begin{matrix}\text{General solution of}\\ \text{homogeneous eqn.}\end{matrix}\right\}$$

The final subsection shows the modern way to study an nth order linear difference equation, rewriting it as a first order system $\mathbf{x}_{k+1} = A\mathbf{x}_k$ $(k = 1,2,\ldots)$. Such systems were introduced in Section 1.9 and they will be discussed further in Sections 4.9 and 5.6.

SOLUTIONS TO EXERCISES

1. If $y_k = (-4)^k$, then $y_{k+1} = (-4)^{k+1}$ and $y_{k+2} = (-4)^{k+2}$. Substitute these formulas into the left side of the equation:

$$y_{k+2} + 2y_{k+1} - 8y_k = (-4)^{k+2} + 2(-4)^{k+1} - 8(-4)^k$$

$$= (-4)^k[(-4)^2 + 2(-4) - 8]$$

$$= (-4)^k[16 - 8 - 8] = 0 \quad \text{for all } k$$

Since the difference equation holds for all k, $(-4)^k$ is a solution. The text answer displays the similar calculations for $y_k = 2^k$.

7. Compute the Casorati matrix for the signals 1^k, 2^k, and $(-2)^k$, setting $k = 0$ for convenience:

$$\begin{bmatrix} 1^0 & 2^0 & (-2)^0 \\ 1^1 & 2^1 & (-2)^1 \\ 1^2 & 2^2 & (-2)^2 \end{bmatrix} = \begin{bmatrix} 1 & 1 & 1 \\ 1 & 2 & -2 \\ 1 & 4 & 4 \end{bmatrix} \sim \begin{bmatrix} 1 & 1 & 1 \\ 0 & 1 & -3 \\ 0 & 3 & 3 \end{bmatrix} \sim \begin{bmatrix} 1 & 1 & 1 \\ 0 & 1 & -3 \\ 0 & 0 & 12 \end{bmatrix}$$

This Casorati matrix has three pivots and hence is invertible, by the IMT. Hence the set of signals $\{1^k, 2^k, (-2)^k\}$ is linearly independent in $\mathbb{S}$. We know (from the text) that these signals are in the solution space H of a third-order difference equation. By Theorem 17, $\dim H = 3$. Since the three signals are linearly independent, they form a basis for H, by the Basis Theorem in Section 4.5.

Warning: Many student papers for Exercise 7 suffer from a lack of precision, often confusing linear independence of the columns of the Casorati matrix with linear independence of the signals in $\mathbb{S}$. There is no need to discuss the columns of the Casorati matrix—just observe that the matrix is invertible. But you must point out that the three *signals* are linearly independent, in order to apply the Basis Theorem (Theorem 12) in Section 4.5 to the vector space H of solutions to the difference equation.

13. The auxiliary equation for $y_{k+2} - y_{k+1} + \frac{2}{9}y_k = 0$ is $r^2 - r + \frac{2}{9} = 0$. By the quadratic formula,

$$r = \frac{1 \pm \sqrt{1 - 8/9}}{2} = \frac{1 \pm 1/3}{2} = \frac{2}{3} \text{ or } \frac{1}{3}$$

Two solutions of the difference equation are $(\frac{2}{3})^k$ and $(\frac{1}{3})^k$. These signals are obviously linearly independent because neither is a multiple of the other. Since the solution space is two-dimensional (Theorem 17), the two signals form a basis for the solution space (Theorem 12).

Study Tip: I think Exercises 7—19 (and 25—28) are good exam questions, because they illustrate how important Theorems 17 and 12 really are. Probably, you do not have to memorize theorem numbers, but your discussion should show that you have two theorems in mind and that you know how to use them. (Check with your instructor.)

19. The auxiliary equation for $y_{k+2} + 4y_{k+1} + y_k = 0$ is $r^2 + 4r + 1 = 0$. By the quadratic formula,

$$r = \frac{-4 \pm \sqrt{16 - 4}}{2} = \frac{-4 \pm 2\sqrt{3}}{2} = -2 \pm \sqrt{3}$$

Two solutions of the difference equation are $(-2 + \sqrt{3})^k$ and $(-2 - \sqrt{3})^k$. They are obviously linearly independent because neither is a multiple of the other. Since the solution space is two-dimensional (Theorem 17), the two signals form a fundamental set of solutions (Theorem 12), and the general solution has the form $c_1(-2 + \sqrt{3})^k + c_2(-2 - \sqrt{3})^k$.

25. To prove that $y_k = k^2$ is a solution of

$$y_{k+2} + 3y_{k+1} - 4y_k = 10k + 7 \qquad (1)$$

show that when k^2, $(k + 1)^2$, and $(k + 2)^2$ are substituted for y_k, y_{k+1}, and y_{k+2}, respectively, the resulting equation is true for all k:

$$\begin{aligned}(k + 2)^2 + 3(k + 1)^2 - 4k^2 &= (k^2 + 4k + 4) + 3(k^2 + 2k + 1) - 4k^2 \\ &= (1 + 3 - 4)k^2 + (4 + 6)k + (4 + 3) \\ &= 10k + 7 \quad \text{for all } k.\end{aligned}$$

So k^2 is a solution of (1). The auxiliary equation for the homogeneous difference equation

$$y_{k+2} + 3y_{k+1} - 4y_k = 0 \quad \text{for all } k \qquad (2)$$

is $r^2 + 3r - 4 = 0$, which factors as $(r - 1)(r + 4) = 0$, so $r = 1, -4$. Thus 1^k and $(-4)^k$ are solutions of (2). The signals are linearly independent (for neither is a multiple of the other), so they form a basis for the two-dimensional solution space. The general solution of (2) is $c_1 \cdot 1^k + c_2(-4)^k$. Add this to a particular solution of (1) and obtain the general solution $k^2 + c_1 + c_2(-4)^k$ of (1).

31. The full explanation is in the text's answer section.

35. For $\{y_k\}$ and $\{z_k\}$ in $\mathbb{S}$, the kth term of $\{y_k\} + \{z_k\}$ is $y_k + z_k$. Hence

$$\begin{aligned} T(\{y_k\} + \{z_k\}) &= (y_{k+2} + z_{k+2}) + a(y_{k+1} + z_{k+1}) + b(y_k + z_k) \\ &= (y_{k+2} + ay_{k+1} + by_k) + (z_{k+2} + az_{k+1} + bz_k) \\ &= T\{y_k\} + T\{z_k\} \end{aligned}$$

For any scalar r, the kth term of $r\{y_k\}$ is ry_k, and so

$$\begin{aligned} T(r\{y_k\}) &= ry_{k+2} + a(ry_{k+1}) + b(ry_k) \\ &= r(y_{k+2} + ay_{k+1} + by_k) = rT\{y_k\} \end{aligned}$$

Thus T has the two properties that define a linear transformation.

Appendix: The Casorati Test

Let $\{\mathbf{y}_1, \ldots, \mathbf{y}_n\}$ be a set of signals in $\mathbb{S}$. For $j = 1, \ldots, n$ and for any k, let $\mathbf{y}_j(k)$ denote the kth entry in the signal $\mathbf{y}_j$ and let

$$C(k) = \begin{bmatrix} \mathbf{y}_1(k) & \cdots & \mathbf{y}_n(k) \\ \mathbf{y}_1(k+1) & \cdots & \mathbf{y}_n(k+1) \\ \vdots & & \vdots \\ \mathbf{y}_1(k+n-1) & \cdots & \mathbf{y}_n(k+n-1) \end{bmatrix} \qquad \text{The Casorati matrix}$$

a. If $C(k)$ is invertible for some k, $\{\mathbf{y}_1, \ldots, \mathbf{y}_n\}$ is linearly independent.

b. If $\mathbf{y}_1, \ldots, \mathbf{y}_n$ all satisfy a homogeneous difference equation of order n,

$$y_{k+n} + a_1 y_{k+n-1} + \cdots + a_n y_k = 0 \quad \text{for all } k \tag{*}$$

(with $a_n \neq 0$), and if the Casorati matrix $C(k)$ is not invertible for some k, then $\{\mathbf{y}_1, \ldots, \mathbf{y}_n\}$ is linearly dependent in $\mathbb{S}$, and for all k, $C(k)$ is not invertible.

Proof. (a) The argument given in the text (page 273) for a set of th signals generalizes immediately to n signals. (b) Suppose that $\mathbf{y}_1$. are in the set H of solutions of (*) and $C(k_0)$ is not invertible k_0. It is readily verified that if $T: H \to \mathbb{R}^n$ is defined by

$$T(\mathbf{y}) = \begin{bmatrix} \mathbf{y}(k_0) \\ \mathbf{y}(k_0+1) \\ \vdots \\ \mathbf{y}(k_0+n-1) \end{bmatrix}$$

then T is a linear transformation. The proof of Theorem 16 is easily modified to show that (*) has a unique solution $\mathbf{y}$ whenever $\mathbf{y}(k_0)$, ..., $\mathbf{y}(k_0+n-1)$ are specified. This means that T is a one-to-one mapping of H onto $\mathbb{R}^n$. Furthermore, the images $T(\mathbf{y}_1)$, ..., $T(\mathbf{y}_n)$ form the columns of the Casorati matrix $C(k_0)$ and hence are linearly dependent, because $C(k_0)$ is not invertible. Since T is one-to-one, $\{\mathbf{y}_1, \ldots, \mathbf{y}_n\}$ is linearly dependent, by Exercise 32 in Section 4.3. In this case, for any k the images $T(\mathbf{y}_1)$, ..., $T(\mathbf{y}_n)$ are linearly dependent, and so $C(k)$ is not invertible. (See Exercise 31 in Section 4.3.)

MATLAB roots

In Exercises 7—16 and 25—28, the polynomial in the auxiliary equation is stored in a row vector **p**, with coefficients in descending order. For instance, if the auxiliary equation is $r^2 + 6r + 9 = 0$, then **p** = [1 6 9].

The MATLAB command **roots(p)** produces a column vector whose entries are the roots of the polynomial described by **p**.

4.9 APPLICATIONS TO MARKOV CHAINS

This section builds on the population movement example in Section 1.9. You should review that example now. Markov chains are widely used in applications and there is a rich theory connected with them. The simple examples and exercises in this section provide a basic foundation on which you can build later as needed. Two of the examples here will be analyzed from a different point of view in Section 5.2.

KEY IDEAS

A probability vector is a list of nonnegative numbers that sum to one. A Markov chain is a sequence of probability vectors $\{\mathbf{x}_k\}$ that satisfy a difference equation $\mathbf{x}_{k+1} = P\mathbf{x}_k$ $(k = 0, 1, \ldots)$ for some stochastic matrix P (whose columns are themselves probability vectors).

The theory of this chapter can be used to show that $P - I$ always has a nontrivial null space when P is a stochastic matrix (Exercise 17). In advanced texts it is shown that the null space of $P - I$ always includes at least one probability vector, which then is a steady-state vector for P, because the equation $(P - I)\mathbf{q} = \mathbf{0}$ is equivalent to $P\mathbf{q} = \mathbf{q}$. (Also, see Exercise 18.)

Our main interest is in a *regular* stochastic matrix P. In this case the steady-state vector is unique, according to Theorem 18. The key to predicting the distant future for a Markov chain associated with such a P is to find the steady-state vector $\mathbf{q}$, since the sequence $\{\mathbf{x}_k\}$ converges to $\mathbf{q}$ no matter what the initial state.

SOLUTIONS TO EXERCISES

1. a. To set up the stochastic matrix P, label the columns N (for news) and M (for music) in some order; use the *same order* for the rows. (Failure to keep the same order is a common source of error in this type of problem.) The data should be arranged so you read *down* a column and then to the right along a row.

From: N	From: M	To:
.7	.6	News
.3	.4	Music

b. You are told that 100% of the listeners are listening to the news at 8:15 a.m., so start the Markov chain then, with $\mathbf{x}_0 = \begin{bmatrix} 1 \\ 0 \end{bmatrix}$.

c. There are two breaks between 8:15 and 9:25, so you need $\mathbf{x}_2$.

$$\mathbf{x}_1 = P\mathbf{x}_0 = \begin{bmatrix} .7 & .6 \\ .3 & .4 \end{bmatrix}\begin{bmatrix} 1 \\ 0 \end{bmatrix} = \begin{bmatrix} .7 \\ .3 \end{bmatrix}$$

$$\mathbf{x}_2 = P\mathbf{x}_1 = \begin{bmatrix} .7 & .6 \\ .3 & .4 \end{bmatrix}\begin{bmatrix} .7 \\ .3 \end{bmatrix} = \begin{bmatrix} .67 \\ .33 \end{bmatrix}$$

The entries in $\mathbf{x}_2$ show that after two station breaks, 67% of the audience is listening to the news and 33% is listening to music.

Study Tip: When you compute a typical probability vector $P\mathbf{x}$, be sure to *compute all* of the entries in the product $P\mathbf{x}$. Then check your work by verifying that the entries sum to 1. (This trick has caught numerous errors of mine!)

7. To find the steady state vector for a regular stochastic matrix P:
(i) set up the matrix $P - I$;
(ii) find the general solution of $(P - I)\mathbf{x} = \mathbf{0}$;
(iii) choose a basis vector for $\text{Nul}(P - I)$ whose entries sum to 1.

$$P = \begin{bmatrix} .7 & 1 & .1 \\ .2 & .8 & .2 \\ .1 & .1 & .7 \end{bmatrix}, \quad P - I = \begin{bmatrix} .7 & 1 & .1 \\ .2 & .8 & .2 \\ .1 & .1 & .7 \end{bmatrix} - \begin{bmatrix} 1 & 0 & 0 \\ 0 & 1 & 0 \\ 0 & 0 & 1 \end{bmatrix} = \begin{bmatrix} -.3 & .1 & .1 \\ .2 & -.2 & .2 \\ .1 & .1 & -.3 \end{bmatrix}$$

Solve $(P - I)\mathbf{x} = \mathbf{0}$:

$$\begin{bmatrix} -.3 & .1 & .1 & 0 \\ .2 & -.2 & .2 & 0 \\ .1 & .1 & -.3 & 0 \end{bmatrix} \sim \begin{bmatrix} 1 & 1 & -3 & 0 \\ 2 & -2 & 2 & 0 \\ -3 & 1 & 1 & 0 \end{bmatrix}$$

Interchange rows 1 and 3
Scale every row by 10

$$\sim \cdots \sim \begin{bmatrix} 1 & 0 & -1 & 0 \\ 0 & 1 & -2 & 0 \\ 0 & 0 & 0 & 0 \end{bmatrix}, \quad \begin{aligned} x_1 \quad - x_3 &= 0 \\ x_2 - 2x_3 &= 0 \; ; \\ x_3 \text{ is free} & \end{aligned} \quad \begin{bmatrix} x_1 \\ x_2 \\ x_3 \end{bmatrix} = \begin{bmatrix} x_3 \\ 2x_3 \\ x_3 \end{bmatrix} = x_3 \begin{bmatrix} 1 \\ 2 \\ 1 \end{bmatrix}$$

The entries in $\begin{bmatrix} 1 \\ 2 \\ 1 \end{bmatrix}$ sum to 4, so $\mathbf{q} = \frac{1}{4}\begin{bmatrix} 1 \\ 2 \\ 1 \end{bmatrix} = \begin{bmatrix} 1/4 \\ 1/2 \\ 1/4 \end{bmatrix}$ or $\begin{bmatrix} .25 \\ .50 \\ .25 \end{bmatrix}$.

Study Tip: Notice that the column sums are all zero for the matrix $P - I$ of Exercise 7. This always happens (see Exercise 17), and so you have a fast way to check your arithmetic for the entries in $P - I$.

13. From Exercise 3, $P = \begin{bmatrix} .50 & .25 & .25 \\ .25 & .50 & .25 \\ .25 & .25 & .50 \end{bmatrix}$. So $P - I = \begin{bmatrix} -.50 & .25 & .25 \\ .25 & -.50 & .25 \\ .25 & .25 & -.50 \end{bmatrix}$.

Solve $(P - I)\mathbf{x} = \mathbf{0}$:

$$\begin{bmatrix} -.50 & .25 & .25 & 0 \\ .25 & -.50 & .25 & 0 \\ .25 & .25 & -.50 & 0 \end{bmatrix} \sim \begin{bmatrix} 1 & 1 & -2 & 0 \\ 1 & -2 & 1 & 0 \\ -2 & 1 & 1 & 0 \end{bmatrix}$$

Scale rows by 4
Interchange rows 1 & 3

$$\sim \cdots \sim \begin{bmatrix} 1 & 0 & -1 & 0 \\ 0 & 1 & -1 & 0 \\ 0 & 0 & 0 & 0 \end{bmatrix} \quad \begin{aligned} x_1 &= x_3 \\ x_2 &= x_3 \\ x_3 &\text{ is free} \end{aligned}$$

A basis for Nul$(P - I)$ is $\left\{ \begin{bmatrix} 1 \\ 1 \\ 1 \end{bmatrix} \right\}$; the steady-state vector is $\mathbf{q} = \begin{bmatrix} 1/3 \\ 1/3 \\ 1/3 \end{bmatrix}$.

So each food will be preferred equally.

Warning: You may have noticed that in Exercises 7 and 13, I scaled rows by 10, to avoid decimals. A common mistake is to do this only to P, before forming $I - P$. That changes P drastically. The scaling I did was permissible because it was applied to all the coefficients in an *equation*

19. a. The product $S\mathbf{x}$ equals the sum of the entries in $\mathbf{x}$. Thus, by definition, $\mathbf{x}$ is a probability vector if and only if its entries are nonnegative and $S\mathbf{x} = 1$.

b. Let $P = [\mathbf{p}_1 \ \ \mathbf{p}_2 \ \ \cdots \ \ \mathbf{p}_n]$, where the $\mathbf{p}_i$ are probability vectors. By matrix multiplication and part (a),

$$SP = [S\mathbf{p}_1 \ \ S\mathbf{p}_2 \ \ \cdots \ \ S\mathbf{p}_n] = [1 \ \ 1 \ \ \cdots \ \ 1] = S$$

c. By part (b), $S(P\mathbf{x}) = (SP)\mathbf{x} = S\mathbf{x} = 1$. The entries in $P\mathbf{x}$ are obviously nonnegative, because P and $\mathbf{x}$ have only nonnegative entries. By (a), the condition $S(P\mathbf{x}) = 1$ shows that $P\mathbf{x}$ is a probability vector.

MATLAB The MATLAB box on page 1-37 contains information that is useful for homework here.

CHAPTER 4 SUPPLEMENTARY EXERCISES

Justifications for the answers to the True/False questions are given below. Other justifications are possible. A solution of Exercise 9(a) is also included.

1. a. True. $\text{Span}\{\mathbf{v}_1, \ldots, \mathbf{v}_p\}$ is a subspace of V, and every subspace is itself a vector space.

b. True. Any linear combination of $\mathbf{v}_1, \ldots, \mathbf{v}_{p-1}$ is also a linear combination of $\mathbf{v}_1, \ldots, \mathbf{v}_{p-1}, \mathbf{v}_p$, using a zero weight on $\mathbf{v}_p$.

c. False. Counterexample: Take $\mathbf{v}_p = 2\mathbf{v}_1$, then $\{\mathbf{v}_1, \ldots, \mathbf{v}_p\}$ is linearly dependent.

d. False. Counterexample: Let $\{\mathbf{e}_1, \mathbf{e}_2, \mathbf{e}_3\}$ be the standard basis for $\mathbb{R}^3$. Then $\{\mathbf{e}_1, \mathbf{e}_2\}$ is a linearly independent set but not a basis for $\mathbb{R}^3$.

e. True, by the Spanning Set Theorem (Section 4.3).

f. True. By the Basis Theorem, S is a basis for V because S contains exactly p vectors. So S must be linearly independent.

g. False. The plane must go through the origin to be a subspace.

h. False. Counterexample: $\begin{bmatrix} 2 & 5 & -2 & 0 \\ 0 & 0 & 7 & 3 \\ 0 & 0 & 0 & 0 \end{bmatrix}$.

i. True. This statement appears before Theorem 13 in Section 4.6. Try to make an example of a specific row operation on some 3×3 matrix.

j. False. Row operations on A do not change the solutions of $A\mathbf{x} = \mathbf{0}$.

k. False. Counterexample: $A = \begin{bmatrix} 1 & 2 \\ 3 & 6 \end{bmatrix}$; two nonzero rows but rank is 1.

l. False. If U has k nonzero rows, then rank A = k and dim Nul A = $n - k$, by the Rank Theorem.

m. True. Row equivalent matrices have the same number of pivot columns.

n. False. Counterexample: $A = \begin{bmatrix} 1 & 0 & 0 \\ 0 & 1 & 0 \end{bmatrix}$. If rank $A = n$ (the number of columns of A), then the transformation $\mathbf{x} \mapsto A\mathbf{x}$ is one-to-one.

o. True. If $\mathbf{x} \mapsto A\mathbf{x}$ is onto, then Col $A = \mathbb{R}^m$ and rank $A = m$. See Theorem 12(a) in Section 1.8.

p. True. See the second paragraph after Theorem 15 in Section 4.7.

q. False. The jth column of $\underset{\mathcal{C} \leftarrow \mathcal{B}}{P}$ is $[\mathbf{b}_j]_{\mathcal{C}}$.

9. a. Any $\mathbf{y}$ in Col AB has the form $(AB)\mathbf{x}$ for some $\mathbf{x}$. Then $\mathbf{y} = A(B\mathbf{x})$, which shows that $\mathbf{y}$ is a linear combination of the columns of A. Thus Col AB is contained in Col A; that is, Col AB is a *subspace* of Col A. By Theorem 11 in Section 4.5, dim Col $AB \le$ dim Col A.; that is, rank $AB \le$ rank A.

b. The solution is in the text.

CHAPTER 4 GLOSSARY CHECKLIST

Check your knowledge by attempting to write definitions of the terms below. Then compare your work with the definitions given in the text's Glossary. Ask your instructor which definitions, if any, might appear on a test.

auxiliary equation: A polynomial equation in a variable r, created from ...

basis (for a subspace H): A set $\mathcal{B} = \{\mathbf{v}_1, \ldots, \mathbf{v}_p\}$ in V such that: ....

$\mathcal{B}$-coordinates of x: *See* coordinates of **x** relative to the basis $\mathcal{B}$.

change-of-coordinates matrix (from a basis $\mathcal{B}$ to a basis $\mathcal{C}$): A matrix $\underset{\mathcal{C}\leftarrow\mathcal{B}}{P}$ that transforms ..., namely, (equation) ....

column space (of an $m \times n$ matrix A): The set Col A of .... In set notation, Col A = { : }.

controllable (pair of matrices): A matrix pair (A, B) where A is $n \times n$, B has n rows, and ....

coordinate mapping (determined by an ordered basis $\mathcal{B}$ in a vector space V): a mapping that associates to each ....

coordinates of x relative to the basis $\mathcal{B} = \{\mathbf{b}_1, \ldots, \mathbf{b}_n\}$: ....

coordinate vector of x relative to $\mathcal{B}$: The vector $[\mathbf{x}]_{\mathcal{B}}$ whose entries ....

dimension (of a vector space V): The number ....

explicit description (of a subspace W of $\mathbb{R}^n$): A parametric representation of W as the set of ....

finite-dimensional (vector space): A vector space that is ....

full rank (matrix): An $m \times n$ matrix whose rank is ....

fundamental set of solutions: A ... for the set of solutions of ....

fundamental subspaces (of A): The ... of A, and ....

implicit description (of a subspace W of $\mathbb{R}^n$): A set of one or more ....

infinite-dimensional (vector space): A nonzero vector space V that ....

isomorphism: A ... mapping from one vector space ....

kernel (of a linear transformation $T: V \to W$): The set of ... such that ....

linear combination: Any sum ....

linear dependence relation: A ... equation where ....

linear filter: A ... equation used to transform ....

linearly dependent (vectors): A set $\{\mathbf{v}_1, \ldots, \mathbf{v}_p\}$ with the property that ....

linearly independent (vectors): A set $\{\mathbf{v}_1, \ldots, \mathbf{v}_p\}$ with the property ....

linear transformation T (from a vector space V into a vector space W): A rule $T: V \rightarrow W$ that to each vector $\mathbf{x}$ in V assigns a unique vector $T(\mathbf{x})$ in W, such that: ....

Markov chain: A sequence of ... vectors $\mathbf{v}_0$, $\mathbf{v}_1$, $\mathbf{v}_2, \ldots$, together with a ... matrix P such that ....

maximal linearly independent set (in V): A linearly independent set $\mathcal{B}$ in V such that if ..., then ...

minimal spanning set (for a subspace H): A set $\mathcal{B}$ that spans H and has the property that if ..., then ....

null space (of an $m \times n$ matrix A): The set Nul A of all .... In set notation, Nul $A = \{ \quad : \quad \}$.

probability vector: A vector in $\mathbb{R}^n$ whose entries ....

proper subspace: Any subspace of a vector space V ....

range (of a linear transformation $T: V \longrightarrow W$): The set of all vectors ....

rank (of a matrix A): ....

regular stochastic matrix: A ... matrix P such that ....

row space (of a matrix A): The set Row A of all ...; also denoted by ....

signal (or **discrete-time signal**): ....

Span $\{\mathbf{v}_1, \ldots, \mathbf{v}_p\}$: The set of .... Also, the ... *spanned* ....

spanning set (for a subspace H): Any set $\{\mathbf{v}_1, \ldots, \mathbf{v}_p\}$ ... such that ....

standard basis: The basis ... for $\mathbb{R}^n$ consisting of ..., or the basis ... for $\mathbb{P}_n$.

state vector: A ... vector. In general, a vector that ... , often in connection with a difference equation ....

steady-state vector (for a stochastic matrix P): A ... vector $\mathbf{v}$ such ....

stochastic matrix: A ... matrix ....

submatrix (of A): Any matrix obtained by ....

subspace: A subset H of some vector space V such that ....

vector space: A set of objects, called vectors, on which ....

zero subspace: The subspace ... consisting of ....

5 EIGENVALUES AND EIGENVECTORS

5.1 EIGENVECTORS AND EIGENVALUES

This section introduces eigenvectors and eigenvalues. A hint about the connection with dynamical systems appears at the end of the section.

KEY IDEAS

In words, a nonzero vector **v** is an eigenvector of a matrix A if and only if the transformed vector $A\mathbf{x}$ points in the same or opposite direction of **v**.

Notice that while an eigenvalue λ might be zero, an eigenvector is never zero (by definition). An eigenspace contains eigenvectors together with the zero vector.

The two equations $A\mathbf{x} = \lambda\mathbf{x}$ and $(A - \lambda I)\mathbf{x} = \mathbf{0}$ are equivalent. See Example 3. The first equation is useful for understanding what eigenvalues and eigenvectors are, and it shows the geometric effect of the linear transformation $\mathbf{x} \mapsto A\mathbf{x}$ on an eigenvector. The second equation shows that the eigenspace is a subspace (because it is the null space of the matrix $A - \lambda I$), and the equation is used to find a basis for the eigenspace, when λ is a known eigenvalue. The second equation will be used again in Section 5.2 for another purpose.

SOLUTIONS TO EXERCISES

1. The number 2 is an eigenvalue of A if and only if the equation $A\mathbf{x} = 2\mathbf{x}$ has a nontrivial solution. This equation is equivalent to $(A - 2I)\mathbf{x} = \mathbf{0}$. Compute

$$A - 2I = \begin{bmatrix} 3 & 2 \\ 3 & 8 \end{bmatrix} - \begin{bmatrix} 2 & 0 \\ 0 & 2 \end{bmatrix} = \begin{bmatrix} 1 & 2 \\ 3 & 6 \end{bmatrix}$$

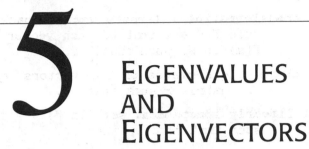

The columns of A are obviously linearly dependent, so $(A - 2I)\mathbf{x} = \mathbf{0}$ has a nontrivial solution, and so 2 is an eigenvalue of A.

7. Proceed as in Exercise 1:

$$A - 4I = \begin{bmatrix} 3 & 0 & -1 \\ 2 & 3 & 1 \\ -3 & 4 & 5 \end{bmatrix} - \begin{bmatrix} 4 & 0 & 0 \\ 0 & 4 & 0 \\ 0 & 0 & 4 \end{bmatrix} = \begin{bmatrix} -1 & 0 & -1 \\ 2 & -1 & 1 \\ -3 & 4 & 1 \end{bmatrix}$$

We need to know whether $A - 4I$ is invertible. This could be checked in several ways, but since we are asked for an eigenvector, in the event that one exists, the best strategy is to row reduce the augmented matrix for $(A - 4I)\mathbf{x} = \mathbf{0}$:

$$\begin{bmatrix} -1 & 0 & -1 & 0 \\ 2 & -1 & 1 & 0 \\ -3 & 4 & 1 & 0 \end{bmatrix} \sim \begin{bmatrix} -1 & 0 & -1 & 0 \\ 0 & -1 & -1 & 0 \\ 0 & 4 & 4 & 0 \end{bmatrix} \sim \begin{bmatrix} -1 & 0 & -1 & 0 \\ 0 & -1 & -1 & 0 \\ 0 & 0 & 0 & 0 \end{bmatrix}$$

Now it is clear that 4 is an eigenvalue of A [because $(A - 4I)\mathbf{x} = \mathbf{0}$ has a nontrivial solution]. The coordinates of an eigenvector satisfy $-x_1 - x_3 = 0$ and $-x_2 - x_3 = 0$. The general solution is not requested, so take any nonzero value for x_3 to produce an eigenvector. If $x_3 = 1$, then $\mathbf{x} = (-1, -1, 1)$.

Checkpoint 1: The answer in the text is different, namely, $(1, 1, -1)$. Why is it also correct?

Helpful Hint: Suppose you think that 4 is an eigenvalue of a matrix, as in Exercise 7, and you row reduce the augmented matrix for $(A - 4I)\mathbf{x} = \mathbf{0}$. If you discover that there are no free variables, then there are only two possibilities: (1) 4 is *not* an eigenvalue of A, or (2) you have made an error.

13. For $\lambda = 1$:

$$A - 1I = \begin{bmatrix} 4 & 0 & 1 \\ -2 & 1 & 0 \\ -2 & 0 & 1 \end{bmatrix} - \begin{bmatrix} 1 & 0 & 0 \\ 0 & 1 & 0 \\ 0 & 0 & 1 \end{bmatrix} = \begin{bmatrix} 3 & 0 & 1 \\ -2 & 0 & 0 \\ -2 & 0 & 0 \end{bmatrix}$$

The equations for $(A - I)\mathbf{x} = \mathbf{0}$ are easy to solve: $\begin{cases} 3x_1 + x_3 = 0 \\ -2x_1 = 0 \end{cases}$

Row operations hardly seem necessary. Obviously x_1 is zero, and hence x_3 is also zero. There are three variables, so x_2 is free. The general solution of $(A - I)\mathbf{x} = \mathbf{0}$ is $x_2\mathbf{e}_2$, where $\mathbf{e}_2 = \begin{bmatrix} 0 \\ 1 \\ 0 \end{bmatrix}$, and so $\mathbf{e}_2$ provides a basis for the eigenspace.

For $\lambda = 2$:

$$A - 2I = \begin{bmatrix} 4 & 0 & 1 \\ -2 & 1 & 0 \\ -2 & 0 & 1 \end{bmatrix} - \begin{bmatrix} 2 & 0 & 0 \\ 0 & 2 & 0 \\ 0 & 0 & 2 \end{bmatrix} = \begin{bmatrix} 2 & 0 & 1 \\ -2 & -1 & 0 \\ -2 & 0 & -1 \end{bmatrix}$$

$$[(A - 2I)\ \mathbf{0}] = \begin{bmatrix} 2 & 0 & 1 & 0 \\ -2 & -1 & 0 & 0 \\ -2 & 0 & -1 & 0 \end{bmatrix} \sim \begin{bmatrix} 2 & 0 & 1 & 0 \\ 0 & -1 & 1 & 0 \\ 0 & 0 & 0 & 0 \end{bmatrix} \sim \begin{bmatrix} 1 & 0 & 1/2 & 0 \\ 0 & 1 & -1 & 0 \\ 0 & 0 & 0 & 0 \end{bmatrix}$$

So $x_1 = -(1/2)x_3$, $x_2 = x_3$, with x_3 free. The general solution of $(A - 2I)\mathbf{x} = \mathbf{0}$ is $x_3\begin{bmatrix} -1/2 \\ 1 \\ 1 \end{bmatrix}$. A nice basis for the eigenspace is $\left\{\begin{bmatrix} -1 \\ 2 \\ 2 \end{bmatrix}\right\}$.

For $\lambda = 3$:

$$A - 3I = \begin{bmatrix} 4 & 0 & 1 \\ -2 & 1 & 0 \\ -2 & 0 & 1 \end{bmatrix} - \begin{bmatrix} 3 & 0 & 0 \\ 0 & 3 & 0 \\ 0 & 0 & 3 \end{bmatrix} = \begin{bmatrix} 1 & 0 & 1 \\ -2 & -2 & 0 \\ -2 & 0 & -2 \end{bmatrix}$$

$$[(A - 3I)\ \mathbf{0}] = \begin{bmatrix} 1 & 0 & 1 & 0 \\ -2 & -2 & 0 & 0 \\ -2 & 0 & -2 & 0 \end{bmatrix} \sim \begin{bmatrix} 1 & 0 & 1 & 0 \\ 0 & -2 & 2 & 0 \\ 0 & 0 & 0 & 0 \end{bmatrix} \sim \begin{bmatrix} 1 & 0 & 1 & 0 \\ 0 & 1 & -1 & 0 \\ 0 & 0 & 0 & 0 \end{bmatrix}$$

So $x_1 = -x_3$, $x_2 = x_3$, with x_3 free. Basis for the eigenspace is $\left\{\begin{bmatrix} -1 \\ 1 \\ 1 \end{bmatrix}\right\}$.

Study Tip: The text's answer to Exercise 15 is likely to be the same as yours, but there are many answers. What should you do if your vectors differ from those in the answer key? Example 2 gives the answer. Whenever you compute an **x** that you think is an eigenvector of A, you can check this simply by computing $A\mathbf{x}$. There is a little more to do in Exercise 15, however. The answer shows a basis of two eigenvectors, which means that the eigenspace is two-dimensional. So your answer must consist of two linearly independent eigenvectors. You can check that they are indeed eigenvectors, and then their linear independence can be checked by inspection.

19. The matrix $\begin{bmatrix} 1 & 2 & 3 \\ 1 & 2 & 3 \\ 1 & 2 & 3 \end{bmatrix}$ is not invertible because its columns are linearly dependent. So the number 0 is an eigenvalue of the matrix. See the discussion following Example 5.

21. a. Carefully read the definition of an eigenvalue.

    b. See the paragraph before the Invertible Matrix Theorem box.

    c. See the discussion of equation (3).

    d. See Example 2 and the paragraph preceding it. Also, see the Numerical Note.

    e. See the Warning after Example 3.

23. If a 2×2 matrix A had three distinct eigenvalues, then by Theorem 2 there would correspond three linearly independent eigenvectors (one for each eigenvalue). This is impossible because the vectors all belong to a two-dimensional vector space in which any set of three vectors is linearly dependent. See Theorem 8 in Section 1.6. In general, if an $n \times n$ matrix has p distinct eigenvalues, then by Theorem 2 there would be a linearly independent set of p eigenvectors (one for each eigenvalue). Since these vectors belong to an n-dimensional vector space, p cannot exceed n.

25. Let $\mathbf{x}$ be a nonzero vector such that $A\mathbf{x} = \lambda\mathbf{x}$. Then $A^{-1}A\mathbf{x} = A^{-1}(\lambda\mathbf{x})$, and $\mathbf{x} = \lambda(A^{-1}\mathbf{x})$. Since $\mathbf{x} \neq \mathbf{0}$ (and since A is invertible), λ cannot be zero. Then $\lambda^{-1}\mathbf{x} = A^{-1}\mathbf{x}$, which shows that λ^{-1} is an eigenvalue of A^{-1}.

Note: The relation between the eigenvalues of A and A^{-1} is important in the so-called *inverse power method* for estimating an eigenvalue of a matrix. (See Section 5.8.)

27. For any λ, $(A - \lambda I)^T = A^T - (\lambda I)^T = A^T - \lambda I$. Since $(A - \lambda I)^T$ is invertible if and only if $A - \lambda I$ is invertible (by Theorem 6(c) in Section 2.2), we conclude that $A^T - \lambda I$ is *not* invertible if and only if $A - \lambda I$ is *not* invertible. That is, λ is an eigenvalue of A^T if and only if λ is an eigenvalue of A.

33. The solution is given in the text. This exercise is important because it sets the stage for work later in the chapter. You ought to spend at least a little time on Exercise 34, too, even if that is not assigned.

    Observe that the set of all solutions of the difference equation $\mathbf{x}_{k+1} = A\mathbf{x}_k$ $(k = 0, 1, \ldots)$ is a subset H of the vector space V of all sequences of vectors in $\mathbb{R}^n$. The zero vector of V (that is, the zero sequence) is in H. So, Exercise 33(b) shows that H is a *subspace* of V.

Answer to Checkpoint: Of course, because the answer in the text is a nonzero multiple of the eigenvector found in the solution to Exercise 7, and any nonzero multiple of an eigenvector is another eigenvector. (The eigenspace is a *vector space* and so is closed under scalar multiplication.)

MATLAB Finding Eigenvectors

The linear algebra Toolbox has a command that will simplify your homework by automatically producing a basis for an eigenspace. For example, if A is a 3×3 matrix with an eigenvalue 7, then the commands

```
C = A - 7*eye(3)
B = nulbasis(C)
```

or, simply, **B = nulbasis(A - 7*eye(3))**, will produce a matrix B whose columns form a basis for the eigenspace for A corresponding to $\lambda = 7$. In general, **eye(k)** is a $k \times k$ identity matrix, and **nulbasis(C)** is a matrix whose columns form a basis for Nul C (the same basis you would get if you started with **ref(C)** and made the calculations by hand).

Remarks:

1. If the numbers in the basis matrix B are messy, try **format rat; B** , which will display the answer to the **nulbasis** command as a matrix with rational entries, if possible.
2. Although MATLAB has more powerful commands for eigenvector calculations, you should not use them yet, because you need to learn basic concepts about eigenvectors and eigenvalues.

5.2 THE CHARACTERISTIC EQUATION

There are several equivalent definitions of the determinant of a matrix. The definition here in terms of the pivots in an echelon form has the advantage that it is easy to state and understand, and in most cases it provides the most efficient way to compute a determinant.

KEY IDEAS

When A is 3×3, the geometric interpretation of det A as a volume explains why det $A = 0$ if and only if A is not invertible:

The determinant of A is zero.
<=> The parallelepiped determined by the columns of A has zero volume.
<=> One column of A is in the subspace spanned by the other columns.
<=> The columns of A are linearly dependent.
<=> The matrix A is not invertible.

If A is $n \times n$, then $\det(A - \lambda I) = 0$ if and only if $A - \lambda I$ is not invertible, and this happens if and only if λ is an eigenvalue of A.

Exercises 1—14 are designed only to provide some basic familiarity with characteristic polynomials. The main use of $\det(A - \lambda I)$ is as a tool for *studying* eigenvalues rather than computing them.

Sometimes the characteristic polynomial is defined as $\det(\lambda I - A)$. A property of determinants implies that $\det(\lambda I - A) = (-1)^n \det(A - \lambda I)$, when A is $n \times n$, so the two polynomials are either the same (when n is even) or they are negatives of one another. The use of $\det(A - \lambda I)$ tends to make hand calculations easier and less prone to copying errors.

SOLUTIONS TO EXERCISES

1. $A = \begin{bmatrix} 2 & 7 \\ 7 & 2 \end{bmatrix}$, $A - \lambda I = \begin{bmatrix} 2 & 7 \\ 7 & 2 \end{bmatrix} - \begin{bmatrix} \lambda & 0 \\ 0 & \lambda \end{bmatrix} = \begin{bmatrix} 2-\lambda & 7 \\ 7 & 2-\lambda \end{bmatrix}$, the characteristic polynomial is

$$\det(A - \lambda I) = (2 - \lambda)^2 - 7^2 = 4 - 4\lambda + \lambda^2 - 49 = \lambda^2 - 4\lambda - 45$$

In factored form, the characteristic equation is $(\lambda - 9)(\lambda + 5) = 0$, so the eigenvalues of A are 9 and -5.

Warning: Don't row reduce a matrix A to find its eigenvalues. Row reduction preserves the null space of A but not the eigenvalues of A.

7. $A = \begin{bmatrix} 5 & 3 \\ -4 & 4 \end{bmatrix}$, $A - \lambda I = \begin{bmatrix} 5-\lambda & 3 \\ -4 & 4-\lambda \end{bmatrix}$, the characteristic polynomial is

$$\det(A - \lambda I) = (5 - \lambda)(4 - \lambda) - (3)(-4) = 20 - 9\lambda + \lambda^2 + 12$$

$$= \lambda^2 - 9\lambda + 32$$

The characteristic polynomial does not factor easily, but the quadratic formula provides the solutions of $\lambda^2 - 9\lambda + 32 = 0$.

$$\lambda = \frac{+9 \pm \sqrt{81 - 4(32)}}{2}$$

These values for λ are not real numbers, so A has no real eigenvalues. There is no nonzero vector $\mathbf{x}$ in $\mathbb{R}^2$ such that $A\mathbf{x} = \lambda\mathbf{x}$ for such a λ. (For any $\mathbf{x} \neq \mathbf{0}$ in $\mathbb{R}^2$, the vector $A\mathbf{x}$ has only real entries and thus could not equal a complex multiple of $\mathbf{x}$.)

Study Tip: If you are asked to work some of Exercises 9—14, you may be tested on them. This is one way of finding out if you know what the characteristic polynomial is and how it is connected with eigenvalues. Also, you can show that you know some elementary properties of determinants.

13. $A - \lambda I = \begin{bmatrix} 6-\lambda & -2 & 0 \\ -2 & 9-\lambda & 0 \\ 5 & 8 & 3-\lambda \end{bmatrix}$

The method using the special 3×3 formula will produce the characteristic polynomial $-\lambda^3 + 18\lambda^2 - 95\lambda + 150$. Factoring such a polynomial to find the eigenvalues requires a little experience. (See the appendix at the end of the exercise solutions.) However, if you use a cofactor expansion down the third column (see Section 3.1), you immediately obtain

$$\det(A - \lambda I) = (3 - \lambda)\cdot\det\begin{bmatrix} 6-\lambda & -2 \\ -2 & 9-\lambda \end{bmatrix}$$

$$= (3 - \lambda)[(6-\lambda)(9-\lambda) - 4]$$

$$= (3 - \lambda)(\lambda^2 - 15\lambda + 50)$$

The characteristic polynomial is already partially factored, and the remaining quadratic factor is itself easily factored. The factored characteristic polynomial is $(3 - \lambda)(\lambda - 10)(\lambda - 5)$ or, equivalently, $-(\lambda - 3)(\lambda - 5)(\lambda - 10)$.

Note: The solutions of Exercises 11—14 are similar to that of Exercise 13. These matrices have the property that if a cofactor expansion is chosen along a column or row that contains two zeros, then the characteristic polynomial appears in a partially factored form.

19. Since the equation $\det(A - \lambda I) = (\lambda_1 - \lambda)(\lambda_2 - \lambda)\cdots(\lambda_n - \lambda)$ holds for all λ, set $\lambda = 0$ and conclude that $\det A = \lambda_1\lambda_2\cdots\lambda_n$.

21. **a.** See Example 1. **b.** See Theorem 3.

c. See Theorem 3. **d.** See the solution of Example 4.

23. If $A = QR$, with Q invertible, and if $A_1 = RQ$, then $A_1 = Q^{-1}QRQ = Q^{-1}AQ$, which shows that A_1 is similar to A.

25. The complete solution is in the text.

29. The Toolbox command **gauss** will produce the echelon form in three steps, without row scaling. See the MATLAB box on p. 1-17. For other matrix programs, see the notes at the back of this *Guide*.

30. In MATLAB, the following commands will produce the graph of one characteristic polynomial:

```
x = linspace(0,3)          Choose 100 points between 0 and 3
grid on                    Include a grid on the graph
hold on                    Keep all graphs on the same axes.
A(3,2)=32; p=poly(A); v=polyval(p,x); plot(x,v,'b')
                           Compute characteristic polynomial, evaluate it
                           at the points in x, plot the curve in blue.
```

Edit the last line to change the value of *a* and the color of the curve.

Appendix: Factoring a Polynomial

In general it is difficult to factor a polynomial of degree 3 or higher (unless you have one of several powerful computer programs available). Fortunately, textbook examples and exercises tend to have small integer solutions. The following observation is helpful.

> *Let* $p(\lambda)$ *be a polynomial with integer coefficients. If* $p(c) = 0$ *for some integer* c*, then* $\lambda - c$ *is a factor of* $p(\lambda)$ *and* c *is a divisor of the constant term of* $p(\lambda)$.

EXAMPLE Find the eigenvalues of the matrix A whose characteristic polynomial is $p(\lambda) = -\lambda^3 + 18\lambda^2 - 96\lambda + 160$.

Solution By the observation above, any integer eigenvalue of A must be a divisor of the constant term 160 in the characteristic polynomial. There are twenty-four such divisors: 1, 2, 4, 5, 8, 10, 16, 20, 32, 40, 80, and 160, together with the negatives of these numbers. We let c be one of these numbers and try to divide $\lambda - c$ into the polynomial by long division. The terms $\lambda \pm 1$ and $\lambda \pm 2$ don't work, and the first successful division is

$$
\begin{array}{r|l}
 & -\lambda^2 + 14\lambda - 40 \\ \hline
\lambda - 4 & -\lambda^3 + 18\lambda^2 - 96\lambda + 160 \\
 & \underline{-\lambda^3 + 4\lambda^2} \\
 & \qquad 14\lambda^2 - 96\lambda \\
 & \qquad \underline{14\lambda^2 - 56\lambda} \\
 & \qquad\qquad -40\lambda + 160 \\
 & \qquad\qquad \underline{-40\lambda + 160}
\end{array}
$$

Thus the characteristic polynomial is $(\lambda - 4)(-\lambda^2 + 14\lambda - 40)$.

The quadratic polynomial factors easily, and the characteristic equation is $(\lambda - 4)(\lambda - 4)(10 - \lambda) = 0$. The eigenvalues of A are 4 (with multiplicity two) and 10. □

MATLAB: You can use the MATLAB command **poly(A)** to check your answers in Exercises 9—14. Note that if A is $n \times n$, **poly(A)** lists the coefficients of the characteristic polynomial of A, in order of decreasing powers of λ, beginning with λ^n. If the polynomial is of odd degree, the coefficients are multiplied by -1, to make +1 the coefficient of λ^n. This corresponds to using $\det(\lambda I - A)$ instead of $\det(A - \lambda I)$.

5.3 DIAGONALIZATION

The factorization $A = PDP^{-1}$ is used to compute powers of A, decouple a dynamical system in Section 5.6, and study symmetric matrices and quadratic forms in Chapter 7.

KEY IDEAS

Example 3 gives the algorithm for diagonalizing a matrix A. After you construct P and D, check your calculations:

1. Compute AP and PD, and check that $AP = PD$.
2. Make sure the columns of P are linearly independent. Use Theorem 7 to save time. You only have to verify that for each eigenvalue, the corresponding eigenvectors are linearly independent. That's easy if the dimension of the eigenspace is 2 or 1.

The key equation in this section is $AP = PD$. It will help you to keep the order of the factors correct when you write $A = PDP^{-1}$, and it also leads immediately to $P^{-1}AP = D$, which you may need occasionally. Possible test question: If $AP = PD$, explain why the first column of P is an eigenvector of A (if the column is nonzero). (Study the proof of Theorem 5.)

Warning: Do not confuse the property of being diagonalizable with the property of being invertible. They are not connected. The matrix in Example 5 is diagonalizable, but it is not invertible because 0 is an eigenvalue. The matrix $\begin{bmatrix} 1 & -2 \\ 0 & 1 \end{bmatrix}$ is invertible, but it is not diagonalizable because the eigenspace for $\lambda = 1$ is only one-dimensional.

SOLUTIONS TO EXERCISES

1. $P = \begin{bmatrix} 5 & 7 \\ 2 & 3 \end{bmatrix}$, $D = \begin{bmatrix} 2 & 0 \\ 0 & 1 \end{bmatrix}$, $A = PDP^{-1}$, and $A^4 = PD^4P^{-1}$. Next, compute

$P^{-1} = \frac{1}{1}\begin{bmatrix} 3 & -7 \\ -2 & 5 \end{bmatrix}$, $D^4 = \begin{bmatrix} 2^4 & 0 \\ 0 & 1 \end{bmatrix} = \begin{bmatrix} 16 & 0 \\ 0 & 1 \end{bmatrix}$. Putting this together,

$$A^4 = \begin{bmatrix} 5 & 7 \\ 2 & 3 \end{bmatrix}\begin{bmatrix} 16 & 0 \\ 0 & 1 \end{bmatrix}\begin{bmatrix} 3 & -7 \\ -2 & 5 \end{bmatrix} = \begin{bmatrix} 80 & 7 \\ 32 & 3 \end{bmatrix}\begin{bmatrix} 3 & -7 \\ -2 & 5 \end{bmatrix} = \begin{bmatrix} 226 & -525 \\ 90 & -209 \end{bmatrix}$$

7. $A = \begin{bmatrix} 1 & 0 \\ 6 & -1 \end{bmatrix}$. The eigenvalues are obviously ±1 (since A is triangular).

<u>For $\lambda = 1$:</u> $A - 1I = \begin{bmatrix} 0 & 0 \\ 6 & -2 \end{bmatrix}$. The equation $(A - I)\mathbf{x} = \mathbf{0}$ amounts to $6x_1 - 2x_2 = 0$. So $x_1 = (1/3)x_2$, with x_2 free. The general solution is $x_2\begin{bmatrix} 1/3 \\ 1 \end{bmatrix}$. A nice basis vector for the eigenspace is $\mathbf{u}_1 = \begin{bmatrix} 1 \\ 3 \end{bmatrix}$.

<u>For $\lambda = -1$:</u> $A - (-1)I = \begin{bmatrix} 2 & 0 \\ 6 & 0 \end{bmatrix}$. The equation $(A + I)\mathbf{x} = \mathbf{0}$ amounts to $2x_1 = 0$, with x_2 free. The general solution is $x_2\begin{bmatrix} 0 \\ 1 \end{bmatrix}$. Take $\mathbf{u}_2 = \begin{bmatrix} 0 \\ 1 \end{bmatrix}$ as a basis vector for the eigenspace.

From $\mathbf{u}_1$ and $\mathbf{u}_2$, construct $P = \begin{bmatrix} \mathbf{u}_1 & \mathbf{u}_2 \end{bmatrix} = \begin{bmatrix} 1 & 0 \\ 3 & 1 \end{bmatrix}$. Then set $D = \begin{bmatrix} 1 & 0 \\ 0 & -1 \end{bmatrix}$, where the eigenvalues in D correspond to $\mathbf{u}_1$ and $\mathbf{u}_2$, respectively.

Warning: The 3×3 matrices in Exercises 12—18 may be diagonalizable even though each matrix has only two distinct eigenvalues. (Theorem 6 gives only a *sufficient* condition for diagonalizability.) You have to check for three linearly independent eigenvectors.

13. $A = \begin{bmatrix} 2 & 2 & -1 \\ 1 & 3 & -1 \\ -1 & -2 & 2 \end{bmatrix}$. The eigenvalues 5 and 1 are given. Because A is 3×3, you need three linearly independent eigenvectors.

<u>For $\lambda = 5$:</u> Solve $(A - 5I)\mathbf{x} = \mathbf{0}$. Form $A - 5I = \begin{bmatrix} -3 & 2 & -1 \\ 1 & -2 & -1 \\ -1 & -2 & -3 \end{bmatrix}$ and compute

$$\begin{bmatrix} -3 & 2 & -1 & 0 \\ 1 & -2 & -1 & 0 \\ -1 & -2 & -3 & 0 \end{bmatrix} \sim \begin{bmatrix} 1 & -2 & -1 & 0 \\ -3 & 2 & -1 & 0 \\ -1 & -2 & -3 & 0 \end{bmatrix} \sim \cdots \sim \begin{bmatrix} 1 & 0 & 1 & 0 \\ 0 & 1 & 1 & 0 \\ 0 & 0 & 0 & 0 \end{bmatrix}$$

So $x_1 = -x_3$, $x_2 = -x_3$, with x_3 free. Take $\mathbf{v}_1 = \begin{bmatrix} -1 \\ -1 \\ 1 \end{bmatrix}$, for instance, as a basis vector for the eigenspace. At this point, you don't know if A is diagonalizable. Your only hope is to find two linearly independent eigenvectors inside the eigenspace for $\lambda = 1$, because there are no other eigenspaces in which to look.

For $\lambda = 1$: Form $A - I = \begin{bmatrix} 1 & 2 & -1 \\ 1 & 2 & -1 \\ -1 & -2 & 1 \end{bmatrix}$. The equation $(A - I)\mathbf{x} = \mathbf{0}$ reduces to $x_1 + 2x_2 - x_3 = 0$. So $x_1 = -2x_2 + x_3$, with x_2 and x_3 free. At this point you know that the eigenspace is two-dimensional (because there are two free variables). So there are the necessary two linearly independent eigenvectors, and hence A is diagonalizable. To produce the eigenvectors, write the solution of $(A - I)\mathbf{x} = \mathbf{0}$ in the form

$$\begin{bmatrix} x_1 \\ x_2 \\ x_3 \end{bmatrix} = \begin{bmatrix} -2x_2+x_3 \\ x_2 \\ x_3 \end{bmatrix} = x_2\begin{bmatrix} -2 \\ 1 \\ 0 \end{bmatrix} + x_3\begin{bmatrix} 1 \\ 0 \\ 1 \end{bmatrix}$$

Set $\mathbf{v}_2 = \begin{bmatrix} -2 \\ 1 \\ 0 \end{bmatrix}$, $\mathbf{v}_3 = \begin{bmatrix} 1 \\ 0 \\ 1 \end{bmatrix}$, and $P = [\mathbf{v}_1 \ \ \mathbf{v}_2 \ \ \mathbf{v}_3] = \begin{bmatrix} -1 & -2 & 1 \\ -1 & 1 & 0 \\ 1 & 0 & 1 \end{bmatrix}$.

The columns of P are linearly independent, by Theorem 7, because the eigenvectors form bases for their respective eigenspaces. So P is invertible. Since the first column of P corresponds to $\lambda = 5$, the first diagonal entry in D must be 5, which means that

$$D = \begin{bmatrix} 5 & 0 & 0 \\ 0 & 1 & 0 \\ 0 & 0 & 1 \end{bmatrix}$$

Warning: A common mistake in Exercise 13 is to build a 3×3 matrix P in the usual way and then take D to be a 2×2 matrix, such as

$$D = \begin{bmatrix} 5 & 0 \\ 0 & 1 \end{bmatrix}$$

Of course this doesn't make sense because PD isn't even defined. Another mistake is to make a 2×2 matrix D as above and build P with only two of the three eigenvectors, say one from each eigenspace. Now $AP = PD$, but P is a 3×2 matrix and is not invertible.

Study Tip: Exercise 18 is a good test question. Try it. Note: the eigenvector (-2,1,2) does not correspond to the eigenvalue $\lambda = 5$. If you have trouble here, review Exercises 5—8 in Section 5.1.

19. $A = \begin{bmatrix} 5 & -3 & 0 & 9 \\ 0 & 3 & 1 & -2 \\ 0 & 0 & 2 & 0 \\ 0 & 0 & 0 & 2 \end{bmatrix}$. The eigenvalues are obviously 5,3, and 2. (Why?)

For $\lambda = 2$: To solve $(A - 2I)\mathbf{x} = \mathbf{0}$, completely reduce $[(A-2I) \;\; \mathbf{0}]$:

$$\begin{bmatrix} 3 & -3 & 0 & 9 & 0 \\ 0 & 1 & 1 & -2 & 0 \\ 0 & 0 & 0 & 0 & 0 \\ 0 & 0 & 0 & 0 & 0 \end{bmatrix} \sim \begin{bmatrix} 3 & 0 & 3 & 3 & 0 \\ 0 & 1 & 1 & -2 & 0 \\ 0 & 0 & 0 & 0 & 0 \\ 0 & 0 & 0 & 0 & 0 \end{bmatrix} \sim \begin{bmatrix} 1 & 0 & 1 & 1 & 0 \\ 0 & 1 & 1 & -2 & 0 \\ 0 & 0 & 0 & 0 & 0 \\ 0 & 0 & 0 & 0 & 0 \end{bmatrix}$$

So $x_1 = -x_3 - x_4$, $x_2 = -x_3 + 2x_4$, with x_3 and x_4 free. The usual calculations produce a basis for the eigenspace:

$$\begin{bmatrix} x_1 \\ x_2 \\ x_3 \\ x_4 \end{bmatrix} = \begin{bmatrix} -x_3 - x_4 \\ -x_3 + 2x_4 \\ x_3 \\ x_4 \end{bmatrix} = x_3\begin{bmatrix} -1 \\ -1 \\ 1 \\ 0 \end{bmatrix} + x_4\begin{bmatrix} -1 \\ 2 \\ 0 \\ 1 \end{bmatrix}. \quad \text{Basis: } \mathbf{v}_1 = \begin{bmatrix} -1 \\ -1 \\ 1 \\ 0 \end{bmatrix}, \; \mathbf{v}_2 = \begin{bmatrix} -1 \\ 2 \\ 0 \\ 1 \end{bmatrix}$$

Checkpoint: If you happened to choose $\lambda = 2$ first, as in this solution, would you have enough information at this point to determine whether A is diagonalizable?

For $\lambda = 3$: To solve $(A - 3I)\mathbf{x} = \mathbf{0}$, completely reduce $[(A-3I) \;\; \mathbf{0}]$:

$$\begin{bmatrix} 2 & -3 & 0 & 9 & 0 \\ 0 & 0 & 1 & -2 & 0 \\ 0 & 0 & -1 & 0 & 0 \\ 0 & 0 & 0 & -1 & 0 \end{bmatrix} \sim \cdots \sim \begin{bmatrix} 1 & -3/2 & 0 & 0 & 0 \\ 0 & 0 & 1 & 0 & 0 \\ 0 & 0 & 0 & 1 & 0 \\ 0 & 0 & 0 & 0 & 0 \end{bmatrix}, \quad \begin{cases} x_1 = (3/2)x_2 \\ x_2 \text{ is free} \\ x_3 = 0 \\ x_4 = 0 \end{cases}$$

Choosing $x_2 = 2$ produces the eigenvector $\mathbf{v}_3 = (3,2,0,0)$.

For $\lambda = 5$: To solve $(A - 3I)\mathbf{x} = \mathbf{0}$, completely reduce $[(A - 5I)\ \ \mathbf{0}]$:

$$\begin{bmatrix} 0 & -3 & 0 & 9 & 0 \\ 0 & -2 & 1 & -2 & 0 \\ 0 & 0 & -3 & 0 & 0 \\ 0 & 0 & 0 & -3 & 0 \end{bmatrix} \sim \cdots \sim \begin{bmatrix} 0 & 1 & 0 & 0 & 0 \\ 0 & 0 & 1 & 0 & 0 \\ 0 & 0 & 0 & 1 & 0 \\ 0 & 0 & 0 & 0 & 0 \end{bmatrix}, \quad \begin{cases} x_1 \text{ is free} \\ x_2 = 0 \\ x_3 = 0 \\ x_4 = 0 \end{cases}$$

A basis vector for the eigenspace is $\mathbf{v}_4 = (1,0,0,0)$. Set

$$P = [\mathbf{v}_1\ \ \mathbf{v}_2\ \ \mathbf{v}_3\ \ \mathbf{v}_4] = \begin{bmatrix} -1 & -1 & 3 & 1 \\ -1 & 2 & 2 & 0 \\ 1 & 0 & 0 & 0 \\ 0 & 1 & 0 & 0 \end{bmatrix}, \quad D = \begin{bmatrix} 2 & 0 & 0 & 0 \\ 0 & 2 & 0 & 0 \\ 0 & 0 & 3 & 0 \\ 0 & 0 & 0 & 5 \end{bmatrix}$$

This answer differs from that in the text. There, $P = [\mathbf{v}_4\ \ \mathbf{v}_3\ \ \mathbf{v}_1\ \ \mathbf{v}_2]$, and the entries in D are rearranged to match the new order of the eigenvectors. According to the Diagonalization Theorem, both answers are correct.

21. **a.** The symbol D does not automatically denote a diagonal matrix.

b. See the remark after the statement of the Diagonalization Theorem.

c. Check out the matrix in Example 4.

d. See Example 4.

23. A is diagonalizable because you know that five linearly independent eigenvectors exist: three in the three-dimensional eigenspace and two in the two-dimensional eigenspace. Theorem 7 guarantees that the set of all five eigenvectors is linearly independent.

25. Let $\{\mathbf{v}_1\}$ be a basis for the one-dimensional eigenspace, let $\mathbf{v}_2$ and $\mathbf{v}_3$ form a basis for the two-dimensional eigenspace, and let $\mathbf{v}_4$ be any eigenvector in the remaining eigenspace. By Theorem 7, $\{\mathbf{v}_1, \mathbf{v}_2, \mathbf{v}_3, \mathbf{v}_4\}$ is linearly independent. Since A is 4×4, the Diagonalization Theorem shows that A is diagonalizable.

27. If A is diagonalizable, then $A = PDP^{-1}$ for some invertible P and diagonal D. Since A is invertible, 0 is not an eigenvalue of A. So the diagonal entries in D (which are eigenvalues of A) are not zero, and D is invertible. By the theorem on the inverse of a product,

$$A^{-1} = (PDP^{-1})^{-1} = (P^{-1})^{-1}D^{-1}P^{-1} = PD^{-1}P^{-1}$$

Since D^{-1} is obviously diagonal, A^{-1} is diagonalizable.

A Second Proof: If A is $n \times n$, it has n linearly independent eigenvectors, say, $\mathbf{v}_1, \ldots, \mathbf{v}_n$. Then $\mathbf{v}_1, \ldots, \mathbf{v}_n$ are also eigenvectors of A^{-1}, by the solution of Exercise 25 in Section 5.1. Hence A^{-1} is diagonalizable, by the Diagonalization Theorem.

29. The diagonal entries in D_1 are reversed from those in D. So interchange the (eigenvector) columns of P to make them correspond properly to the eigenvalues in D_1. In this case,

$$P_1 = \begin{bmatrix} 1 & 1 \\ -2 & -1 \end{bmatrix} \quad \text{and} \quad D_1 = \begin{bmatrix} 3 & 0 \\ 0 & 5 \end{bmatrix}$$

Although the first column of P must be an eigenvector corresponding to the eigenvalue 3, there is nothing to prevent us from selecting some multiple of $\begin{bmatrix} 1 \\ -2 \end{bmatrix}$, say $\begin{bmatrix} -3 \\ 6 \end{bmatrix}$, and letting $P_2 = \begin{bmatrix} -3 & 1 \\ 6 & -1 \end{bmatrix}$. We now have three different factorizations or "diagonalizations" of A:

$$A = PDP^{-1} = P_1D_1P_1^{-1} = P_2D_1P_2^{-1}$$

Answer to Checkpoint: Yes. In Exercise 19, the fact that the eigenspace for $\lambda = 2$ is two-dimensional guarantees that A is diagonalizable, because each of the other two eigenvalues will produce at least one eigenvector, and the resulting set of four eigenvectors will be linearly independent, by Theorem 7. So A is diagonalizable, by the Diagonalization Theorem. (Note that since one eigenspace is two-dimensional, the other eigenspaces must be one-dimensional, because there could not possibly be *more* than four linearly independent eigenvectors in $\mathbb{R}^4$.)

Mastering Linear Algebra Concepts: Eigenvalue, Eigenvector, Eigenspace

I suggest that you take time now to prepare review sheets for the three terms listed above. Begin with their definitions, of course.

Eigenvalue:

- equivalent description — Sec. 5.1: Equation (3); Box on page 307
- geometric interpretation — Sec. 5.1: Fig. 2
- special cases — Theorems 1, 6
- typical computations — Sec. 5.1: Exer. 7, 19; Sec. 5.2: Exer. 7, 15
- connections with other concepts — The IMT; Theorem 4; Sec. 5.2: Exer. 19

Eigenvector:

- special cases — Theorem 2
- typical computations — Sec. 5.1: Examples 2, 4, Exer. 5, 15
- connections with other concepts — Sec. 5.1: Exer. 33; Sec. 5.2: Example 5; Sec. 5.3: Theorem 5, Exer. 5, 18

Eigenspace:

- ▶ equivalent description — Sec. 5.1: Equation (3)
- ▶ geometric interpretation — Sec. 5.1: Fig. 2, 3
- ▶ typical computations — Sec. 5.1: Example 4
- ▶ connections with other concepts — Theorem 7

MATLAB eig

The command **ev = eig(A)** produces a vector **ev** that lists the eigenvalues of A. To learn the diagonalization procedure, you should use the method of Section 5.1 to produce the eigenvectors. For instance, if A is 5×5, the command

```
V = nulbasis(A - ev(1)*eye(5))
```

creates a matrix whose column(s) give a basis for the eigenspace corresponding to the first eigenvalue of A listed in **ev**.

In later work, you can automate the diagonalization process. The command **[P D] = eig(A)** produces two square matrices P and D (or any other names you choose) such that $AP = PD$, with D diagonal. If A happens to be diagonalizable, then P will be invertible. In any case, P is likely to be quite different from what you construct for the homework. The columns of P are scaled so they are "unit" vectors (such that the sum of squares of the entries is 1). Dividing P by one of its entries may make some of the new entries recognizable, because the problems in the text usually involve simple numbers.

5.4 EIGENVALUES AND LINEAR TRANSFORMATIONS

This section introduces the matrix of a linear transformation relative to specified bases for vector spaces, and then uses this concept to give a new interpretation of the matrix factorization $A = PDP^{-1}$.

STUDY NOTES

The exercises will help you learn the definition of the matrix representation of a transformation relative to specified bases, say $\mathcal{B}$ and $\mathcal{C}$. Compare this definition with the standard matrix of a transformation from $\mathbb{R}^n$ into $\mathbb{R}^m$. What are $\mathcal{B}$ and $\mathcal{C}$ in this case?

After Example 1 (and Exercises 1-6 and 9-10), the section focuses on the case when $T: V \to V$ and $\mathcal{B} = \mathcal{C}$. The following algorithm and the solution to Exercise 7 describe two different ways to construct $[T]_{\mathcal{B}}$.

> **Algorithm for Finding the B-matrix of T: V ⟶ V**
>
> 1. Compute the images of the basis vectors:
>
> $$T(\mathbf{b}_1),\ \ldots,\ T(\mathbf{b}_n)$$
>
> 2. Convert these images into $\mathcal{B}$-coordinate vectors:
>
> $$[T(\mathbf{b}_1)]_{\mathcal{B}},\ \ldots,\ [T(\mathbf{b}_n)]_{\mathcal{B}}$$
>
> 3. Place the $\mathcal{B}$-coordinate vectors into the columns of $[T]_{\mathcal{B}}$.

The sentence before Theorem 8 summarizes the main idea of the theorem. Studying the proof should help you understand the theorem and review important concepts. The subsection on Similarity of Matrix Representations could have contained more ideas. Take notes carefully if your instructor decides to expand this material somewhat.

SOLUTIONS TO EXERCISES

1. $T(\mathbf{b}_1) = 3\mathbf{d}_1 - 5\mathbf{d}_2$, $T(\mathbf{b}_2) = -\mathbf{d}_1 + 6\mathbf{d}_2$, $T(\mathbf{b}_3) = 4\mathbf{d}_2$

$$[T(\mathbf{b}_1)]_{\mathcal{D}} = \begin{bmatrix} 3 \\ -5 \end{bmatrix} \qquad [T(\mathbf{b}_2)]_{\mathcal{D}} = \begin{bmatrix} -1 \\ 6 \end{bmatrix} \qquad [T(\mathbf{b}_3)]_{\mathcal{D}} = \begin{bmatrix} 0 \\ 4 \end{bmatrix}$$

Matrix for T relative to $\mathcal{B}$ and $\mathcal{D}$: $\begin{bmatrix} 3 & -1 & 0 \\ -5 & 6 & 4 \end{bmatrix}$

7. The method of Example 2 works. But, perhaps a faster way is to realize that the information given provides the general form of $T(\mathbf{p})$ as shown in the figure below:

$$a_0 + a_1 t + a_2 t^2 \xrightarrow{\ T\ } 3a_0 + (5a_0 - 2a_1)t + (4a_1 + a_2)t^2$$

coordinate mapping ↓ (on both sides)

$$\begin{bmatrix} a_0 \\ a_1 \\ a_2 \end{bmatrix} \xrightarrow[\text{by } [T]_B]{\text{multiplication}} \begin{bmatrix} 3a_0 \\ 5a_0 - 2a_1 \\ 4a_1 + a_2 \end{bmatrix}$$

The matrix that implements the multiplication along the bottom of the figure is easily filled in by inspection:

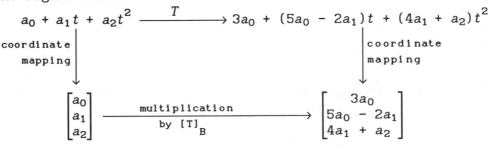

$$\begin{bmatrix} ? & ? & ? \\ ? & ? & ? \\ ? & ? & ? \end{bmatrix}\begin{bmatrix} a_0 \\ a_1 \\ a_2 \end{bmatrix} = \begin{bmatrix} 3a_0 \\ 5a_0 - 2a_1 \\ 4a_1 + a_2 \end{bmatrix} \text{ implies that } [T]_{\mathcal{B}} = \begin{bmatrix} 3 & 0 & 0 \\ 5 & -2 & 0 \\ 0 & 4 & 1 \end{bmatrix}$$

Study Tip: See the Study Guide notes for Section 1.8 (particularly for Exercise 19). This method allows you to find a matrix that implements the mapping without assuming that T is linear. In fact, this derivation *proves* that T is linear, because it is now represented as a matrix transformation. Why not try this method on Example 2?

13. Start by diagonalizing $A = \begin{bmatrix} 0 & 1 \\ -3 & 4 \end{bmatrix}$. The characteristic polynomial is $\lambda^2 - 4\lambda + 3 = (\lambda - 1)(\lambda - 3)$, so the eigenvalues are 1 and 3.

For $\lambda = 1$: $A - I = \begin{bmatrix} -1 & 1 \\ -3 & 3 \end{bmatrix}$. The equation $(A - I)\mathbf{x} = \mathbf{0}$ amounts to $-x_1 + x_2 = 0$. So $x_1 = x_2$, with x_2 free. As a basis vector, take $\mathbf{u}_1 = \begin{bmatrix} 1 \\ 1 \end{bmatrix}$.

For $\lambda = 3$: $A - 3I = \begin{bmatrix} -3 & 1 \\ -3 & 1 \end{bmatrix}$. The equation $(A - I)\mathbf{x} = \mathbf{0}$ amounts to $-3x_1 + x_2 = 0$. So $x_1 = x_2/3$, with x_2 free. Take $\mathbf{u}_2 = \begin{bmatrix} 1 \\ 3 \end{bmatrix}$ as a basis vector. The vectors $\mathbf{u}_1, \mathbf{u}_2$ can form the columns of a matrix P that diagonalizes A. By Theorem 8, the basis $\mathcal{B} = \{\mathbf{u}_1, \mathbf{u}_2\}$ has the property that the $\mathcal{B}$-matrix of the transformation $\mathbf{x} \mapsto A\mathbf{x}$ is a diagonal matrix.

19. If A is similar to B, then there exists an invertible matrix P such that $P^{-1}AP = B$. Thus B is invertible because it is the product of invertible matrices. By a theorem about inverses of products, $B^{-1} = P^{-1}A^{-1}(P^{-1})^{-1} = P^{-1}A^{-1}P$, which shows that A^{-1} is similar to B^{-1}.

21. By hypothesis, there exist invertible P and Q such that $P^{-1}BP = A$ and $Q^{-1}CQ = A$. Then $P^{-1}BP = Q^{-1}CQ$. Left-multiplying by Q, and right-multiplying by Q^{-1}, we have $QP^{-1}BPQ^{-1} = QQ^{-1}CQQ^{-1}$. So $C = QP^{-1}BPQ^{-1} = (PQ^{-1})^{-1}B(PQ^{-1})$, which shows that B is similar to C.

23. If $A\mathbf{x} = \lambda\mathbf{x}$, $\mathbf{x} \neq \mathbf{0}$, then $P^{-1}A\mathbf{x} = \lambda P^{-1}\mathbf{x}$. If $B = P^{-1}AP$, then

$$B(P^{-1}\mathbf{x}) = P^{-1}AP(P^{-1}\mathbf{x}) = P^{-1}A\mathbf{x} = \lambda P^{-1}\mathbf{x} \qquad (*)$$

by the first calculation. Note that $P^{-1}\mathbf{x} \neq \mathbf{0}$, because $\mathbf{x} \neq \mathbf{0}$ and P^{-1} is invertible. Hence (*) shows that $P^{-1}\mathbf{x}$ is an eigenvector of B corresponding to λ. (Of course, λ is an eigenvalue of both A and B because the matrices are similar, by Theorem 4 in Section 5.2.)

25. If $A = PBP^{-1}$, then

$$\text{tr}(A) = \text{tr}((PB)P^{-1}) = \text{tr}(P^{-1}(PB)) \quad \text{By the trace property}$$

$$= \text{tr}(P^{-1}PB) = \text{tr}(IB) = \text{tr}(B)$$

If B is diagonal, then the diagonal entries of B must be the eigenvalues of A, by the Diagonalization Theorem (Theorem 5 in Section 5.3). So $\text{tr}\,A = \text{tr}\,B = \{\text{sum of the eigenvalues of } A\}$.

Study Tip: It can be shown that for *any* square matrix A, the trace of A is the sum of the eigenvalues of A, counted according to multiplicities. You can use this fact to provide a quick check on your eigenvalue calculations. The sum of the eigenvalues must match the sum of the diagonal entries in A (even if A is not diagonalizable).

29. If $\mathcal{B} = \{\mathbf{b}_1, \ldots, \mathbf{b}_n\}$, then the $\mathcal{B}$-coordinate vector of $\mathbf{b}_j$ is $\mathbf{e}_j$, the standard basis vector for $\mathbb{R}^n$. For instance,

$$\mathbf{b}_1 = 1\cdot\mathbf{b}_1 + 0\cdot\mathbf{b}_2 + \cdots + 0\cdot\mathbf{b}_n$$

Thus $[I(\mathbf{b}_j)]_{\mathcal{B}} = [\mathbf{b}_j]_{\mathcal{B}} = \mathbf{e}_j$, and

$$[I]_{\mathcal{B}} = [[I(\mathbf{b}_1)]_{\mathcal{B}} \quad \cdots \quad [I(\mathbf{b}_n)]_{\mathcal{B}}] = [\mathbf{e}_1 \quad \cdots \quad \mathbf{e}_n] = I$$

Appendix: The Characteristic Polynomial of a 2×2 Matrix

By Exercise 19 in Section 5.2, $\det A$ is the product of the eigenvalues of A. From the Study Tip above, $\text{tr}\,A$ is the sum of the eigenvalues. Thus, if a 2×2 matrix A has eigenvalues λ_1 and λ_2, the characteristic polynomial of A has the form

$$(\lambda_1 - \lambda)(\lambda_2 - \lambda) = \lambda^2 - (\lambda_1 + \lambda_2)\lambda + \lambda_1\lambda_2$$

$$= \lambda^2 - (\text{tr}\,A)\lambda + (\det A)$$

5.5 COMPLEX EIGENVALUES

If the characteristic equation of an $n \times n$ real matrix A has a complex eigenvalue λ and if $\mathbf{v}$ is a nonzero vector in $\mathbb{C}^n$ such that $A\mathbf{v} = \lambda\mathbf{v}$, then both λ and $\mathbf{v}$ provide useful information about A.

KEY IDEAS

Only matrices with real entries are considered here. If λ is a complex eigenvalue of A, with $\mathbf{v}$ a corresponding eigenvector, then $\bar{\lambda}$ is also an eigenvalue of A, with $\bar{\mathbf{v}}$ an eigenvector. Find out if you should know how to prove this fact.

Example 6 describes the prototype for all 2×2 matrices with a complex eigenvalue λ. Only λ is needed if you want to know the angle φ of rotation and the scale factor $|\lambda|$, but an associated eigenvector $\mathbf{v}$ is also needed if you want to factor A as PCP^{-1}, as in Example 7.

SOLUTIONS TO EXERCISES

1. $A = \begin{bmatrix} 1 & -2 \\ 1 & 3 \end{bmatrix}$, $A - \lambda I = \begin{bmatrix} 1-\lambda & -2 \\ 1 & 3-\lambda \end{bmatrix}$

$$\det(A - \lambda I) = (1 - \lambda)(3 - \lambda) - (-2) = \lambda^2 - 4\lambda + 5$$

Use the quadratic formula to find the eigenvalues: $\lambda = \dfrac{4 \pm \sqrt{16 - 20}}{2} = 2 \pm i$. Example 2 gives a shortcut for finding one eigenvector, and Example 5 shows how to write the other eigenvector with no effort.

<u>For $\lambda = 2 + i$:</u> $A - (2+i)I = \begin{bmatrix} -1-i & -2 \\ 1 & 1-i \end{bmatrix}$. The equation $(A - \lambda I)\mathbf{x} = \mathbf{0}$ gives

$$\begin{aligned} (-1 - i)x_1 - \quad\quad 2x_2 &= 0 \\ x_1 + (1 - i)x_2 &= 0 \end{aligned}$$

As in Example 2, the two equations are equivalent—each determines the same relation between x_1 and x_2. So use the second equation to obtain $x_1 = -(1 - i)x_2$, with x_2 free. The general solution is $x_2\begin{bmatrix} -1 + i \\ 1 \end{bmatrix}$, and the vector $\mathbf{v}_1 = \begin{bmatrix} -1 + i \\ 1 \end{bmatrix}$ provides a basis for the eigenspace.

<u>For $\lambda = 2 - i$:</u> Let $\mathbf{v}_2 = \overline{\mathbf{v}_1} = \begin{bmatrix} -1 - i \\ 1 \end{bmatrix}$. The remark prior to Example 5 shows that $\mathbf{v}_2$ is automatically an eigenvector for $\overline{2 + i}$. In fact,

calculations similar to those above would show that $\{\mathbf{v}_2\}$ is a basis for the eigenspace. (In general, for a real matrix A, it can be shown that the set of complex conjugates of the vectors in a basis of the eigenspace for λ is a basis of the eigenspace for $\bar{\lambda}$.)

7. $A = \begin{bmatrix} \sqrt{3} & -1 \\ 1 & \sqrt{3} \end{bmatrix}$. The eigenvalues are $\sqrt{3} \pm i$. Ask your instructor if you are permitted to write down the eigenvalues of $\begin{bmatrix} a & -b \\ b & a \end{bmatrix}$ from memory, or if you are expected to find them via the characteristic equation. Note that the eigenvectors are easy to remember, too. See the Practice Problem.

The scale factor associated with the transformation $\mathbf{x} \mapsto A\mathbf{x}$ is simply $r = |\lambda| = ((\sqrt{3})^2 + 1^2)^{1/2} = 2$. For the angle of the rotation, plot the point $(a,b) = (\sqrt{3},1)$ in the xy-plane and use trigonometry.

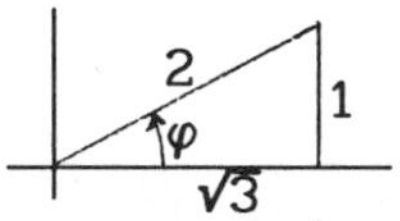

$\varphi = \pi/6$ radians

13. From Exercise 1, $\lambda = 2 \pm i$, and the eigenvector $\mathbf{v} = \begin{bmatrix} -1-i \\ 1 \end{bmatrix}$ corresponds to $\lambda = 2 - i$. Since $\text{Re}\,\mathbf{v} = \begin{bmatrix} -1 \\ 1 \end{bmatrix}$ and $\text{Im}\,\mathbf{v} = \begin{bmatrix} -1 \\ 0 \end{bmatrix}$, take $P = \begin{bmatrix} -1 & -1 \\ 1 & 0 \end{bmatrix}$. Then compute

$$C = P^{-1}AP = \begin{bmatrix} 0 & 1 \\ -1 & -1 \end{bmatrix}\begin{bmatrix} 1 & -2 \\ 1 & 3 \end{bmatrix}\begin{bmatrix} -1 & -1 \\ 1 & 0 \end{bmatrix} = \begin{bmatrix} 0 & 1 \\ -1 & -1 \end{bmatrix}\begin{bmatrix} -3 & -1 \\ 2 & -1 \end{bmatrix} = \begin{bmatrix} 2 & -1 \\ 1 & 2 \end{bmatrix}$$

Actually, Theorem 9 gives the formula for C. Note that the eigenvector $\mathbf{v}$ corresponds to $a - bi$ instead of $a + bi$. If, for instance, you use the eigenvector for $2 + i$, your C will be $\begin{bmatrix} 2 & 1 \\ -1 & 2 \end{bmatrix}$. The imaginary part of the eigenvalue is the (1,2)-entry in C.

So there are two possible choices for C (depending on the vector used to produce P). On an exam, if you are not sure of the form of C, you can always compute it quickly from the formula $C = P^{-1}AP$, as in the solution above.

Note: Because there are two possibilities for C in the factorization of a 2×2 matrix as in Exercise 13, the measure of rotation φ associated with the transformation $\mathbf{x} \mapsto A\mathbf{x}$ is determined only up to a change of sign. The "orientation" of the angle is determined by the change of variable $\mathbf{x} = P\mathbf{u}$. See Fig. 4 in the text.

19. $A = \begin{bmatrix} 1.52 & -.7 \\ .56 & .4 \end{bmatrix}$, $\det(A - \lambda I) = \lambda^2 - 1.92\lambda + 1$.

Use the quadratic formula to solve $\lambda^2 - 1.92\lambda + 1 = 0$:

$$\lambda = \frac{1.92 \pm \sqrt{-.3136}}{2} = .96 \pm .28i$$

To find the eigenvector for $\lambda = .96 - .28i$, solve

$$\begin{aligned} (1.52 - .96 + .28i)x_1 \qquad\qquad -.7x_2 &= 0 \\ .56x_1 + (.4 - .96 + .28i)x_2 &= 0 \end{aligned}$$

There is a nonzero solution (because $.96 - .28i$ is an eigenvalue), so you can use either equation to find the solution. From the second equation,

$$x_1 = \frac{.56 - .28i}{.56}x_2 = (1 - .5i)x_2$$

Setting $x_2 = 2$ produces the (complex) eigenvector $\mathbf{v} = \begin{bmatrix} 2 - i \\ 2 \end{bmatrix}$.

Since $\text{Re}\,\mathbf{v} = \begin{bmatrix} 2 \\ 2 \end{bmatrix}$ and $\text{Im}\,\mathbf{v} = \begin{bmatrix} -1 \\ 0 \end{bmatrix}$, take $P = \begin{bmatrix} 2 & -1 \\ 2 & 0 \end{bmatrix}$. Finally, compute

$$P^{-1}AP = \frac{1}{2}\begin{bmatrix} 0 & 1 \\ -2 & 2 \end{bmatrix}\begin{bmatrix} 1.52 & -.7 \\ .56 & .4 \end{bmatrix}\begin{bmatrix} 2 & -1 \\ 2 & 0 \end{bmatrix} = \begin{bmatrix} .96 & -.28 \\ .28 & .96 \end{bmatrix}$$

This final matrix, which has the proper form, is C.

25. Writing $\mathbf{x} = \text{Re}\,\mathbf{x} + i(\text{Im}\,\mathbf{x})$, we have $A\mathbf{x} = A(\text{Re}\,\mathbf{x}) + i\,A(\text{Im}\,\mathbf{x})$. Since A is real, so are $A(\text{Re}\,\mathbf{x})$ and $A(\text{Im}\,\mathbf{x})$. Thus $A(\text{Re}\,\mathbf{x})$ is the real part of $A\mathbf{x}$ and $A(\text{Im}\,\mathbf{x})$ is the imaginary part of $A\mathbf{x}$.

MATLAB Complex Eigenvalues

The command **[V D] = eig(A)** (mentioned in Section 5.3) works for matrices with complex eigenvalues. In this case V and the diagonal matrix D have some complex entries. For a 2×2 real matrix with a complex eigenvalue, MATLAB tends to place the eigenvalue $a - bi$ (where $b > 0$) as the (2,2)-entry of D. MATLAB does not produce matrices for a factorization $A = PCP^{-1}$ of the sort described in this section.

For any matrix V, the commands **real(V)** and **imag(V)** produce the real and imaginary parts of the entries in V, displayed as matrices the same size as V.

5.6 DISCRETE DYNAMICAL SYSTEMS

This section presents the climax to a crescendo of ideas that began in Section 1.9 and flowed through parts of Chapters 4 and 5. You need not have read all the material to appreciate the interesting applications in this section, but you will profit from a review of page 302 and Example 5 in Section 5.2.

KEY IDEAS

A **solution** of a first order homogeneous difference equation

$$\mathbf{x}_{k+1} = A\mathbf{x}_k \quad (k = 0, 1, 2, \ldots) \tag{1}$$

is a sequence $\{\mathbf{x}_k\}$ that satisfies (1) and is described by a formula for each $\mathbf{x}_k$ that does not depend on the preceding terms in the sequence other than the initial term $\mathbf{x}_0$. In Section 5.1, you saw how a solution can be constructed when $\mathbf{x}_0$ is an eigenvector. When $\mathbf{x}_0$ is *not* an eigenvector, look for an eigenvector decomposition of $\mathbf{x}_0$:

$$\mathbf{x}_0 = c_1\mathbf{v}_1 + \cdots + c_n\mathbf{v}_n \qquad \text{Each } \mathbf{v}_i \text{ is an eigenvector.} \tag{2}$$

To make (2) possible for any $\mathbf{x}_0$ in $\mathbb{R}^n$, the section assumes that the $n \times n$ matrix A has n linearly independent eigenvectors. If $\mathbf{x}_k = A^k\mathbf{x}_0$, then

$$\mathbf{x}_k = c_1(\lambda_1)^k\mathbf{v}_1 + \cdots + c_n(\lambda_n)^k\mathbf{v}_n$$

When $\{\mathbf{x}_k\}$ describes the "state" of a system at discrete times (denoted by $k = 0,1,2,\ldots$), the *long-term behavior* of this dynamical system is a description of what happens to $\mathbf{x}_k$ as $k \to \infty$. The text focuses on the following important situation.

> Let A be an $n \times n$ matrix with n linearly independent eigenvectors, corresponding to eigenvalues such that $|\lambda_1| \geq 1 > |\lambda_j|$ for $j = 2, \ldots, n$. If $\mathbf{x}_0$ is given by (2) with $c_1 \neq 0$, then for all sufficiently large k,
>
> $\mathbf{x}_{k+1} \approx \lambda_1\mathbf{x}_k$ — Each entry in $\mathbf{x}_k$ grows by a factor of λ_1.
>
> $\mathbf{x}_k \approx c_1(\lambda_1)^k\mathbf{v}_1$ — $\mathbf{x}_k$ is approximately a multiple of $\mathbf{v}_1$, and so the ratio between any two entries in $\mathbf{x}_k$ is nearly the same as the corresponding ratio for $\mathbf{v}_1$.

STUDY NOTES

When the two approximations above are true in an application, the eigenvalue λ_1 and the eigenvector $\mathbf{v}_1$ have interesting physical interpretations. Make sure you can describe these on an exam. (See the last four sentences in the solution of Example 1, for instance.)

The predator-prey model is rather primitive and provides only a starting point for more refined models. Still, you might enjoy considering what the model in Example 1 predicts if $\mathbf{x}_0$ happens to be a multiple of $\mathbf{v}_2 = (5,1)$, or if initially there are *more* than 5 owls for every 1 thousand rats, assuming $p = .104$.

The graphical descriptions of solutions to difference equations should help you understand what can happen to $\mathbf{x}_k$ as $k \to \infty$. I hope you enjoy studying the figures even if your class does not have time to cover this part of the section. Only the simplest cases are shown, but these cases form the foundation for studying *nonlinear* dynamical systems which are widely used (but require calculus techniques not covered here). Even for nonlinear systems, eigenvalues and eigenvectors of certain matrices play an important role.

SOLUTIONS TO EXERCISES

1. a. The eigenvectors $\mathbf{v}_1 = \begin{bmatrix} 1 \\ 1 \end{bmatrix}$ and $\mathbf{v}_2 = \begin{bmatrix} -1 \\ 1 \end{bmatrix}$ form a basis for $\mathbb{R}^2$. To find the action of A on $\mathbf{x}_0 = \begin{bmatrix} 9 \\ 1 \end{bmatrix}$, express $\mathbf{x}_0$ in terms of $\mathbf{v}_1$ and $\mathbf{v}_2$. That is, find c_1, c_2 such that $\mathbf{x}_0 = c_1\mathbf{v}_1 + c_2\mathbf{v}_2$:

$$\begin{bmatrix} 1 & -1 & 9 \\ 1 & 1 & 1 \end{bmatrix} \sim \begin{bmatrix} 1 & 0 & 5 \\ 0 & 1 & -4 \end{bmatrix} \Rightarrow \mathbf{x}_0 = 5\mathbf{v}_1 - 4\mathbf{v}_2$$

Since $\mathbf{v}_1$, $\mathbf{v}_2$ are eigenvectors (for the eigenvalues 3 and 1/3):

$$\mathbf{x}_1 = A\mathbf{x}_0 = 5A\mathbf{v}_1 - 4A\mathbf{v}_2 = 5\cdot 3\mathbf{v}_1 - 4\cdot(1/3)\mathbf{v}_2$$

$$= \begin{bmatrix} 15 \\ 15 \end{bmatrix} - \begin{bmatrix} -4/3 \\ 4/3 \end{bmatrix} = \begin{bmatrix} 49/3 \\ 41/3 \end{bmatrix}$$

b. Each time A acts on a linear combination of $\mathbf{v}_1$ and $\mathbf{v}_2$, the $\mathbf{v}_1$ term is multiplied by the eigenvalue 3 and the $\mathbf{v}_2$ term is multiplied by the eigenvalue 1/3.

$$\mathbf{x}_2 = A\mathbf{x}_1 = A[5(3)\mathbf{v}_1 - 4(1/3)\mathbf{v}_2] = 5(3)^2\mathbf{v}_1 - 4(1/3)^2\mathbf{v}_2$$

In general, $\mathbf{x}_k = 5(3)^k\mathbf{v}_1 - 4(1/3)^k\mathbf{v}_2$ for $k \geq 0$.

7. **a**. The matrix A in Exercise 1 has eigenvalues 3 and 1/3. Since $|3| > 1$ and $|1/3| < 1$, the origin is a saddle point.

   b. The direction of greatest attraction is determined by $\mathbf{v}_2 = \begin{bmatrix} -1 \\ 1 \end{bmatrix}$, the eigenvector corresponding to the eigenvalue with absolute value less than 1. The direction of greatest repulsion is determined by $\mathbf{v}_1 = \begin{bmatrix} 1 \\ 1 \end{bmatrix}$, the eigenvector corresponding to the eigenvalue greater than 1.

   c. The drawing below shows: (1) lines through the eigenvectors and the origin, (2) arrows toward the origin (showing attraction) on the line through $\mathbf{v}_2$ and arrows away from the origin (showing repulsion) on the line through $\mathbf{v}_1$, (3) several typical trajectories (with arrows) that show the general flow of points. No specific points other than $\mathbf{v}_1$ and $\mathbf{v}_2$ were computed. This type of drawing is about all that one can make without using a computer to plot points.

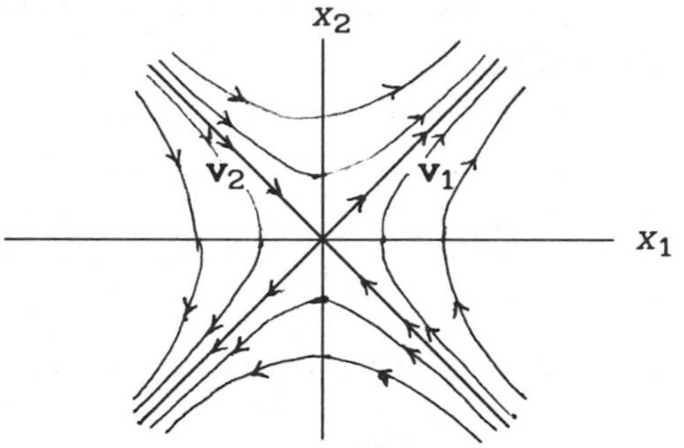

Remark: Sketching trajectories for a dynamical system in which the origin is an attractor or a repellor is more difficult than the sketch in Exercise 7. There has been no discussion of the direction in which the trajectories "bend" as they move toward or away from the origin. For instance, if you rotate Figure 1 of Section 5.6 through a quarter-turn and relabel the axes so that x_1 is on the horizontal axis, then the new figure corresponds to the matrix A with the diagonal entries .8 and .64 interchanged. In general, if A is a diagonal matrix, with positive diagonal entries a and d, unequal to 1, then the trajectories lie on the axes or on curves whose equations have the form $x_2 = r(x_1)^s$, where $s = (\ln d)/(\ln a)$ and r depends on the initial point $\mathbf{x}_0$. (See *Encounters with Chaos*, by Denny Gulick, New York: McGraw-Hill, 1992, pp. 147-150.)

Study Tip: If your instructor wants you to graph trajectories when the origin is an attractor or repellor, there will need to be some class discussion of exactly how to do this.

13. $A = \begin{bmatrix} .8 & .3 \\ -.4 & 1.5 \end{bmatrix}$. First find the eigenvalues:

$$\begin{aligned} \det(A - \lambda I) &= (.8 - \lambda)(1.5 - \lambda) - (.3)(-.4) = \lambda^2 - 2.3 + 1.32 \\ &= (\lambda - 1.2)(\lambda - 1.1) \quad \text{Use the quadratic formula, if needed.} \\ &= 0 \end{aligned}$$

Since both eigenvalues, 1.2 and 1.1, are greater than 1, the origin is a repellor. For the direction of greatest repulsion, find the eigenvector for the larger eigenvalue, 1.2:

$$[(A - 1.2I) \quad \mathbf{0}] = \begin{bmatrix} -.4 & .3 & 0 \\ -.4 & .3 & 0 \end{bmatrix} \sim \begin{bmatrix} 1 & -3/4 & 0 \\ 0 & 0 & 0 \end{bmatrix}, \quad \mathbf{x} = x_2\begin{bmatrix} 3/4 \\ 1 \end{bmatrix}$$

Any multiple of $\begin{bmatrix} 3/4 \\ 1 \end{bmatrix}$, such as $\begin{bmatrix} 3 \\ 4 \end{bmatrix}$, determines the direction of greatest repulsion.

MATLAB Plotting Trajectories

Given a vector **x**, the command **x = A*x** will compute the "next" point on the trajectory. Use the up-arrow (↑) and <Enter> to repeat the command, over and over.

The following steps create a "trajectory" matrix T whose columns are the points $\mathbf{x}$, $A\mathbf{x}$, $A^2\mathbf{x}, \ldots, A^{15}\mathbf{x}$. (Change 15 to any integer you wish.)

```
T = x                 Put x in the first column of T.
for j=1:15            This loop repeats the next two lines 15 times.
  x = A*x;            Compute the next point on the trajectory.
  T = [T  x]          Store the new point in T.
end                   End of the loop.
```

After you type the line beginning with "for", MATLAB will suspend all calculations (while you type additional lines) until you type "end".

If you want MATLAB itself to plot the points in T, use the commands:

```
plot(T(1,:),T(2,:),'ow'), grid
```

The 'o' produces a small circle at each point on the trajectory. The 'w' makes the circle white. If you have the data for another trajectory stored in a matrix S, you can plot both trajectories on the same graph:

```
plot(T(1,:),T(2,:),'ow',S(1,:),S(2,:),'*g'), grid     g is for green.
```

Each new "plot" command erases the previous graph.

5.7 APPLICATIONS TO DIFFERENTIAL EQUATIONS

If you plan to take a course in differential equations, the material in this section will be a valuable reference. If you have already studied differential equations, you may gain new understanding as you work through this section.

KEY IDEAS

A basic solution of the differential equation $\mathbf{x}' = A\mathbf{x}$ is an *eigenfunction* $\mathbf{x}(t) = \mathbf{v}e^{\lambda t}$, where λ is an eigenvalue of A and $\mathbf{v}$ is a corresponding eigenvector. In all examples and exercises in this section, every solution of $\mathbf{x}' = A\mathbf{x}$ is a linear combination of eigenfunctions. (This is because A is diagonalizable. The general case is usually handled in a full course in differential equations.) An initial condition, $\mathbf{x}(0) = \mathbf{x}_0$, determines the weights for the linear combination of eigenfunctions.

The eigenvalues of A determine the nature of the origin for the dynamical system described by $\mathbf{x}' = A\mathbf{x}$. Most of the discussion involves the case when A is 2×2. If one eigenvalue is negative and one is positive, the origin is a saddle point. If the real parts of the eigenvalues are negative (this includes the case when both eigenvalues are real and negative), the origin is an *attractor* of the dynamical system. If the real parts are positive, the origin is a *repellor*. If an eigenvalue is complex, then the trajectory of a corresponding eigenfunction forms a *spiral*—either toward the origin, away from the origin, or on an ellipse around the origin, depending on the real part of the eigenvalue.

Study Tip: Note that conditions on eigenvalues here differ from those in Section 5.6. For differential equations, the real parts of the eigenvalues determine the nature of the trajectories; for difference equations, the absolute values of the eigenvalues are important. You can remember this if you note that a basic solution $\mathbf{v}e^{\lambda t}$ of $\mathbf{x}' = A\mathbf{x}$ tends to $\mathbf{0}$ (as $t \to \infty$) only if the real part of λ is negative. In contrast, a basic solution $\lambda^k\mathbf{v}$ of $\mathbf{x}_{k+1} = A\mathbf{x}_k$ tends to $\mathbf{0}$ (as $k \to \infty$) only if the absolute value of λ is less than 1.

SOLUTIONS TO EXERCISES

1. The eigenfunctions for $\mathbf{x}' = A\mathbf{x}$ are $\mathbf{v}_1e^{4t}$ and $\mathbf{v}_2e^{2t}$. The general solution of $\mathbf{x}' = A\mathbf{x}$ has the form

$$c_1\begin{bmatrix}-3\\1\end{bmatrix}e^{4t} + c_2\begin{bmatrix}-1\\1\end{bmatrix}e^{2t}$$

The initial condition $\mathbf{x}(0) = (-6,1)$ determines c_1 and c_2:

$$c_1\begin{bmatrix}-3\\1\end{bmatrix}e^{4(0)} + c_2\begin{bmatrix}-1\\1\end{bmatrix}e^{2(0)} = \begin{bmatrix}-6\\1\end{bmatrix}$$

$$\begin{bmatrix}-3 & -1 & -6\\1 & 1 & 1\end{bmatrix} \sim \begin{bmatrix}1 & 1 & 1\\-3 & -1 & -6\end{bmatrix} \sim \cdots \sim \begin{bmatrix}1 & 0 & 5/2\\0 & 1 & -3/2\end{bmatrix}$$

Thus $c_1 = 5/2$, $c_2 = -3/2$, and $\mathbf{x}(t) = \frac{5}{2}\begin{bmatrix}-3\\1\end{bmatrix}e^{4t} - \frac{3}{2}\begin{bmatrix}-1\\1\end{bmatrix}e^{2t}$.

Checkpoint: Let A be the matrix $\begin{bmatrix}1 & 3/2\\-1/2 & -1\end{bmatrix}$, obtained by dividing the matrix in Exercise 3 by 2. Now the eigenvalues of A are .5 and -.5. Is the origin an attractor or a saddle point for the equation $\mathbf{x}' = A\mathbf{x}$?

7. Use the eigenvectors $\mathbf{v}_1 = \begin{bmatrix}1\\3\end{bmatrix}$ and $\mathbf{v}_2 = \begin{bmatrix}1\\1\end{bmatrix}$ (found in Exercise 5) to create $P = [\mathbf{v}_1 \;\; \mathbf{v}_2]$. Match the eigenvectors with the eigenvalues in $D = \begin{bmatrix}4 & 0\\0 & 6\end{bmatrix}$. The details of the substitution of $\mathbf{x} = P\mathbf{y}$ into $\mathbf{x}' = A\mathbf{x}$ are given in the answer section of the text.

Helpful Hint: The idea of changing variables to uncouple a differential equation is fairly common in engineering texts. Exercises 7 and 8 test your understanding of the value of a diagonalization. (You might see such a question on an exam.)

13. An eigenvalue of A is $\lambda = 1 + 3i$, with eigenvector $\mathbf{v} = (1 + i, 2)$. The complex eigenfunctions $\mathbf{v}e^{\lambda t}$ and $\bar{\mathbf{v}}e^{\bar{\lambda} t}$ provide a basis for the solution space of all complex solutions of $\mathbf{x}' = A\mathbf{x}$. The general (complex) solution is

$$c_1\begin{bmatrix}1 + i\\2\end{bmatrix}e^{(1+3i)t} + c_2\begin{bmatrix}1 - i\\2\end{bmatrix}e^{(1-3i)t} \quad (c_1 \text{ and } c_2 \text{ are complex})$$

Use the real and imaginary parts of $\mathbf{v}e^{(1+3i)t}$ to build the general real solution. Rewrite $\mathbf{v}e^{(1+3i)t}$ as:

$$\begin{bmatrix}1 + i\\2\end{bmatrix}e^{(1+3i)t} = \begin{bmatrix}1 + i\\2\end{bmatrix}(\cos 3t + i \sin 3t)e^{t}$$

$$= \begin{bmatrix} \cos 3t - \sin 3t \\ 2\cos 3t \end{bmatrix} e^t + i \begin{bmatrix} \sin 3t + \cos 3t \\ 2\sin 3t \end{bmatrix} e^t$$

The general real solution has the form

$$c_1 \begin{bmatrix} \cos 3t - \sin 3t \\ 2\cos 3t \end{bmatrix} e^t + c_2 \begin{bmatrix} \sin 3t + \cos 3t \\ 2\sin 3t \end{bmatrix} e^t \quad (c_1 \text{ and } c_2 \text{ are real})$$

The trajectories are spirals because the eigenvalues are complex. The spirals tend away from the origin because the real parts of the eigenvalues are positive.

19. Substitute $R_1 = 5$, $R_2 = 3$, $C_1 = 4$, and $C_2 = 3$ into the formula for A given in Example 1, and use a matrix program to find the eigenvalues and eigenvectors:

$$A = \begin{bmatrix} -2 & 3/4 \\ 1 & -1 \end{bmatrix}; \quad \lambda_1 = -.5: \ \mathbf{v}_1 = \begin{bmatrix} 1 \\ 2 \end{bmatrix}; \quad \lambda_2 = -2.5: \ \mathbf{v}_2 = \begin{bmatrix} -3 \\ 2 \end{bmatrix}$$

General solution: $\mathbf{x}(t) = c_1 \begin{bmatrix} 1 \\ 2 \end{bmatrix} e^{-.5t} + c_2 \begin{bmatrix} -3 \\ 2 \end{bmatrix} e^{-2.5t}$.

The condition $\mathbf{x}(0) = \begin{bmatrix} 4 \\ 4 \end{bmatrix}$ implies that $\begin{bmatrix} 1 & -3 \\ 2 & 2 \end{bmatrix} \begin{bmatrix} c_1 \\ c_2 \end{bmatrix} = \begin{bmatrix} 4 \\ 4 \end{bmatrix}$. By a matrix program, $c_1 = 5/2$ and $c_2 = -1/2$, so that

$$\begin{bmatrix} v_1(t) \\ v_2(t) \end{bmatrix} = \mathbf{x}(t) = \frac{5}{2} \begin{bmatrix} 1 \\ 2 \end{bmatrix} e^{-.5t} - \frac{1}{2} \begin{bmatrix} -3 \\ 2 \end{bmatrix} e^{-2.5t}$$

Helpful Hint: (for 21 and 22) Find the general real solution before you use the initial condition to find the constants c_1 and c_2. Otherwise, your c_1 and c_2 will probably be complex, and you will have to do unnecessary complex arithmetic to write the solution using only real scalars.

Answer to Checkpoint: One eigenvalue is positive and one is negative, so the origin is a saddle point. If you consider the *difference* equation $\mathbf{x}_{k+1} = A\mathbf{x}_k$ (with the same matrix A), then the origin is an attractor, because both eigenvalues are less than 1 in absolute value. Be careful if you have a test that covers both Sections 5.6 and 5.7.

5.8 ITERATIVE ESTIMATES FOR EIGENVALUES

The algorithms in this section illustrate another use of the eigenvector decomposition described in Section 5.6 (on page 337). Other methods for eigenvalue estimation were mentioned in Section 5.2 (on page 311).

STUDY NOTES

Throughout the section, we suppose that the initial vector $\mathbf{x}_0$ can be written as $\mathbf{x}_0 = c_1\mathbf{v}_1 + \cdots + c_n\mathbf{v}_n$, where $\mathbf{v}_1, \ldots, \mathbf{v}_n$ are eigenvectors of A and $c_1 \neq 0$. (In practice, you will not know $c_1\mathbf{v}_1, \ldots, c_n\mathbf{v}_n$. This eigenvector decomposition is used only to explain why the power method works.)

The Power Method: Assume the eigenvalue λ_1 for $\mathbf{v}_1$ is a strictly dominant eigenvalue (so that $|\lambda_1| > |\lambda_j|$ for $j = 2, \ldots, n$). Then, for large k, the line through $A^k\mathbf{x}_0$ and $\mathbf{0}$ nearly coincides with the line through $\mathbf{v}_1$ and $\mathbf{0}$. The vector $A^k\mathbf{x}_0$ itself may never approach a multiple of $\mathbf{v}_1$ (see Exercise 21), but if each $A^k\mathbf{x}_0$ is scaled so its largest entry is 1, then the scaled vectors approach an eigenvector (a multiple of $\mathbf{v}_1$) as $k \to \infty$.

The Inverse Power Method: You must start with an initial estimate α for a particular eigenvalue, say λ_2, and α must be closer to λ_2 than to any other eigenvalue of A. In this case, $1/(\lambda_2 - \alpha)$ is a strictly dominant eigenvalue of the matrix $B = (A - \alpha I)^{-1}$. The inverse power method avoids computing B. Instead of multiplying $\mathbf{x}_k$ by B to get $\mathbf{x}_{k+1}$ (suitably scaled), you solve the equation $(A - \alpha I)\mathbf{y}_k = \mathbf{x}_k$ for $\mathbf{y}_k$ and then scale $\mathbf{y}_k$ to produce $\mathbf{x}_{k+1}$.

SOLUTIONS TO EXERCISES

1. The vectors in the sequence $\begin{bmatrix} 1 \\ 0 \end{bmatrix}, \begin{bmatrix} 1 \\ .25 \end{bmatrix}, \begin{bmatrix} 1 \\ .3158 \end{bmatrix}, \begin{bmatrix} 1 \\ .3298 \end{bmatrix}, \begin{bmatrix} 1 \\ .3326 \end{bmatrix}$ approach an eigenvector $\mathbf{v}_1$. Of these vectors, the last one, $\mathbf{x}_4$, is probably the best estimate of $\mathbf{v}_1$. To compute an estimate of λ_1, multiply one of the vectors by A and examine its entries. Again, the best information probably comes from $A\mathbf{x}_4 = \begin{bmatrix} 4.9978 \\ 1.6652 \end{bmatrix}$, whose entries are approximately λ_1 times the entries in $\mathbf{x}_4$. From the first entry, the estimate of λ_1 is 4.9978.

The computed value of $A\mathbf{x}_4$ can be used as an estimate of the direction of the eigenspace, but this vector is not necessarily a better estimate for an eigenvector than $\mathbf{x}_4$. For that, you should use $\mathbf{x}_5$, the *scaled* version of $A\mathbf{x}_4$. With the given data, the distance from $A\mathbf{x}_4$ to the eigenspace is about 7.0×10^{-4}, but the distance from $\mathbf{x}_5$ to the eigenspace is less than 1.4×10^{-4}.

Study Tip: Exercises 1-6 make good exam questions because they test your understanding of the power method without requiring extensive calculation.

7. The data in the table below and the tables in Exercise 19 were produced by MATLAB, which carried more decimal places than shown here.

k	0	1	2	3	4	5
$\mathbf{x}_k$	$\begin{bmatrix}1\\0\end{bmatrix}$	$\begin{bmatrix}.75\\1\end{bmatrix}$	$\begin{bmatrix}1\\.9565\end{bmatrix}$	$\begin{bmatrix}.9932\\1\end{bmatrix}$	$\begin{bmatrix}1\\.9990\end{bmatrix}$	$\begin{bmatrix}.9998\\1\end{bmatrix}$
$A\mathbf{x}_k$	$\begin{bmatrix}6\\8\end{bmatrix}$	$\begin{bmatrix}11.5\\11.0\end{bmatrix}$	$\begin{bmatrix}12.70\\12.78\end{bmatrix}$	$\begin{bmatrix}12.959\\12.946\end{bmatrix}$	$\begin{bmatrix}12.9927\\12.9948\end{bmatrix}$	$\begin{bmatrix}12.9990\\12.9987\end{bmatrix}$
μ_k	8	11.5	12.78	12.959	12.9948	12.9990

The exact eigenvalues are 13 and -2. The subspaces determined by $A^k\mathbf{x}$ are lines whose slopes alternate above and below the slope of the eigenspace (the eigenspace is the line $x_2 = x_1$).

13. If the eigenvalues close to 4 and -4 have different absolute values, then one of these eigenvalues is a strictly dominant eigenvalue, so the power method will work. But the power method depends on powers of the quotients λ_2/λ_1 and λ_3/λ_1 going to zero. If $|\lambda_2/\lambda_1|$ is close to 1, its powers will go to zero slowly.

15. Suppose $A\mathbf{x} = \lambda\mathbf{x}$, with $\mathbf{x} \neq \mathbf{0}$. For any α, $A\mathbf{x} - \alpha I\mathbf{x} = (\lambda - \alpha)\mathbf{x}$, and $(A - \alpha I)\mathbf{x} = (\lambda - \alpha)\mathbf{x}$. If α is *not* an eigenvalue of A, then $A - \alpha I$ is invertible and $\lambda - \alpha$ is not 0; hence

$$\mathbf{x} = (A - \alpha I)^{-1}(\lambda - \alpha)\mathbf{x} \quad \text{and} \quad (\lambda - \alpha)^{-1}\mathbf{x} = (A - \alpha I)^{-1}\mathbf{x}$$

This last equation shows that $\mathbf{x}$ is an eigenvector of $(A - \alpha I)^{-1}$ corresponding to the eigenvalue $(\lambda - \alpha)^{-1}$.

19. a. The data in the table on the next page show that $\mu_6 = 30.2887 = \mu_7$ to four decimal places. Actually, to six places, the largest eigenvalue is 30.288685, with eigenvector (.957629, .688937, 1, .943782).

k	0	1	2	3	4	5	6	7
$\mathbf{x}_k$	$\begin{bmatrix}1\\0\\0\\0\end{bmatrix}$	$\begin{bmatrix}1\\.7\\.8\\.7\end{bmatrix}$	$\begin{bmatrix}.99\\.71\\1\\.93\end{bmatrix}$	$\begin{bmatrix}.961\\.691\\1\\.942\end{bmatrix}$	$\begin{bmatrix}.9581\\.6893\\1\\.9436\end{bmatrix}$	$\begin{bmatrix}.9577\\.6890\\1\\.9438\end{bmatrix}$	$\begin{bmatrix}.957637\\.688942\\1\\.943778\end{bmatrix}$	$\begin{bmatrix}.957630\\.688938\\1\\.943781\end{bmatrix}$
$A\mathbf{x}_k$	$\begin{bmatrix}10\\7\\8\\7\end{bmatrix}$	$\begin{bmatrix}26.2\\18.8\\26.5\\24.7\end{bmatrix}$	$\begin{bmatrix}29.4\\21.1\\30.6\\28.8\end{bmatrix}$	$\begin{bmatrix}29.05\\20.90\\30.32\\28.61\end{bmatrix}$	$\begin{bmatrix}29.01\\20.87\\30.29\\28.59\end{bmatrix}$	$\begin{bmatrix}29.006\\20.868\\30.289\\28.586\end{bmatrix}$	$\begin{bmatrix}29.0054\\20.8671\\30.2887\\28.5859\end{bmatrix}$	$\begin{bmatrix}29.0053\\20.8670\\30.2887\\28.5859\end{bmatrix}$
μ_k	10	26.5	30.6	30.32	30.29	30.2892	30.2887	30.2887

b. The inverse power method (with $\alpha = 0$) produces $\nu_1 = \mu_1^{-1} = .010141$, and $\nu_2 = .0101501$, which seems to be accurate to at least four places. Actually, ν_2 is accurate to six places, ν_3 is accurate to eight places , and ν_4 is accurate to ten places. The convergence is so rapid because the next-to-smallest eigenvalue is near .85, which is much farther away from 0 than .0101501. The vector $\mathbf{x}_4$ gives an estimate for the eigenvector that is accurate to seven places in each entry.

k	0	1	2	3	4
$\mathbf{x}_k$	$\begin{bmatrix}1\\0\\0\\0\end{bmatrix}$	$\begin{bmatrix}-.6098\\1\\-.2439\\.1463\end{bmatrix}$	$\begin{bmatrix}-.60401\\1\\-.25105\\.14890\end{bmatrix}$	$\begin{bmatrix}-.603973\\1\\-.251134\\.148953\end{bmatrix}$	$\begin{bmatrix}-.6039723\\1\\-.2511351\\.1489534\end{bmatrix}$
$\mathbf{y}_k$	$\begin{bmatrix}25\\-41\\10\\-6\end{bmatrix}$	$\begin{bmatrix}-59.56\\98.61\\-24.76\\14.68\end{bmatrix}$	$\begin{bmatrix}-59.5041\\98.5211\\-24.7420\\14.6750\end{bmatrix}$	$\begin{bmatrix}-59.5044\\98.5217\\-24.7423\\14.6751\end{bmatrix}$	$\begin{bmatrix}-59.50438\\98.52170\\-24.74226\\14.67515\end{bmatrix}$
ν_k	-.024	.010141	.0101501	.010150049	.0101500484

MATLAB Power Method and Inverse Power Method

Use **format long** to display 15 decimal digits in your data. The algorithms below assume that A has a strictly dominant eigenvalue, and the initial vector is **x**, with largest entry 1 (in magnitude). (If your initial vector is called **x0**, rename it by entering **x = x0** .)

The Power Method When the following sequence of commands is performed over and over, the values of **x** approach (in many cases) an eigenvector for a strictly dominant eigenvalue:

```
y = A*x                                                              (1)
[t  r] = max(abs(y)); mu = y(r)    mu = estimate for eigenvalue      (2)
x = y/y(r)                         Estimate for the eigenvector      (3)
```

In (2), t is the absolute value of the largest entry in **y** and r is the index of that entry. As these commands are repeated, the numbers that appear in **y**(r) are the μ_k that approach the dominant eigenvalue.

Recall that MATLAB commands can be recalled by the up-arrow key (↑). After entering (1)-(3), your keystrokes can be

↑ ↑ ↑ <Enter> ↑ ↑ ↑ <Enter> ↑ ↑ ↑ <Enter>

and so on. Alternatively, you could enclose lines (1)-(3) in a loop (see page 5-25).

The Inverse Power Method Store the initial estimate of the eigenvalue in the variable **a**, and enter the command **C = A - a*eye(n)**, where n is the number of columns of A. Then enter the commands

```
y = C\x                            Solves the equation (A - aI)y = x   (1)
[t  r] = max(abs(y)); nu = a + 1/y(r)                                  (2)
                                   nu = estimated eigenvalue
x = y/y(r)                         Estimate for the eigenvector        (3)
```

As these commands are repeated (using ↑↑↑ <Enter> each time), lines (2) and (3) produce the sequences $\{\nu_k\}$ and $\{\mathbf{x}_k\}$ described in the text.

Displaying Data If your computer screen displays only 24 or 25 lines, vectors in the sequence $\{\mathbf{x}_k\}$ tend to scroll off the screen soon after you compute them. To see more vectors at once, and to compare their entries more easily, you can *display* them as row vectors. Change (1) to **y = A*x; y'** (power method) or **y = C\x; y'** (inverse power), and for both methods, change (3) to **x = y/y(r); x'** .

For even more data on your screen, use the command **format compact** , which removes extra lines between data displays. The simple command **format** returns everything to normal.

CHAPTER 5 SUPPLEMENTARY EXERCISES

Justifications for the answers to the True/False questions are given below. Other justifications are possible.

1. a. True. If A is invertible and if $A\mathbf{x} = 1\cdot\mathbf{x}$ for some nonzero $\mathbf{x}$, then $\mathbf{x} = A^{-1}(1\cdot\mathbf{x})$, and so $A^{-1}\mathbf{x} = 1\cdot\mathbf{x}$, which shows that 1 is an eigenvalue for A^{-1}.

   b. False. Take $A = \begin{bmatrix} 1 & 2 \\ 0 & 1 \end{bmatrix}$, or let A be the matrix in Example 4 of Section 5.3. In both cases A is not diagonalizable, but A is invertible and hence is row equivalent to an identity matrix.

   c. True. If A contains a row or column of zeros, then A is not invertible, and hence 0 is an eigenvalue of A, by the Invertible Matrix Theorem (stated in Section 5.1).

   d. True. The rotation matrix from Example 1 in Section 5.5 has no eigenvectors in $\mathbb{R}^2$.

   e. False. Zero is an eigenvalue of every singular square matrix.

   f. True. By definition, an eigenvector must be nonzero.

   g. False. Let $\mathbf{v}$ be an eigenvector for A. Then $\mathbf{v}$ and $2\mathbf{v}$ are distinct eigenvectors, but $\mathbf{v}$ and $2\mathbf{v}$ are linearly dependent.

   h. False. Let A be a matrix with a two-dimensional eigenspace. Then there exist two linearly independent eigenvectors of A.

   i. True. This follows from Theorem 4 in Section 5.2.

   j. False. Let A be the 3×3 matrix in Example 3 of Section 5.3. Then A is similar to a diagonal matrix D. The eigenvectors of D are the columns of I_3, but the eigenvectors of A are entirely different.

   k. False. Let $A = \begin{bmatrix} 2 & 0 \\ 0 & 3 \end{bmatrix}$. Then $\mathbf{e}_1 = \begin{bmatrix} 1 \\ 0 \end{bmatrix}$ and $\mathbf{e}_2 = \begin{bmatrix} 0 \\ 1 \end{bmatrix}$ are eigenvectors of A, but $\mathbf{e}_1 + \mathbf{e}_2$ is not. (Actually, it can be shown that if two eigenvectors of A correspond to distinct eigenvalues, then their sum cannot be an eigenvector.)

   l. False. *All* the diagonal entries of an upper triangular matrix are the eigenvalues of the matrix. (Theorem 1 in Section 5.1.)

m. True, because A and A^T have the same characteristic polynomial. By the determinant transpose property, $\det(A^T - \lambda I) = \det(A - \lambda I)^T = \det(A - \lambda I)$.

n. False. Counterexample: Let A be the 5×5 identity matrix.

o. False. If A is a diagonal matrix with 0 on the diagonal, then the columns of A are not linearly independent.

p. True. If $A\mathbf{u} = \lambda_1\mathbf{u}$ and $A\mathbf{u} = \lambda_2\mathbf{u}$, then $\lambda_1\mathbf{u} = \lambda_2\mathbf{u}$ and $(\lambda_1 - \lambda_2)\mathbf{u} = \mathbf{0}$. If $\mathbf{u} \neq \mathbf{0}$, then λ_1 must equal λ_2.

q. False. Let A be a singular matrix that is diagonalizable. (For instance, let A be a diagonal matrix with 0 on the diagonal.) Then, by Theorem 8 in Section 5.4, the transformation $\mathbf{x} \mapsto A\mathbf{x}$ is represented by a diagonal matrix relative to a coordinate system determined by eigenvectors of A.

8. To show that A^k tends to the zero matrix, it suffices to show that each column of A^k can be made as close to the zero vector as desired by taking k sufficiently large. The jth column of A is $A\mathbf{e}_j$, where $\mathbf{e}_j$ is the jth column of the identity matrix. Since A is diagonalizable, there is a basis for $\mathbb{R}^n$ consisting of eigenvectors $\mathbf{v}_1, \ldots, \mathbf{v}_n$, corresponding to eigenvalues $\lambda_1, \ldots, \lambda_n$. So there exist scalars $c_1, \ldots, c_n$, such that

$$\mathbf{e}_j = c_1\mathbf{v}_1 + \cdots + c_n\mathbf{v}_n \quad \text{(an eigenvector decomposition of } \mathbf{e}_j\text{)}$$

Then, for $k = 1, 2, \ldots,$

$$A^k\mathbf{e}_j = c_1(\lambda_1)^k\mathbf{v}_1 + \cdots + c_n(\lambda_n)^k\mathbf{v}_n \qquad (*)$$

If the eigenvalues are all less than 1 in absolute value, then their kth powers all tend to zero. So (*) shows that $A^k\mathbf{e}_j$ tends to the zero vector, as desired.

CHAPTER 5 GLOSSARY CHECKLIST

Check your knowledge by attempting to write definitions of the terms below. Then compare your work with the definitions given in the text's Glossary. Ask your instructor which definitions, if any, might appear on a test.

algebraic multiplicity: The multiplicity of an eigenvalue as ....

attractor (of a dynamical system in $\mathbb{R}^2$): The origin of $\mathbb{R}^2$ when all trajectories tend ....

$\mathcal{B}$-matrix (for T): A matrix $[T]_{\mathcal{B}}$ for a linear transformation $T: V \to V$ relative to a basis $\mathcal{B}$ for V, with the property that ....

characteristic equation (of A): ....

characteristic polynomial (of A): ....

companion matrix: A special form of matrix whose characteristic ... is ....

complex eigenvalue: A nonreal root of the characteristic equation of an $n \times n$ matrix A, when ....

complex eigenvector: A nonzero vector $\mathbf{x}$ in $\mathbb{C}^n$ such that ..., where ....

decoupled system: A difference equation $\mathbf{y}_{k+1} = A\mathbf{y}_k$, or a differential equation, in which A is ....

determinant (of a square matrix A): A number $\det A$ computed from A; equal to ....

diagonalizable (matrix): A matrix that may be written in ... form as ....

difference equation (or **linear recurrence relation**): An equation of the form ... whose solution is ....

discrete linear dynamical system (or briefly, a **dynamical system**): A difference equation of the form ... that describes ....

eigenfunction (of the equation $\mathbf{x}'(t) = A\mathbf{x}(t)$): A function of the form ....

eigenspace (of A corresponding to λ): The set of ... solutions of ....

eigenvalue (of A): A scalar λ such that ....

eigenvector (of A): A ... vector $\mathbf{x}$ such that ....

eigenvector basis: A basis consisting entirely of ....

eigenvector decomposition (of $\mathbf{x}$): An equation $\mathbf{x} = ...$

fundamental set of solutions (for $\mathbf{x}' = A\mathbf{x}$): A basis for ....

Im x: The vector in $\mathbb{R}^n$ formed from ....

invariant subspace (for A): A subspace H such that ....

inverse power method: An algorithm for estimating ....

matrix for T relative to bases $\mathcal{B}$ and $\mathcal{C}$: A matrix M for a linear transformation $T: V \to W$ with the property that .... When $W = V$ and $\mathcal{C} = \mathcal{B}$, the matrix M is called ... and is denoted by ....

power method: An algorithm for estimating ....

repellor (of a dynamical system in $\mathbb{R}^2$) : The origin in $\mathbb{R}^2$ when all trajectories ... tend ......

Rayleigh quotient: $R(\mathbf{x})$ = ... An estimate of ....

Re x: The vector in $\mathbb{R}^n$ formed from ....

saddle point (of a dynamical system in $\mathbb{R}^2$): The origin in $\mathbb{R}^2$ when ....

similar (matrices): Matrices A and B such that ....

spiral point (of a dynamical system in $\mathbb{R}^2$): The origin in $\mathbb{R}^2$ when ....

stage-matrix model: A difference equation $\mathbf{x}_{k+1} = A\mathbf{x}_k$ where $\mathbf{x}_k$ lists ....

strictly dominant eigenvalue: An eigenvalue λ_1 of a matrix A with the property that ...

trace (of a square matrix A): The ..., denoted by $\text{tr}\, A$.

trajectory: The graph of a solution $\{\mathbf{x}_0, \mathbf{x}_1, \mathbf{x}_2, \ldots\}$ of a .... Also, the graph of $\mathbf{x}(t)$ for $t \geq 0$, when ....

6 ORTHOGONALITY AND LEAST-SQUARES

6.1 INNER PRODUCT, LENGTH, AND ORTHOGONALITY

The concepts of length, distance, and orthogonality introduced in this section are essential for many geometric descriptions in the rest of the text.

STUDY NOTES

The first half of the section is computational and easily learned. The second half, however, requires more attention. Read it carefully. The concepts of orthogonality and orthogonal complements are the foundation for the rest of the chapter. In fact, Theorem 3 is sometimes called the Fundamental Theorem of Linear Algebra.

In Figure 8, when a subspace such as Col *A* is referred to as a fundamental subspace of *A*, the phrase "of *A*" points to the important connection between Col *A* and the matrix; the phrase does not imply that Col *A* in some sense lies inside the matrix.

SOLUTIONS TO EXERCISES

1. $\mathbf{u} = \begin{bmatrix} -1 \\ 2 \end{bmatrix}$, $\mathbf{v} = \begin{bmatrix} 4 \\ 6 \end{bmatrix}$, $\mathbf{u \cdot u} = (-1)^2 + 2^2 = 5$,
$\mathbf{v \cdot u} = 4(-1) + 6(2) = 8$, $\dfrac{\mathbf{v \cdot u}}{\mathbf{u \cdot u}} = \dfrac{8}{5}$

7. $\mathbf{w} = (3, -1, -5)$, $\|\mathbf{w}\|^2 = \mathbf{w \cdot w} = 3^2 + (-1)^2 + (-5)^2 = 35$. So $\|\mathbf{w}\| = \sqrt{35}$.

13. $\mathbf{x} = (10, -3)$, $\mathbf{y} = (-1, -5)$
$\|\mathbf{x} - \mathbf{y}\|^2 = [10 - (-1)]^2 + [-3 - (-5)]^2 = 121 + 4 = 125$
$\text{dist}(\mathbf{x}, \mathbf{y}) = \|\mathbf{x} - \mathbf{y}\| = \sqrt{125}$, or $5\sqrt{5}$

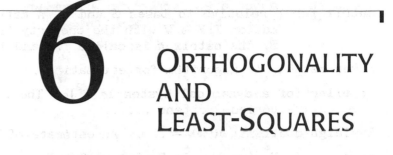

19. **a.** See the definition of $\|\mathbf{v}\|$. **b.** See Theorem 1(c).

c. See the discussion of Fig. 5.

d. Think about a 2×2 matrix of zeros and ones, and see Theorem 3 for statements that are always true.

e. See the box following Example 6.

21. Theorem 1(b): $(\mathbf{u} + \mathbf{v})\cdot\mathbf{w} = (\mathbf{u} + \mathbf{v})^T\mathbf{w} = (\mathbf{u}^T + \mathbf{v}^T)\mathbf{w} = \mathbf{u}^T\mathbf{w} + \mathbf{v}^T\mathbf{w} = \mathbf{u}\cdot\mathbf{w} + \mathbf{v}\cdot\mathbf{w}$. The second and third equalities used Theorems 3(b) and 2(c), respectively, from Section 2.1.

Theorem 1(c): $(c\mathbf{u})\cdot\mathbf{v} = (c\mathbf{u})^T\mathbf{v} = (c\mathbf{u}^T)\mathbf{v} = c(\mathbf{u}^T\mathbf{v}) = c(\mathbf{u}\cdot\mathbf{v})$, by Theorems 3(c) and 2(d) in Section 2.1. Also, $\mathbf{u}\cdot(c\mathbf{v}) = \mathbf{u}^T(c\mathbf{v}) = c\mathbf{u}^T\mathbf{v} = c(\mathbf{u}\cdot\mathbf{v})$.

27. If $\mathbf{y}$ is orthogonal to $\mathbf{u}$ and $\mathbf{v}$, then $\mathbf{y}\cdot\mathbf{u} = 0$ and $\mathbf{y}\cdot\mathbf{v} = 0$, and hence by a property of the inner product, $\mathbf{y}\cdot(\mathbf{u} + \mathbf{v}) = \mathbf{y}\cdot\mathbf{u} + \mathbf{y}\cdot\mathbf{v} = 0 + 0 = 0$. So $\mathbf{y}$ is orthogonal to $\mathbf{u} + \mathbf{v}$.

29. Take a typical vector $\mathbf{w} = c_1\mathbf{v}_1 + \cdots + c_p\mathbf{v}_p$ in W. If $\mathbf{x}$ is orthogonal to each $\mathbf{v}_j$, then using the linearity of the inner product (Theorem 1(b) and 1(c)), $\mathbf{w}\cdot\mathbf{x} = (c_1\mathbf{v}_1 + \cdots + c_p\mathbf{v}_p)\cdot\mathbf{x} = c_1\mathbf{v}_1\cdot\mathbf{x} + \cdots + c_p\mathbf{v}_p\cdot\mathbf{x} = 0$. So $\mathbf{x}$ is orthogonal to each $\mathbf{w}$ in W.

31. Suppose $\mathbf{x}$ is in W and $W^\perp$. Then, since $\mathbf{x}$ is in $W^\perp$, $\mathbf{x}$ is orthogonal to every vector in W, including $\mathbf{x}$ itself. So $\mathbf{x}\cdot\mathbf{x} = 0$. This is true only if $\mathbf{x} = \mathbf{0}$. This problem shows that $W \cap W^\perp$ is the zero subspace. (See Exercise 32 in Section 4.1.)

> **MATLAB** The inner product of real column vectors **u** and **v** is **u'*v** (and **v'*u**); the length of **v** is **norm(v)** . See the MATLAB note on page 2-4.

6.2 ORTHOGONAL SETS

Orthogonal sets and orthogonal bases are used throughout the chapter. The "orthogonal projection" discussed in this section is an important special case of the orthogonal projections studied in Section 6.3.

STUDY NOTES

The proofs of Theorems 4 and 5 are worth studying because they involve a calculation you will see and use several times.

The subsection entitled *An Orthogonal Projection* is simple but extremely important. Also, the geometric interpretation of Theorem 5 on page 382 will be helpful when you study Theorem 8 in the next section.

The attention paid to Theorems 6 and 7 will depend on what your instructor plans to do later in the course. In some cases, an instructor may discuss Theorems 6 and 7 only for square matrices. The $m \times n$ case is needed later, for Theorems 10, 12 and 15. Remember: the term *orthogonal matrix* applies only to a square matrix. Also, the columns of an orthogonal matrix must be ortho*normal*, not simply orthogonal.

SOLUTIONS TO EXERCISES

1. $\mathbf{u} = \begin{bmatrix} -1 \\ 4 \\ -3 \end{bmatrix}$, $\mathbf{v} = \begin{bmatrix} 5 \\ 2 \\ 1 \end{bmatrix}$, $\mathbf{w} = \begin{bmatrix} 3 \\ -4 \\ -7 \end{bmatrix}$, $\mathbf{u}\cdot\mathbf{v} = -5 + 8 - 3 = 0$,

$\mathbf{u}\cdot\mathbf{w} = -3 - 16 + 21 = 2 \neq 0$. The set $\{\mathbf{u}, \mathbf{v}, \mathbf{w}\}$ is not orthogonal. There is no need to check $\mathbf{v}\cdot\mathbf{w}$.

7. $\mathbf{u}_1 = \begin{bmatrix} 2 \\ -3 \end{bmatrix}$, $\mathbf{u}_2 = \begin{bmatrix} 6 \\ 4 \end{bmatrix}$, $\mathbf{x} = \begin{bmatrix} 9 \\ -7 \end{bmatrix}$. $\mathbf{u}_1\cdot\mathbf{u}_2 = 12 - 12 = 0$, so $\{\mathbf{u}_1, \mathbf{u}_2\}$ is an orthogonal set. Since the vectors are nonzero, $\mathbf{u}_1$ and $\mathbf{u}_2$ are linearly independent, by Theorem 4. But two such vectors in $\mathbb{R}^2$ automatically form a basis for $\mathbb{R}^2$. So $\{\mathbf{u}_1, \mathbf{u}_2\}$ is an orthogonal basis for $\mathbb{R}^2$. By Theorem 5,

$$\mathbf{x} = \frac{\mathbf{x}\cdot\mathbf{u}_1}{\mathbf{u}_1\cdot\mathbf{u}_1}\mathbf{u}_1 + \frac{\mathbf{x}\cdot\mathbf{u}_2}{\mathbf{u}_2\cdot\mathbf{u}_2}\mathbf{u}_2 = \frac{18 + 21}{4 + 9}\mathbf{u}_1 + \frac{54 - 28}{36 + 16}\mathbf{u}_2 = 3\begin{bmatrix} 2 \\ -3 \end{bmatrix} + \frac{1}{2}\begin{bmatrix} 6 \\ 4 \end{bmatrix}$$

13. $\mathbf{y} = \begin{bmatrix} 2 \\ 3 \end{bmatrix}$, $\mathbf{u} = \begin{bmatrix} 4 \\ -7 \end{bmatrix}$. The orthogonal projection of $\mathbf{y}$ onto $\mathbf{u}$ is

$$\hat{\mathbf{y}} = \frac{\mathbf{y}\cdot\mathbf{u}}{\mathbf{u}\cdot\mathbf{u}}\mathbf{u} = \frac{8 - 21}{16 + 49}\mathbf{u} = \frac{-13}{65}\mathbf{u} = \frac{-1}{5}\begin{bmatrix} 4 \\ -7 \end{bmatrix} = \begin{bmatrix} -4/5 \\ 7/5 \end{bmatrix}$$

The component of $\mathbf{y}$ orthogonal to $\mathbf{u}$ is $\mathbf{y} - \hat{\mathbf{y}} = \begin{bmatrix} 2 \\ 3 \end{bmatrix} - \begin{bmatrix} -4/5 \\ 7/5 \end{bmatrix} = \begin{bmatrix} 14/5 \\ 8/5 \end{bmatrix}$.

Thus, $\mathbf{y} = \hat{\mathbf{y}} + (\mathbf{y} - \hat{\mathbf{y}}) = \begin{bmatrix} -4/5 \\ 7/5 \end{bmatrix} + \begin{bmatrix} 14/5 \\ 8/5 \end{bmatrix}$.

19. $\mathbf{u} = \begin{bmatrix} -.6 \\ .8 \end{bmatrix}$, $\mathbf{v} = \begin{bmatrix} .8 \\ .6 \end{bmatrix}$, $\mathbf{u}\cdot\mathbf{v} = -.48 + .48 = 0$, so $\{\mathbf{u}, \mathbf{v}\}$ is an orthogonal set. Also, $\|\mathbf{u}\|^2 = \mathbf{u}\cdot\mathbf{u} = (-.6)^2 + (.8)^2 = .36 + .64 = 1$. Similarly, $\|\mathbf{v}\|^2 = \mathbf{v}\cdot\mathbf{v} = 1$. Thus $\{\mathbf{u}, \mathbf{v}\}$ is an orthonormal set.

23. **a.** See Example 3, for instance. **b.** See Theorem 5.
c. See the paragraph after Example 5.
d. See the paragraph before Example 7. **e.** See Example 4.

25. $(U\mathbf{x})\cdot(U\mathbf{y}) = (U\mathbf{x})^T(U\mathbf{y}) = \mathbf{x}^T U^T U\mathbf{y} = \mathbf{x}^T\mathbf{y} = \mathbf{x}\cdot\mathbf{y}$ (because $U^TU = I$).
If $\mathbf{y} = \mathbf{x}$, Theorem 7(b) says that $\|U\mathbf{x}\|^2 = \|\mathbf{x}\|^2$, which implies part (a). Part (c) of Theorem 7 follows immediately from part (b).

Study Tip: If your instructor emphasizes orthogonal matrices, work Exercises 27—29. (They make good test questions.) In each case, mention explicitly how you use the fact that the matrices are square. Don't read the solutions below until you have first *written* down your own solution.

27. If U has orthonormal columns, then $U^TU = I$, by Theorem 6. If U is also *square*, then the equation $U^TU = I$ implies that U is invertible, by the Invertible Matrix Theorem.

29. Since U and V are orthogonal, each is invertible. By Theorem 6 in Section 2.2, UV is invertible and $(UV)^{-1} = V^{-1}U^{-1} = V^TU^T = (UV)^T$ (by Theorem 3 in Section 2.1). Thus UV is an orthogonal matrix.

Mastering Linear Algebra Concepts: Orthogonal Basis

To the review sheet(s) you have on "basis", add the concepts of an orthogonal basis and an orthonormal basis for a subspace. You need to know what special properties they posses.

▶ basic definitions	Pages 379 and 383
▶ equivalent descriptions	Theorems 4 and 6
▶ geometric interpretation	Figs. 1 and 6
▶ special cases	Matrix with orthonormal columns
▶ examples and counterexamples	Examples 2 and 5
▶ algorithms and computations	Example 2
▶ connections with other concepts	Orthogonal matrix

MATLAB In Exercises 1—9 and 17—22, the fastest way (counting the keystrokes) in MATLAB to test a set such as $\{\mathbf{u}_1, \mathbf{u}_2, \mathbf{u}_3\}$ for orthogonality is to use a matrix **U = [u1 u2 u3]** whose columns are the vectors from the set. See the proof of Theorem 6.

For column vectors **y** and **u**, the orthogonal projection of **y** onto **u** is

```
(y'*u)/(u'*u)*u
```

The parentheses (and the final *****) are essential. MATLAB computes the scalar quotient $(\mathbf{y}^T\mathbf{u})/(\mathbf{u}^T\mathbf{u})$ and then multiplies **u** by this scalar.

6.3 ORTHOGONAL PROJECTIONS

A familiar idea in Euclidean geometry is to construct a line segment perpendicular to a line or plane. This section treats an analogous situation in $\mathbb{R}^n$, namely, the orthogonal projection of a vector (a point in $\mathbb{R}^n$) onto a subspace. The case when the subspace is a line through the origin was already examined in Section 6.2.

KEY IDEAS

If $\mathbf{y}$ is in $\mathbb{R}^n$ and if W is a subspace of $\mathbb{R}^n$, then the orthogonal projection of $\mathbf{y}$ onto W, denoted by $\hat{\mathbf{y}}$ or $\text{proj}_W \mathbf{y}$, has two important properties:

(i) $\mathbf{y} - \hat{\mathbf{y}}$ is orthogonal to W (so $\mathbf{y}$ is the sum of a vector $\hat{\mathbf{y}}$ in W and a vector $\mathbf{y} - \hat{\mathbf{y}}$ in $W^\perp$), and

(ii) $\hat{\mathbf{y}}$ is the closest point in W to $\mathbf{y}$.

Properties (i) and (ii) are described in the Orthogonal Decomposition Theorem and the Best Approximation Theorem. You should learn the statements of both theorems. (By now you probably know that whenever a theorem has an official name, an instructor has an easy time asking test questions about it.) When you need one of these theorems in a discussion (homework or test question), you should mention the theorem by name.

If your class covers Theorem 10, then the paragraph following the theorem will help you understand the difference between an *orthogonal matrix* (which must be square) and a rectangular matrix with orthonormal columns.

SOLUTIONS TO EXERCISES

1. $\mathbf{u}_1 = \begin{bmatrix} 0 \\ 1 \\ -4 \\ -1 \end{bmatrix}$, $\mathbf{u}_2 = \begin{bmatrix} 3 \\ 5 \\ 1 \\ 1 \end{bmatrix}$, $\mathbf{u}_3 = \begin{bmatrix} 1 \\ 0 \\ 1 \\ -4 \end{bmatrix}$, $\mathbf{u}_4 = \begin{bmatrix} 5 \\ -3 \\ -1 \\ 1 \end{bmatrix}$, $\mathbf{x} = \begin{bmatrix} 10 \\ -8 \\ 2 \\ 0 \end{bmatrix}$. You could calculate all the inner products in the decomposition:

$$\mathbf{x} = \underbrace{\frac{\mathbf{x}\cdot\mathbf{u}_1}{\mathbf{u}_1\cdot\mathbf{u}_1}\mathbf{u}_1 + \frac{\mathbf{x}\cdot\mathbf{u}_2}{\mathbf{u}_2\cdot\mathbf{u}_2}\mathbf{u}_2 + \frac{\mathbf{x}\cdot\mathbf{u}_3}{\mathbf{u}_3\cdot\mathbf{u}_3}\mathbf{u}_3}_{\text{in Span}\{\mathbf{u}_1,\mathbf{u}_2,\mathbf{u}_3\}} + \underbrace{\frac{\mathbf{x}\cdot\mathbf{u}_4}{\mathbf{u}_4\cdot\mathbf{u}_4}\mathbf{u}_4}_{\text{in Span}\{\mathbf{u}_4\}} \qquad (*)$$

However, once you know the vector in Span$\{\mathbf{u}_4\}$, the vector in Span$\{\mathbf{u}_1, \mathbf{u}_2, \mathbf{u}_3\}$ is determined completely by (*). So all you need is

$$\frac{\mathbf{x}\cdot\mathbf{u}_4}{\mathbf{u}_4\cdot\mathbf{u}_4}\mathbf{u}_4 = \frac{50 + 24 - 2 + 0}{25 + 9 + 1 + 1}\mathbf{u}_4 = 2\mathbf{u}_4 = \begin{bmatrix} 10 \\ -6 \\ -2 \\ 2 \end{bmatrix}$$

The vector in Span$\{\mathbf{u}_1, \mathbf{u}_2, \mathbf{u}_3\}$ is $\mathbf{x} - 2\mathbf{u}_4 = \begin{bmatrix} 10 \\ -8 \\ 2 \\ 0 \end{bmatrix} - \begin{bmatrix} 10 \\ -6 \\ -2 \\ 2 \end{bmatrix} = \begin{bmatrix} 0 \\ -2 \\ 4 \\ -2 \end{bmatrix}$.

Study Tip: One way to check whether $\text{proj}_W \mathbf{y}$ is computed correctly is to verify that $\mathbf{y} - \text{proj}_W \mathbf{y}$ is orthogonal to each vector in the orthogonal basis $\{\mathbf{u}_1, \ldots, \mathbf{u}_p\}$ for W. A faster check that will catch most errors (but not all) is to verify that $\mathbf{y} - \text{proj}_W \mathbf{y}$ is orthogonal to $\text{proj}_W \mathbf{y}$.

7. $\mathbf{y} = \begin{bmatrix} 1 \\ 3 \\ 5 \end{bmatrix}$, $\mathbf{u}_1 = \begin{bmatrix} 1 \\ 3 \\ -2 \end{bmatrix}$, $\mathbf{u}_2 = \begin{bmatrix} 5 \\ 1 \\ 4 \end{bmatrix}$. First, make sure that $\{\mathbf{u}_1, \mathbf{u}_2\}$ is an orthogonal basis for Span$\{\mathbf{u}_1, \mathbf{u}_2\}$. This is easy, since $\mathbf{u}_1$ and $\mathbf{u}_2$ are nonzero and $\mathbf{u}_1\cdot\mathbf{u}_2 = 0$. Next, by the Orthogonal Decomposition Theorem, $\mathbf{y}$ is the sum of $\text{proj}_W \mathbf{y}$ and $\mathbf{y} - \text{proj}_W \mathbf{y}$, where $W = \text{Span}\{\mathbf{u}_1, \mathbf{u}_2\}$.

$$\text{proj}_W \mathbf{y} = \frac{\mathbf{y}\cdot\mathbf{u}_1}{\mathbf{u}_1\cdot\mathbf{u}_1}\mathbf{u}_1 + \frac{\mathbf{y}\cdot\mathbf{u}_2}{\mathbf{u}_2\cdot\mathbf{u}_2}\mathbf{u}_2 = \frac{1 + 9 - 10}{1 + 9 + 4}\mathbf{u}_1 + \frac{5 + 3 + 20}{25 + 1 + 16}\mathbf{u}_2$$

$$= 0\mathbf{u}_1 + \frac{2}{3}\mathbf{u}_2 = \begin{bmatrix} 10/3 \\ 2/3 \\ 8/3 \end{bmatrix}$$

and

$$\mathbf{y} - \text{proj}_W \mathbf{y} = \begin{bmatrix} 1 \\ 3 \\ 5 \end{bmatrix} - \begin{bmatrix} 10/3 \\ 2/3 \\ 8/3 \end{bmatrix} = \begin{bmatrix} -7/3 \\ 7/3 \\ 7/3 \end{bmatrix}$$

As a check, scale $\mathbf{y} - \text{proj}_W \mathbf{y}$ to $\begin{bmatrix} -1 \\ 1 \\ 1 \end{bmatrix}$, and observe that the scaled vector is obviously orthogonal to $\mathbf{u}_1$ and $\mathbf{u}_2$. Thus $\mathbf{y} - \text{proj}_W \mathbf{y}$ is in $W^\perp$, as it should be.

Warning: The formula for $\text{proj}_W \mathbf{y}$ applies only if $\{\mathbf{u}_1, \ldots, \mathbf{u}_p\}$ is an *orthogonal* basis for W. That's why you should check orthogonality, as in Exercise 7, if you are not sure that the basis is orthogonal. If an orthogonal basis is not available, then other methods can be used to compute $\hat{\mathbf{y}}$. (See Exercise 23 in Section 6.5, for example.)

13. $\mathbf{z} = \begin{bmatrix} 3 \\ -7 \\ 2 \\ 3 \end{bmatrix}$, $\mathbf{v}_1 = \begin{bmatrix} 2 \\ -1 \\ -3 \\ 1 \end{bmatrix}$, $\mathbf{v}_2 = \begin{bmatrix} 1 \\ 1 \\ 0 \\ -1 \end{bmatrix}$. Note that $\mathbf{v}_1$ and $\mathbf{v}_2$ are orthogonal. By the Best Approximation Theorem, the closest point in Span$\{\mathbf{v}_1, \mathbf{v}_2\}$ to $\mathbf{z}$ is the orthogonal projection $\hat{\mathbf{z}}$, where

$$\hat{\mathbf{z}} = \frac{\mathbf{z}\cdot\mathbf{v}_1}{\mathbf{v}_1\cdot\mathbf{v}_1}\mathbf{v}_1 + \frac{\mathbf{z}\cdot\mathbf{v}_2}{\mathbf{v}_2\cdot\mathbf{v}_2}\mathbf{v}_2 = \frac{10}{15}\mathbf{v}_1 + \frac{-7}{3}\mathbf{v}_2 = \frac{2}{3}\begin{bmatrix} 2 \\ -1 \\ -3 \\ 1 \end{bmatrix} - \frac{7}{3}\begin{bmatrix} 1 \\ 1 \\ 0 \\ -1 \end{bmatrix} = \begin{bmatrix} -1 \\ -3 \\ -2 \\ 3 \end{bmatrix}$$

Check: $\mathbf{z} - \hat{\mathbf{z}} = \begin{bmatrix} 4 \\ -4 \\ 4 \\ 0 \end{bmatrix}$. The vector $\begin{bmatrix} 1 \\ -1 \\ 1 \\ 0 \end{bmatrix}$ is orthogonal to both $\mathbf{v}_1$ and $\mathbf{v}_2$.

19. By the Orthogonal Decomposition Theorem, $\mathbf{u}_3$ is the sum of a vector in W = Span$\{\mathbf{u}_1, \mathbf{u}_2\}$ and a vector $\mathbf{v}$ orthogonal to W. First,

$$\text{proj}_W\,\mathbf{u}_3 = \frac{-2}{6}\mathbf{u}_1 + \frac{2}{30}\mathbf{u}_2 = \begin{bmatrix} -2/6 \\ -2/6 \\ 4/6 \end{bmatrix} + \begin{bmatrix} 10/30 \\ -2/30 \\ 4/30 \end{bmatrix} = \begin{bmatrix} 0 \\ -2/5 \\ 4/5 \end{bmatrix}$$

Then

$$\mathbf{v} = \mathbf{u}_3 - \text{proj}_W\,\mathbf{u}_3 = \begin{bmatrix} 0 \\ 0 \\ 1 \end{bmatrix} - \begin{bmatrix} 0 \\ -2/5 \\ 4/5 \end{bmatrix} = \begin{bmatrix} 0 \\ 2/5 \\ 1/5 \end{bmatrix}$$

Not only is $\mathbf{v}$ orthogonal to W, but also any multiple of $\mathbf{v}$ is in $W^{\perp}$.

Study Tip: It would be a good idea to try Exercise 20 and compare the result with Exercise 19. Then think about the following problem: Suppose that $\{\mathbf{u}_1, \mathbf{u}_2\}$ is an orthogonal set of nonzero vectors in $\mathbb{R}^3$. How would you find an orthogonal basis of $\mathbb{R}^3$ that contains $\mathbf{u}_1$ and $\mathbf{u}_2$? You might discuss this with your instructor.

21. **a.** See the calculations for $\mathbf{z}_2$ in Example 1 or the box after Example 6 in Section 6.1.

b. See the Orthogonal Decomposition Theorem.

c. See the the second paragraph after the statement of Theorem 9.

d. See the box before The Best Approximation Theorem.

e. See the paragraph after Theorem 10.

23. By the Orthogonal Decomposition Theorem, each $\mathbf{x}$ in $\mathbb{R}^n$ can be written uniquely as $\mathbf{x} = \mathbf{p} + \mathbf{u}$, with $\mathbf{p}$ in Row A and $\mathbf{u}$ in $(\text{Row}\,A)^\perp$. By Theorem 3 in Section 6.1, $\mathbf{u}$ is in Nul A.

Next, suppose that $A\mathbf{x} = \mathbf{b}$ is consistent. Let $\mathbf{x}$ be a solution, and write $\mathbf{x} = \mathbf{p} + \mathbf{u}$, as above. Then $A\mathbf{p} = A(\mathbf{x} - \mathbf{u}) = A\mathbf{x} - A\mathbf{u} = \mathbf{b} - \mathbf{0} = \mathbf{b}$. So the equation $A\mathbf{x} = \mathbf{b}$ has at least one solution $\mathbf{p}$ in Row A.

Finally, suppose that $\mathbf{p}$ and $\mathbf{p}_1$ are both in Row A and satisfy $A\mathbf{x} = \mathbf{b}$. Then $\mathbf{p} - \mathbf{p}_1$ is in Nul A because

$$A(\mathbf{p} - \mathbf{p}_1) = A\mathbf{p} - A\mathbf{p}_1 = \mathbf{b} - \mathbf{b} = \mathbf{0}$$

The equations $\mathbf{p} = \mathbf{p}_1 + (\mathbf{p} - \mathbf{p}_1)$ and $\mathbf{p} = \mathbf{p} + \mathbf{0}$ both decompose $\mathbf{p}$ as the sum of a vector in Row A and a vector in $(\text{Row}\,A)^\perp$. By the uniqueness of the orthogonal decomposition (Theorem 8), $\mathbf{p}_1 = \mathbf{p}$, so $\mathbf{p}$ is unique.

Warning: I had to work hard to make the arithmetic simple in the exercises for this section, to avoid distractions for you and to save you time. You might not be so lucky on an exam. Even if a problem is designed to be numerically simple, there is always a chance that a minor error will make the calculations messy. In such a case, don't despair. Carry out the arithmetic as best you can, showing the details of your work (patterned after the solutions in this *Study Guide*). Chances are that you will get substantial credit for showing that you understand the concepts.

MATLAB Orthogonal Projections

The orthogonal projection of $\mathbf{y}$ onto a single vector was described in the MATLAB note for Section 6.2. The orthogonal projection onto the set spanned by an orthogonal set of vectors is the sum of the one-dimensional projections. Another way to construct this projection is to normalize the orthogonal vectors, place them in the columns of a matrix U, and use Theorem 10. For instance, if $\{\mathbf{y}_1, \mathbf{y}_2, \mathbf{y}_3\}$ is an orthogonal set of nonzero vectors, then the matrix

```
U = [y1/norm(y1)  y2/norm(y2)  y3/norm(y3)]
```

has orthonormal columns, and **U*(U'*y)** produces the orthogonal projection of $\mathbf{y}$ onto the subspace spanned by $\{\mathbf{y}_1, \mathbf{y}_2, \mathbf{y}_3\}$. (The parentheses around **U'*y** speeds up the computation of **U*U'*y**.)

6.4 THE GRAM-SCHMIDT PROCESS

This section has a nice geometric appeal. The Gram-Schmidt process is well-liked by students and faculty because it is easily learned. Although the process is seldom used in practical work, it has important generalizations to spaces other than $\mathbb{R}^n$ (to be discussed briefly in Section 6.7).

KEY IDEAS

When the Gram-Schmidt process is applied to $\{\mathbf{x}_1,\dots,\mathbf{x}_p\}$, the first step is to set $\mathbf{v}_1 = \mathbf{x}_1$. For $k = 2,\dots,n$, the kth step consists of subtracting from $\mathbf{x}_k$ its projection onto the subspace spanned by the previous **x**'*s*. At each step the projection is easy to compute because an orthogonal basis for the appropriate subspace has already been constructed.

The QR factorization of a matrix *A* encapsulates the result of applying the Gram-Schmidt process to the columns of *A*, just as the LU factorization of a matrix encodes the row operations that reduce a matrix to echelon form. Also, just as the LU factorization can be implemented via multiplication by elementary matrices, so can the QR factorization be constructed via multiplication by orthogonal matrices.

SOLUTIONS TO EXERCISES

1. $\mathbf{x}_1 = \begin{bmatrix} 3 \\ 0 \\ -1 \end{bmatrix}$, $\mathbf{x}_2 = \begin{bmatrix} 8 \\ 5 \\ -6 \end{bmatrix}$. Set $\mathbf{v}_1 = \mathbf{x}_1$ and compute

$$\mathbf{v}_2 = \mathbf{x}_2 - \frac{\mathbf{x}_2\cdot\mathbf{v}_1}{\mathbf{v}_1\cdot\mathbf{v}_1}\mathbf{v}_1 = \begin{bmatrix} 8 \\ 5 \\ -6 \end{bmatrix} - \frac{30}{10}\begin{bmatrix} 3 \\ 0 \\ -1 \end{bmatrix} = \begin{bmatrix} -1 \\ 5 \\ -3 \end{bmatrix}$$

Check: $\mathbf{v}_2\cdot\mathbf{v}_1 = -3 + 0 + 3 = 0$. So an orthogonal basis is $\left\{\begin{bmatrix} 3 \\ 0 \\ -1 \end{bmatrix}, \begin{bmatrix} -1 \\ 5 \\ -3 \end{bmatrix}\right\}$.

7. $\mathbf{x}_1 = \begin{bmatrix} 2 \\ -5 \\ 1 \end{bmatrix}$, $\mathbf{x}_2 = \begin{bmatrix} 4 \\ -1 \\ 2 \end{bmatrix}$. From Exercise 3, use $\mathbf{v}_1 = \begin{bmatrix} 2 \\ -5 \\ 1 \end{bmatrix}$ and $\mathbf{v}_2 = \begin{bmatrix} 3 \\ 3/2 \\ 3/2 \end{bmatrix}$ as an orthogonal basis for $W = \text{Span}\{\mathbf{x}_1, \mathbf{x}_2\}$. Scale $\mathbf{v}_2$ to (2,1,1) before normalizing, and then obtain

$$\mathbf{u}_1 = \frac{1}{\sqrt{30}}\begin{bmatrix} 2 \\ -5 \\ 1 \end{bmatrix} = \begin{bmatrix} 2/\sqrt{30} \\ -5/\sqrt{30} \\ 1/\sqrt{30} \end{bmatrix}, \quad \mathbf{u}_2 = \frac{1}{\sqrt{6}}\begin{bmatrix} 2 \\ 1 \\ 1 \end{bmatrix} = \begin{bmatrix} 2/\sqrt{6} \\ 1/\sqrt{6} \\ 1/\sqrt{6} \end{bmatrix}$$

Study Tip: If you need to normalize a vector by hand, first consider scaling the entries in the vector to make them small integers, if possible.

13. $A = \begin{bmatrix} 5 & 9 \\ 1 & 7 \\ -3 & -5 \\ 1 & 5 \end{bmatrix}$, $Q = \begin{bmatrix} 5/6 & -1/6 \\ 1/6 & 5/6 \\ -3/6 & 1/6 \\ 1/6 & 3/6 \end{bmatrix}$. Let

$$R = Q^TA = \begin{bmatrix} 5/6 & 1/6 & -3/6 & 1/6 \\ -1/6 & 5/6 & 1/6 & 3/6 \end{bmatrix}\begin{bmatrix} 5 & 9 \\ 1 & 7 \\ -3 & -5 \\ 1 & 5 \end{bmatrix} = \begin{bmatrix} 36/6 & 72/6 \\ 0 & 36/6 \end{bmatrix} = \begin{bmatrix} 6 & 12 \\ 0 & 6 \end{bmatrix}$$

As a check, compute $QR = \begin{bmatrix} 5/6 & -1/6 \\ 1/6 & 5/6 \\ -3/6 & 1/6 \\ 1/6 & 3/6 \end{bmatrix}\begin{bmatrix} 6 & 12 \\ 0 & 6 \end{bmatrix} = \begin{bmatrix} 5 & 54/6 \\ 1 & 42/6 \\ -3 & -30/6 \\ 1 & 30/6 \end{bmatrix} = A.$

Remark: The reason the R in Exercise 13 works is that the columns of Q form an orthonormal basis for Col A (since they were obtained by the Gram-Schmidt process). Thus $QQ^T\mathbf{y} = \mathbf{y}$ for all $\mathbf{y}$ in Col A, by Theorem 10 in Section 6.3. In particular, $QQ^TA = A$. So if R is Q^TA, then $QR = Q(Q^TA) = A$.

17. **a.** See the remark after Example 5 in Section 6.2, and the reference there to Exercise 32.

b. See (1) in the statement of Theorem 11.

c. See the solution of Example 4.

19. The full solution is in the text.

21. The solution in the text is complete, except for the details of extending an orthonormal basis for Span$\{\mathbf{q}_1, \ldots, \mathbf{q}_n\}$ to an orthonormal basis for $\mathbb{R}^m$. Here is one algorithm. Let $\mathbf{e}_1, \ldots, \mathbf{e}_m$ be the standard basis for $\mathbb{R}^m$. Let $\mathbf{f}_1$ be the first vector in this basis that is *not* in W_n = Span$\{\mathbf{q}_1, \ldots, \mathbf{q}_n\}$, and let $\mathbf{u}_1 = \mathbf{f}_1 - \text{proj}_{W_n}\mathbf{f}_1$. Then $\{\mathbf{q}_1, \ldots, \mathbf{q}_n, \mathbf{u}_1\}$ is an orthogonal basis for W_{n+1} = Span$\{\mathbf{q}_1, \ldots, \mathbf{q}_n, \mathbf{u}_1\}$. Let $\mathbf{f}_2$ be the first vector in $\{\mathbf{e}_1, \ldots, \mathbf{e}_m\}$ that is not in W_{n+1}. (Of course $\mathbf{f}_2$ occurs later than $\mathbf{f}_1$ in the list $\mathbf{e}_1, \ldots, \mathbf{e}_m$.) Form $\mathbf{u}_2 = \mathbf{f}_2 - \text{proj}_{W_{n+1}}\mathbf{f}_2$ and W_{n+2} = Span$\{\mathbf{q}_1, \ldots, \mathbf{q}_n, \mathbf{u}_1, \mathbf{u}_2\}$. This process will continue until $m - n$ vectors have been added to the original n vectors. Normalizing the new vectors produces an orthonormal basis for $\mathbb{R}^m$.

MATLAB The Gram-Schmidt Process

If A has only two columns, then the Gram-Schmidt process is

```
v1 = A(:,1)
v2 = A(:,2) - (A(:,2)'*v1)/(v1'*v1)*v1
```

If A has three columns, add the command

```
v3 = A(:,3) - (A(:,3)'*v1)/(v1'*v1)*v1 - (A(:,3)'*v2)/(v2'*v2)*v2
```

You should use these commands for a while, to learn the general procedure. After that, you can use the Toolbox command **proj(x,V)** , which computes the projection of a vector **x** onto the subspace spanned by the columns of a matrix (or vector) V. For example,

```
v2 = A(:,2) - proj(A(:,2),v1)          V = v1
v3 = A(:,3) - proj(A(:,3),[v1  v2])    V = [v1  v2]
```

The columns of V in **proj(x,V)** need not be orthogonal for the command to work, but if they are, the entries in **proj(x,V)** will usually agree with those computed via Theorem 10 in Section 6.3, to twelve or more decimal places. Enter **help proj** to learn more about **proj**.

To check your work or to save time, enter **Q = gs(A)** , which uses the Gram-Schmidt process to construct the columns of Q. See **help gs** .

Although you should construct the QR factorization of a matrix using the approach in the text, you might like to see what MATLAB does. The command **[Q1 R1] = qr(A)** creates a modified QR factorization of an $m \times n$ matrix A as described in Exercise 21 in the text. If rank $A = r$, then the first r rows of R_1 are nonzero and the first r columns of Q_1 form an orthonormal basis for Col A.

6.5 LEAST-SQUARES PROBLEMS

The basic geometric principles in this section provide the foundation for all the applications in Sections 6.6—6.8.

KEY IDEAS

The material up to and including Figure 2 needs to be read carefully several times, so you understand what the term "least-squares solution" means. Be careful to distinguish between $\hat{\mathbf{x}}$ and $\hat{\mathbf{b}}$. The vector $\hat{\mathbf{b}}$ is unique since it is the closest point in Col A to $\mathbf{b}$. The least-squares solution $\hat{\mathbf{x}}$ is unique if and only if A has linearly independent columns (see Theorems 13 and 14).

Warning: A common mistake is to think that $\hat{\mathbf{x}}$ itself somehow has the least-squares norm or is the closest point to $\mathbf{b}$. Look at Fig. 2 again. The vector closest to $\mathbf{b}$ is $A\hat{\mathbf{x}}$, not $\hat{\mathbf{x}}$.

Theorem 13 provides a common way to find least-squares solutions. One way to remember the normal equations is to observe that they are the same as $A\mathbf{x} = \mathbf{b}$, with A^T left-multiplied on each side of the equation. It is completely wrong, however, to try to *derive* the normal equations from $A\mathbf{x} = \mathbf{b}$ by multiplying by A^T. If the equation $A\mathbf{x} = \mathbf{b}$ has no solution, then the equation itself is a false statement about every $\mathbf{x}$. Matrix algebra on such a false statement is meaningless.

SOLUTIONS TO EXERCISES

1. $A = \begin{bmatrix} -1 & 2 \\ 2 & -3 \\ -1 & 3 \end{bmatrix}$, $\mathbf{b} = \begin{bmatrix} 4 \\ 1 \\ 2 \end{bmatrix}$

$$A^TA = \begin{bmatrix} -1 & 2 & -1 \\ 2 & -3 & 3 \end{bmatrix}\begin{bmatrix} -1 & 2 \\ 2 & -3 \\ -1 & 3 \end{bmatrix} = \begin{bmatrix} 6 & -11 \\ -11 & 22 \end{bmatrix}, \quad A^T\mathbf{b} = \begin{bmatrix} -1 & 2 & -1 \\ 2 & -3 & 3 \end{bmatrix}\begin{bmatrix} 4 \\ 1 \\ 2 \end{bmatrix} = \begin{bmatrix} -4 \\ 11 \end{bmatrix}$$

a. The normal equations: $\begin{bmatrix} 6 & -11 \\ -11 & 22 \end{bmatrix}\begin{bmatrix} x_1 \\ x_2 \end{bmatrix} = \begin{bmatrix} -4 \\ 11 \end{bmatrix}$

b. Since A^TA is only 2×2, $(A^TA)^{-1}$ is easy to compute, and

$$\hat{\mathbf{x}} = \begin{bmatrix} 6 & -11 \\ -11 & 22 \end{bmatrix}^{-1}\begin{bmatrix} -4 \\ 11 \end{bmatrix} = \frac{1}{11}\begin{bmatrix} 22 & 11 \\ 11 & 6 \end{bmatrix}\begin{bmatrix} -4 \\ 11 \end{bmatrix} = \frac{1}{11}\begin{bmatrix} 33 \\ 22 \end{bmatrix} = \begin{bmatrix} 3 \\ 2 \end{bmatrix}$$

Warning: It is important to distinguish between the normal equations $A^TA\hat{\mathbf{x}} = A^T\mathbf{b}$ and the formula $\hat{\mathbf{x}} = (A^TA)^{-1}A^T\mathbf{b}$. Both equations describe $\hat{\mathbf{x}}$ (implicitly or explicitly), but the formula for $\hat{\mathbf{x}}$ holds only when A has linearly independent columns. Note that the expression $(A^TA)^{-1}A^T$ *cannot* be simplified when A is not invertible.

7. $A = \begin{bmatrix} 1 & -2 \\ -1 & 2 \\ 0 & 3 \\ 2 & 5 \end{bmatrix}$, $\mathbf{b} = \begin{bmatrix} 3 \\ 1 \\ -4 \\ 2 \end{bmatrix}$, $A^TA = \begin{bmatrix} 1 & -1 & 0 & 2 \\ -2 & 2 & 3 & 5 \end{bmatrix}\begin{bmatrix} 1 & -2 \\ -1 & 2 \\ 0 & 3 \\ 2 & 5 \end{bmatrix} = \begin{bmatrix} 6 & 6 \\ 6 & 42 \end{bmatrix}$

$$A^T\mathbf{b} = \begin{bmatrix} 1 & -1 & 0 & 2 \\ -2 & 2 & 3 & 5 \end{bmatrix}\begin{bmatrix} 3 \\ 1 \\ -4 \\ 2 \end{bmatrix} = \begin{bmatrix} 6 \\ -6 \end{bmatrix}$$

The normal equations: $\begin{bmatrix} 6 & 6 \\ 6 & 42 \end{bmatrix}\begin{bmatrix} x_1 \\ x_2 \end{bmatrix} = \begin{bmatrix} 6 \\ -6 \end{bmatrix}$

The particular numbers in A^TA suggest that the normal equations might be solved easily via row operations:

$$\begin{bmatrix} 6 & 6 & 6 \\ 6 & 42 & -6 \end{bmatrix} \sim \begin{bmatrix} 6 & 6 & 6 \\ 0 & 36 & -12 \end{bmatrix} \sim \begin{bmatrix} 1 & 1 & 1 \\ 0 & 1 & -1/3 \end{bmatrix} \sim \begin{bmatrix} 1 & 0 & 4/3 \\ 0 & 1 & -1/3 \end{bmatrix}$$

Thus $\hat{\mathbf{x}} = \begin{bmatrix} 4/3 \\ -1/3 \end{bmatrix}$. The least-squares error is $\|A\hat{\mathbf{x}} - \mathbf{b}\|$, so compute

$$A\hat{\mathbf{x}} - \mathbf{b} = \begin{bmatrix} 1 & -2 \\ -1 & 2 \\ 0 & 3 \\ 2 & 5 \end{bmatrix}\begin{bmatrix} 4/3 \\ -1/3 \end{bmatrix} - \begin{bmatrix} 3 \\ 1 \\ -4 \\ 2 \end{bmatrix} = \begin{bmatrix} 2 \\ -2 \\ -1 \\ 1 \end{bmatrix} - \begin{bmatrix} 3 \\ 1 \\ -4 \\ 2 \end{bmatrix} = \begin{bmatrix} -1 \\ -3 \\ 3 \\ -1 \end{bmatrix}$$

$$\|A\hat{\mathbf{x}} - \mathbf{b}\|^2 = 1 + 9 + 9 + 1 = 20, \text{ and } \|A\hat{\mathbf{x}} - \mathbf{b}\| = \sqrt{20} = 2\sqrt{5}$$

Study Tip: A good way to check your work is to verify that $A\hat{\mathbf{x}} - \mathbf{b}$ is orthogonal to each column of A.

13. $A\mathbf{u} = \begin{bmatrix} 3 & 4 \\ -2 & 1 \\ 3 & 4 \end{bmatrix}\begin{bmatrix} 5 \\ -1 \end{bmatrix} = \begin{bmatrix} 11 \\ -11 \\ 11 \end{bmatrix}$, $\mathbf{b} - A\mathbf{u} = \begin{bmatrix} 11 \\ -9 \\ 5 \end{bmatrix} - \begin{bmatrix} 11 \\ -11 \\ 11 \end{bmatrix} = \begin{bmatrix} 0 \\ 2 \\ -6 \end{bmatrix}$, $\|\mathbf{b} - A\mathbf{u}\| = \sqrt{40}$

$A\mathbf{v} = \begin{bmatrix} 3 & 4 \\ -2 & 1 \\ 3 & 4 \end{bmatrix}\begin{bmatrix} 5 \\ -2 \end{bmatrix} = \begin{bmatrix} 7 \\ -12 \\ 7 \end{bmatrix}$, $\mathbf{b} - A\mathbf{v} = \begin{bmatrix} 11 \\ -9 \\ 5 \end{bmatrix} - \begin{bmatrix} 7 \\ -12 \\ 7 \end{bmatrix} = \begin{bmatrix} 4 \\ 3 \\ -2 \end{bmatrix}$, $\|\mathbf{b} - A\mathbf{u}\| = \sqrt{29}$

Obviously, $A\mathbf{u}$ is not the closest point of Col A to $\mathbf{b}$, because $A\mathbf{v}$ is closer. Hence $\mathbf{u}$ is *not* the least-squares solution of $A\mathbf{x} = \mathbf{b}$.

17. **a.** See the beginning of the section.

b. See the comments about equation (1).

c. Read the definition of a least-squares solution.

d. See Theorem 13. **e.** See Theorem 14.

19. The full solution is in the text.

21. **a.** If A has linearly independent columns, then the equation $A\mathbf{x} = \mathbf{0}$ has only the trivial solution. By Exercise 17, $A^TA\mathbf{x} = \mathbf{0}$ also has only the trivial solution. Since A^TA is *square*, it must be invertible, by the Invertible Matrix Theorem.

b. Since the n linearly independent columns of A belong to $\mathbb{R}^m$, m could not be less than n.

c. The n linearly independent columns of A form a basis for Col A, so the rank of A is n.

MATLAB The Backslash Command

For Exercises 15 and 16, see the Numerical Note on page 410 in the text. The MATLAB "backslash" command **R\(Q'*b)** will solve $R\hat{\mathbf{x}} = Q^T\mathbf{b}$ by row reduction. When A is nonsquare with linearly independent columns, this procedure of first forming Q and R and then solving $R\hat{\mathbf{x}} = Q^T\mathbf{b}$ is what MATLAB itself does (internally) when the MATLAB command **A\b** is used to solve $A\mathbf{x} = \mathbf{b}$. Compute **A\b** for the data in Exercises 15 and 16, and compare the results with your calculations involving Q and R.

For Exercise 26, the command **A = [A1;A2]** creates a (partitioned) matrix whose top block is A1 and bottom block is A2. This command works as long as A1 and A2 have the same number of columns.

6.6 APPLICATIONS TO LINEAR MODELS

This section of the text will be a valuable reference for any person who works with data that requires statistical analysis. Many graduate fields require such work, often in connection with doctoral research. Even most undergraduates will take a course where least-squares lines are used.

KEY IDEA

Linear algebra unifies the study of many problems in statistics and data analysis. All the examples in this section, from ordinary linear regression (using a least-squares line) to multiple regression, concern just one idea: find a least-squares solution of $X\boldsymbol{\beta} = \mathbf{y}$. Only the design matrix X varies. The exercises help you practice choosing X. The least-squares solution $\hat{\boldsymbol{\beta}}$ always satisfies the normal equations $X^TX\hat{\boldsymbol{\beta}} = X^T\mathbf{y}$.

STUDY NOTES

Don't confuse the least-squares line in Fig. 1 with the lines and planes in Section 6.5 onto which we projected various vectors **b**. The line is nothing more than a special case of the curves in Figures 2—5. In each case, the "linearity" of the model lies not in the curve, but rather in the fact that the unknown parameters (or *weights*) β_0, $\beta_1, \ldots$ occur linearly in the formula for the curve, just as the variables x_1, $x_2, \ldots$ occur in an ordinary linear equation.

Any 4×4 submatrix of the design matrix in Example 3 is called a Vandermonde matrix. Using Exercise 11 on page 178, one can show that if at least four of the values $x_1, \ldots, x_n$ are distinct, then the least-squares solution $\hat{\boldsymbol{\beta}}$ will be unique, by Theorem 14 in Section 6.5.

FURTHER READING

An important generalization of the discussion here is to *multivariate* analysis, which involves several **y** vectors rather than just one. In this case the basic equation is $XB = Y$, where each column of Y is a data set for one dependent variable, and each column of B is a set of parameters to be determined. That is, $X[\boldsymbol{\beta}_1 \cdots \boldsymbol{\beta}_p] = [\mathbf{y}_1 \cdots \mathbf{y}_p]$. For more information, see the classic text by T. W. Anderson, *An Introduction to Multivariate Statistical Analysis*, John Wiley & Sons, New York, 1984 (and 1958). The preface of the text says, "A knowledge of matrix algebra is a prerequisite [for understanding the text]." Most modern multivariate statistics texts rely heavily on matrix notation and matrix algebra.

SOLUTIONS TO EXERCISES

1. Place the x-coordinates of the data in the second column of X and the y-coordinates in the vector $\mathbf{y}$. So $X = \begin{bmatrix} 1 & 0 \\ 1 & 1 \\ 1 & 2 \\ 1 & 3 \end{bmatrix}$ and $\mathbf{y} = \begin{bmatrix} 1 \\ 1 \\ 2 \\ 2 \end{bmatrix}$. Compute

$$\underset{X^T}{\begin{bmatrix} 1 & 1 & 1 & 1 \\ 0 & 1 & 2 & 3 \end{bmatrix}} \underset{X}{\begin{bmatrix} 1 & 0 \\ 1 & 1 \\ 1 & 2 \\ 1 & 3 \end{bmatrix}} = \begin{bmatrix} 4 & 6 \\ 6 & 14 \end{bmatrix}, \qquad \underset{X^T}{\begin{bmatrix} 1 & 1 & 1 & 1 \\ 0 & 1 & 2 & 3 \end{bmatrix}} \underset{\mathbf{y}}{\begin{bmatrix} 1 \\ 1 \\ 2 \\ 2 \end{bmatrix}} = \begin{bmatrix} 6 \\ 11 \end{bmatrix}$$

The matrix normal equation and its solution are:

$$\begin{bmatrix} 4 & 6 \\ 6 & 14 \end{bmatrix} \begin{bmatrix} \beta_0 \\ \beta_1 \end{bmatrix} = \begin{bmatrix} 6 \\ 11 \end{bmatrix}$$

$$\begin{bmatrix} \beta_0 \\ \beta_1 \end{bmatrix} = \begin{bmatrix} 4 & 6 \\ 6 & 14 \end{bmatrix}^{-1} \begin{bmatrix} 6 \\ 11 \end{bmatrix} = \frac{1}{20} \begin{bmatrix} 14 & -6 \\ -6 & 4 \end{bmatrix} \begin{bmatrix} 6 \\ 11 \end{bmatrix} = \frac{1}{20} \begin{bmatrix} 18 \\ 8 \end{bmatrix} = \begin{bmatrix} .9 \\ .4 \end{bmatrix}$$

The least-squares line, $y = \beta_0 + \beta_1 x$, is $y = .9 + .4x$

7. a. $\mathbf{y} = X\boldsymbol{\beta} + \boldsymbol{\varepsilon}$, where $\mathbf{y} = \begin{bmatrix} 1.8 \\ 2.7 \\ 3.4 \\ 3.8 \\ 3.9 \end{bmatrix}$, $X = \begin{bmatrix} 1 & 1 \\ 2 & 4 \\ 3 & 9 \\ 4 & 16 \\ 5 & 25 \end{bmatrix}$, $\boldsymbol{\beta} = \begin{bmatrix} \beta_1 \\ \beta_2 \end{bmatrix}$, $\boldsymbol{\varepsilon} = \begin{bmatrix} \varepsilon_1 \\ \varepsilon_2 \\ \varepsilon_3 \\ \varepsilon_4 \\ \varepsilon_5 \end{bmatrix}$.

b. In this problem, X^TX is invertible, and you can use your matrix program to solve the normal equations via row reduction or the inverse of X^TX. MATLAB offers two additional possibilities: the command **β = (X'*X)\(X'*y)** solves the equation $(X^TX)\boldsymbol{\beta} = X^T\mathbf{y}$, and the command **β = X\y** immediately computes the least-squares solution of $X\boldsymbol{\beta} = \mathbf{y}$. In any case,

$$\hat{\boldsymbol{\beta}} = \begin{bmatrix} \beta_1 \\ \beta_2 \end{bmatrix} = \begin{bmatrix} 1.76 \\ -.20 \end{bmatrix} \quad \text{(to two decimal places)}$$

The desired least-squares equation is $y = 1.76x - .20x^2$.

13. Let **1** be the vector in $\mathbb{R}^{13}$ with 1 in each entry, let $\mathbf{t} = (0, \ldots, 12)$, and for $k = 2$ and 3, let **t**^.k denote the vector whose entries are the kth powers of the entries in **t**. (See the MATLAB box at the end of this *Study Guide* section.) Then the design matrix is

X = [**1** **t** **t**^.2 **t**^.3]

The observation vector **y** lists the measured positions of the plane.

a. Numerical solution of the normal equations yields

$$\boldsymbol{\beta} = (-.8558,\ 4.7025,\ 5.5554,\ -.0274)$$

The least-squares polynomial (position of the plane at time t) is

$$y = -.8558 + 4.7025t + 5.5554t^2 - .0274t^3$$

b. The velocity is the derivative of the position function:

$$v(t) = 4.7025 + 11.1108t - .0822t^2$$

When $t = 4.5$ seconds, $v(4.5) = 53.0$ ft/sec.

15. From equation (1) on page 415,

$$X^TX = \begin{bmatrix} 1 & \cdots & 1 \\ x_1 & \cdots & x_n \end{bmatrix} \begin{bmatrix} 1 & x_1 \\ \vdots & \vdots \\ 1 & x_n \end{bmatrix} = \begin{bmatrix} n & \Sigma x \\ \Sigma x & \Sigma x^2 \end{bmatrix}$$

$$X^T\mathbf{y} = \begin{bmatrix} 1 & \cdots & 1 \\ x_1 & \cdots & x_n \end{bmatrix} \begin{bmatrix} y_1 \\ \vdots \\ y_n \end{bmatrix} = \begin{bmatrix} \Sigma y \\ \Sigma xy \end{bmatrix}$$

The equations (7) in the text follow immediately from the usual matrix normal equation $X^T X\boldsymbol{\beta} = X^T\mathbf{y}$.

16. *Note*: The formulas you should derive are

$$\hat{\beta}_0 = \frac{(\Sigma x^2)(\Sigma y) - (\Sigma x)(\Sigma xy)}{n\Sigma x^2 - (\Sigma x)^2}, \qquad \hat{\beta}_1 = \frac{n\Sigma xy - (\Sigma x)(\Sigma y)}{n\Sigma x^2 - (\Sigma x)^2}$$

Some statistics texts present other equivalent formulas for $\hat{\beta}_0$ and $\hat{\beta}_1$.

19. The equation to be proved is $\|\mathbf{y}\|^2 = \|X\hat{\boldsymbol{\beta}}\|^2 + \|\mathbf{y} - X\hat{\boldsymbol{\beta}}\|^2$. This follows from the Pythagorean Theorem (in Section 6.1) and the figure below.

Appendix: The Geometry of a Linear Model

The column space of the design matrix X is sometimes called the **design subspace**. If $\hat{\boldsymbol{\beta}}$ is the least-squares solution of $\mathbf{y} = X\boldsymbol{\beta}$, then the residual vector $\boldsymbol{\varepsilon} = \mathbf{y} - X\hat{\boldsymbol{\beta}}$ is orthogonal to the design subspace, and the equation $\mathbf{y} = X\hat{\boldsymbol{\beta}} + \boldsymbol{\varepsilon}$ is an orthogonal decomposition of the observed $\mathbf{y}$ into the sum of the least-squares predicted $\hat{\mathbf{y}}$ and the residual vector $\boldsymbol{\varepsilon}$.

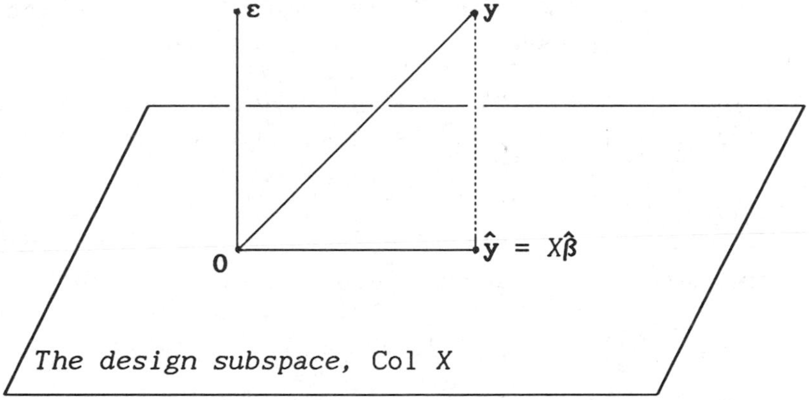

MATLAB Functions of Vectors

If **x** is a vector and k is a positive integer, then **x^.k** is a vector the same size as **x** whose entries are the kth powers of the entries in **x**. The function **cos(x).^k** was mentioned in the MATLAB note on page 4-12. The exponential function, **exp(x)** , and natural logarithm function, **log(x)** , also act on each entry in **x**. For instance, the entries in the vector **exp(-.02*x)** are computed by applying the function $e^{-.02x}$ to the corresponding entries in **x**.

6.7 INNER PRODUCT SPACES

Three examples of inner product spaces are described here, in Examples 1, 2, and 7. Corresponding applications appear in the next section. Material in Sections 6.7 and 6.8 will be useful for many careers, particularly science, engineering, and mathematics. If your course does not cover this now, the text and *Study Guide* can help you learn it later on your own.

KEY IDEAS

The concepts of length and orthogonality in $\mathbb{R}^n$ have analogues in a number of other vector spaces. The definition of an inner product identifies the basic properties needed for a theory that parallels the familiar theory for $\mathbb{R}^n$. Two useful facts, the Cauchy-Schwarz inequality and the triangle inequality, were not developed earlier, but they are important for applications both in $\mathbb{R}^n$ and in other inner product spaces. Every mathematics major will need to know these facts in other undergraduate courses.

The inner product in Example 1 is used in Section 6.8 to describe weighted least-squares problems. The inner product in Examples 2—6 provides a more sophisticated approach to the least-squares curve fitting discussed in Section 6.6. See the "trend analysis" in Section 6.8.

Be sure to read the paragraph preceding Example 6. The idea of "best approximation" to a function is of fundamental importance in mathematics. The most common applications of best approximation (such as Fourier series, introduced in Section 6.8) involve the inner product in Example 7.

SOLUTIONS TO EXERCISES

1. The inner product is $\langle \mathbf{x}, \mathbf{y} \rangle = 4x_1y_1 + 5x_2y_2$. Let $\mathbf{x} = (1,1)$, $\mathbf{y} = (5,-1)$.

   a. $\|\mathbf{x}\|^2 = 4\cdot1\cdot1 + 5\cdot1\cdot1 = 9$, $\|\mathbf{x}\| = 3$

   $\|\mathbf{y}\|^2 = 4\cdot5\cdot5 + 5(-1)(-1) = 105$, $\|\mathbf{y}\| = \sqrt{105}$

   $|\langle \mathbf{x}, \mathbf{y} \rangle|^2 = |4\cdot1\cdot5 + 5\cdot1(-1)|^2 = |15|^2 = 225$

   b. A vector $\mathbf{z} = (z_1, z_2)$ is orthogonal to $\mathbf{y}$ if and only if $\langle \mathbf{z}, \mathbf{y} \rangle = 0$, that is,

   $$4\cdot z_1\cdot5 + 5\cdot z_2\cdot(-1) = 0, \quad 20z_1 - 5z_2 = 0, \quad \text{and} \quad z_2 = 4z_1$$

   Thus (z_1, z_2) is orthogonal to $\mathbf{y}$ if and only if $z_2 = 4z_1$.

7. Given $p(t) = 4 + t$ and $q(t) = 5 - 4t^2$. The orthogonal projection $\hat{q}$ of q onto the subspace spanned by p is $[\langle q,p\rangle/\langle p,p\rangle]p$. The notation of Example 5 organizes the calculations nicely:

$$\begin{array}{lccl} \text{Polynomial:} & p & q & \\ \text{Vector of values:} & \begin{bmatrix}3\\4\\5\end{bmatrix}, & \begin{bmatrix}1\\5\\1\end{bmatrix} & \begin{array}{l}\leftarrow \text{value at } -1\\ \leftarrow \text{value at } 0\\ \leftarrow \text{value at } 1\end{array} \end{array}$$

The inner product $\langle q,p\rangle$ equals the (standard) inner product of the two corresponding vectors in $\mathbb{R}^3$: $\langle q,p\rangle = 3\cdot1 + 4\cdot5 + 5\cdot1 = 28$. Similarly, $\langle p,p\rangle = 3^2 + 4^2 + 5^2 = 50$. Thus

$$\hat{q}(t) = \frac{28}{50}(4 + t) = \frac{56}{25} + \frac{14}{25}t$$

13. Suppose A is invertible and $\langle \mathbf{u},\mathbf{v}\rangle = (A\mathbf{u})\cdot(A\mathbf{v})$, for $\mathbf{u},\mathbf{v}$ in $\mathbb{R}^n$. Note that $\langle \mathbf{u},\mathbf{v}\rangle$ is in $\mathbb{R}$, and check each axiom in the definition on page 422:

i. $\langle \mathbf{u},\mathbf{v}\rangle = (A\mathbf{u})\cdot(A\mathbf{v}) = (A\mathbf{v})\cdot(A\mathbf{u})$ — Property of dot product
$= \langle \mathbf{v},\mathbf{u}\rangle$

ii. $\langle \mathbf{u}+\mathbf{v},\mathbf{w}\rangle = [A(\mathbf{u}+\mathbf{v})]\cdot(A\mathbf{w}) = [A\mathbf{u} + A\mathbf{v}]\cdot(A\mathbf{w})$ — Matrix multiplication
$= (A\mathbf{u})\cdot(A\mathbf{w}) + (A\mathbf{v})\cdot(A\mathbf{w})$ — Property of dot product
$= \langle \mathbf{u},\mathbf{w}\rangle + \langle \mathbf{v},\mathbf{w}\rangle$

iii. $\langle c\mathbf{u},\mathbf{v}\rangle = [A(c\mathbf{u})]\cdot(A\mathbf{v}) = [c(A\mathbf{u})]\cdot(A\mathbf{v})$ — Matrix multiplication
$= c(A\mathbf{u})\cdot(A\mathbf{v})$ — Property of dot product
$= c\langle \mathbf{u},\mathbf{v}\rangle$

iv. $\langle \mathbf{u},\mathbf{u}\rangle = (A\mathbf{u})\cdot(A\mathbf{u}) = \|A\mathbf{u}\|^2 \geq 0$, and this quantity is zero if and only if the vector $A\mathbf{u}$ is $\mathbf{0}$. But $A\mathbf{u} = \mathbf{0}$ if and only if $\mathbf{u} = \mathbf{0}$, because A is invertible.

Another method for verifying the axioms is to use properties of the transpose operation. The calculations are similar. However, for (i), you need to use the fact that the transpose of a scalar (which is a 1×1 matrix) is the scalar itself: $\langle \mathbf{u},\mathbf{v}\rangle = \langle \mathbf{u},\mathbf{v}\rangle^T = [(A\mathbf{u})^T(A\mathbf{v})]^T = (A\mathbf{v})^T(A\mathbf{u})^{TT} = (A\mathbf{v})^T(A\mathbf{u}) = \langle \mathbf{v},\mathbf{u}\rangle$.

17. $\|\mathbf{u}+\mathbf{v}\|^2 = \langle \mathbf{u}+\mathbf{v},\mathbf{u}+\mathbf{v}\rangle = \langle \mathbf{u},\mathbf{u}+\mathbf{v}\rangle + \langle \mathbf{v},\mathbf{u}+\mathbf{v}\rangle$ — Axiom ii
$= \langle \mathbf{u}+\mathbf{v},\mathbf{u}\rangle + \langle \mathbf{u}+\mathbf{v},\mathbf{v}\rangle$ — Axiom i
$= \langle \mathbf{u},\mathbf{u}\rangle + \langle \mathbf{v},\mathbf{u}\rangle + \langle \mathbf{u},\mathbf{v}\rangle + \langle \mathbf{v},\mathbf{v}\rangle$ — Axiom ii
$= \langle \mathbf{u},\mathbf{u}\rangle + 2\langle \mathbf{u},\mathbf{v}\rangle + \langle \mathbf{v},\mathbf{v}\rangle$ — Axiom i

Next, replace $\mathbf{v}$ by $-\mathbf{v}$ and use the fact that $\mathbf{u} - \mathbf{v} = \mathbf{u} + (-1)\mathbf{v}$.

$$\begin{aligned}\|\mathbf{u} - \mathbf{v}\|^2 &= \langle \mathbf{u}, \mathbf{u}\rangle + 2\langle \mathbf{u}, -\mathbf{v}\rangle + \langle -\mathbf{v}, -\mathbf{v}\rangle && \text{Replacing v above by -v.}\\ &= \langle \mathbf{u}, \mathbf{u}\rangle - 2\langle \mathbf{u}, \mathbf{v}\rangle + (-1)^2\langle \mathbf{v}, \mathbf{v}\rangle && \text{Axiom iii and Exercise 15}\end{aligned}$$

Subtracting, $\|\mathbf{u} + \mathbf{v}\|^2 - \|\mathbf{u} - \mathbf{v}\|^2 = 2\langle \mathbf{u}, \mathbf{v}\rangle - (-2\langle \mathbf{u}, \mathbf{v}\rangle) = 4\langle \mathbf{u}, \mathbf{v}\rangle$. Division by 4 gives the desired identity.

19. The full solution is in the text.

25. In the space $C[-1, 1]$ with the integral inner product, the polynomials t and 1 are orthogonal, because

$$\langle t, 1\rangle = \int_{-1}^{1} t \cdot 1\, dt = \frac{1}{2}t^2\Big|_{-1}^{1} = \frac{1}{2}(1)^2 - \frac{1}{2}(-1)^2 = 0$$

So 1 and t can be in an orthogonal basis for Span$\{1, t, t^2\}$. Next, compute $\text{proj}_W\, t^2$, the orthogonal projection of the vector t^2 onto the subspace W spanned by 1 and t.

$$\langle t^2, 1\rangle = \int_{-1}^{1} t^2 \cdot 1\, dt = \frac{1}{3}t^3\Big|_{-1}^{1} = \frac{1}{3}(1)^3 - \frac{1}{3}(-1)^3 = \frac{2}{3}$$

$$\langle 1, 1\rangle = \int_{-1}^{1} 1 \cdot 1\, dt = t\Big|_{-1}^{1} = 1 - (-1) = 2$$

$$\langle t^2, t\rangle = \int_{-1}^{1} t^2 \cdot t\, dt = \frac{1}{4}t^4\Big|_{-1}^{1} = \frac{1}{4}(1)^4 - \frac{1}{4}(-1)^4 = 0$$

There is no need to compute $\langle t, t\rangle$, because t^2 is orthogonal to t. Thus

$$\text{proj}_W\, t^2 = \frac{\langle t^2, 1\rangle}{\langle 1, 1\rangle}1 + \frac{\langle t^2, t\rangle}{\langle t, t\rangle}t = \frac{2/3}{2}1 + 0 = \frac{1}{3}$$

A polynomial orthogonal to W is $t^2 - \text{proj}_W\, t^2 = t^2 - \frac{1}{3}$. Another choice is this polynomial scaled by 3, namely, $3t^2 - 1$. Thus, the polynomials, 1, t^2, and $3t^2 - 1$ form an orthogonal basis for Span$\{1, t, t^2\}$.

Can you find the next Legendre polynomial, a cubic polynomial that is orthogonal to each of the first three Legendre polynomials?

6.8 APPLICATIONS OF INNER PRODUCT SPACES

Of the three applications in this section, the discussion of Fourier series is by far the most important. Such series have great practical value, particularly in mathematics, engineering and the physical sciences. Calculations with Fourier series are simple because sine and cosine functions are orthogonal. This fact is often overlooked in undergraduate courses that do not assume a linear algebra background.

KEY IDEAS

The text gives the normal equations for the weighted least-squares solution of $A\mathbf{x} = \mathbf{y}$. When applied to a least-squares line problem, the most common situation, the normal equations are usually written as

$$(WX)^T WX\hat{\boldsymbol{\beta}} = (WX)^T \mathbf{y}$$

where W is the (diagonal) weighting matrix, X is the design matrix, $\hat{\boldsymbol{\beta}}$ is the least-squares parameter vector, and $\mathbf{y}$ is the observation vector.

Trend analysis is really a least-squares regression problem of the type described in Section 6.6, with data points $(x_1, y_1), \ldots, (x_n, y_n)$ fitted by a curve of the form

$$y = \beta_0 f_0(x) + \beta_1 f_1(x) + \cdots + \beta_k f_k(x)$$

where the functions $f_0, \ldots, f_k$ are polynomials that are orthogonal with respect to an inner product on $\mathbb{P}_{n-1}$ defined by

$$\langle p, q \rangle = p(x_1)q(x_1) + \cdots + p(x_n)q(x_n)$$

Usually, $x_1, \ldots, x_n$ are arranged to be evenly spaced and sum to zero, and the functions $f_1, \ldots, f_k$ are of degree 3 or 4 or less.

In $C[0, 2\pi]$ with the integral inner product, the set

$$\{1, \cos t, \cos 2t, \ldots, \cos nt, \sin t, \sin 2t, \ldots, \sin nt\} \qquad (*)$$

is orthogonal. The nth order Fourier approximation to some f in $C[0, 2\pi]$ is simply the orthogonal projection of f onto the subspace W of trigonometric polynomials spanned by the functions in (*). The Fourier coefficients of f are the weights in the usual formula for the orthogonal projection of f onto W.

If an application involves an interval $[0, T]$ instead of $[0, 2\pi]$, then the inner product requires an integral over $[0, T]$, and the appropriate orthogonal set is obtained by replacing t in each function in (*) with $2\pi t/T$.

SOLUTIONS TO EXERCISES

1. For the data (-2,0), (-1,0), (0,2), (1,4), (2,4), construct

$$X = \begin{bmatrix} 1 & -2 \\ 1 & -1 \\ 1 & 0 \\ 1 & 1 \\ 1 & 2 \end{bmatrix} \text{ Design matrix} \qquad \boldsymbol{\beta} = \begin{bmatrix} \beta_0 \\ \beta_1 \end{bmatrix} \text{ Parameter vector} \qquad \mathbf{y} = \begin{bmatrix} 0 \\ 0 \\ 2 \\ 4 \\ 4 \end{bmatrix} \text{ Observation vector}$$

Since the first and last data points are about half as reliable as the other points, a suitable weighting matrix is

$$W = \begin{bmatrix} 1 & 0 & 0 & 0 & 0 \\ 0 & 2 & 0 & 0 & 0 \\ 0 & 0 & 2 & 0 & 0 \\ 0 & 0 & 0 & 2 & 0 \\ 0 & 0 & 0 & 0 & 1 \end{bmatrix}, \text{ so } WX = \begin{bmatrix} 1 & -2 \\ 2 & -2 \\ 2 & 0 \\ 2 & 2 \\ 1 & 2 \end{bmatrix} \text{ and } W\mathbf{y} = \begin{bmatrix} 0 \\ 0 \\ 4 \\ 8 \\ 4 \end{bmatrix}$$

The remaining calculations are the same as in ordinary least-squares, except that the *weighted* design matrix WX and the *weighted* observation vector $W\mathbf{y}$ appear in place of X and $\mathbf{y}$, respectively.

$$(WX)^T(WX) = \begin{bmatrix} 1 & 2 & 2 & 2 & 1 \\ -2 & -2 & 0 & 2 & 2 \end{bmatrix} \begin{bmatrix} 1 & -2 \\ 2 & -2 \\ 2 & 0 \\ 2 & 2 \\ 1 & 2 \end{bmatrix} = \begin{bmatrix} 14 & 0 \\ 0 & 16 \end{bmatrix}$$

$$(WX)^T(W\mathbf{y}) = \begin{bmatrix} 1 & 2 & 2 & 2 & 1 \\ -2 & -2 & 0 & 2 & 2 \end{bmatrix} \begin{bmatrix} 0 \\ 0 \\ 4 \\ 8 \\ 4 \end{bmatrix} = \begin{bmatrix} 28 \\ 24 \end{bmatrix}$$

The normal equations and solution are

$$\begin{bmatrix} 14 & 0 \\ 0 & 16 \end{bmatrix} \begin{bmatrix} \beta_0 \\ \beta_1 \end{bmatrix} = \begin{bmatrix} 28 \\ 24 \end{bmatrix}, \qquad \begin{bmatrix} \beta_0 \\ \beta_1 \end{bmatrix} = \begin{bmatrix} 1/14 & 0 \\ 0 & 1/16 \end{bmatrix} \begin{bmatrix} 28 \\ 24 \end{bmatrix} = \begin{bmatrix} 2 \\ 3/2 \end{bmatrix}$$

The equation of the least-squares line is $y = 2 + (3/2)x$

7. $\|\cos kt\|^2 = \int_0^{2\pi} \cos kt \cdot \cos kt\,dt = \int_0^{2\pi} \frac{1 + \cos 2kt}{2}\,dt$

$$= \left[\frac{1}{2}t + \frac{\sin 2kt}{4k}\right]\Bigg|_0^{2\pi} = (\tfrac{1}{2}\cdot 2\pi + 0) - 0 = \pi \quad (\text{if } k \neq 0)$$

$\|\sin kt\|^2 = \int_0^{2\pi} \sin kt \cdot \sin kt\,dt = \int_0^{2\pi} \frac{1 - \cos 2kt}{2}\,dt$

$$= \left[\frac{1}{2}t - \frac{\sin 2kt}{4k}\right]\Bigg|_0^{2\pi} = (\tfrac{1}{2}\cdot 2\pi - 0) - 0 = \pi \quad (\text{if } k \neq 0)$$

9. $f(t) = 2\pi - t$. The definite integrals of $t\cos kt$ and $t\sin kt$, shown below, were computed in Example 4. The Fourier coefficients of f are:

$$\frac{a_0}{2} = \frac{1}{2}\cdot\frac{1}{\pi}\int_0^{2\pi}(2\pi - t)\,dt = \frac{1}{2\pi}(-\tfrac{1}{2})(2\pi - t)^2\Big|_0^{2\pi} = 0 + \frac{1}{4\pi}(2\pi)^2 = \pi$$

and for $k > 0$,

$$a_k = \frac{1}{\pi}\int_0^{2\pi}(2\pi - t)\cos kt\,dt = \frac{1}{\pi}\int_0^{2\pi} 2\pi\cos kt\,dt - \frac{1}{\pi}\int_0^{2\pi} t\cos kt\,dt$$

$$= 0 - 0 = 0$$

$$b_k = \frac{1}{\pi}\int_0^{2\pi}(2\pi - t)\sin kt\,dt = \frac{1}{\pi}\int_0^{2\pi} 2\pi\sin kt\,dt - \frac{1}{\pi}\int_0^{2\pi} t\sin kt\,dt$$

$$= 0 - (-\tfrac{2}{k}) = \frac{2}{k}$$

The third-order Fourier approximation to f is

$$\pi + 2\sin t + \sin 2t + \frac{2}{3}\sin 3t$$

13. Take f and g in $C[0,2\pi]$ and let m be a nonnegative integer. Then, the linearity of the inner product shows that

$$\langle (f + g), \cos mt\rangle = \langle f, \cos mt\rangle + \langle g, \cos mt\rangle$$

Dividing each term in this equality by $\langle \cos mt, \cos mt\rangle$, we conclude that the Fourier coefficient a_m of $f + g$ is the sum of the corresponding Fourier coefficients of f and of g. Similarly, the Fourier coefficient b_m of $f + g$ is the sum of the corresponding Fourier coefficients of f and of g.

Appendix: The Linearity of an Orthogonal Projection

The argument for Exercise 13 is a special case of a general principle. In any inner product space, the mapping $\mathbf{y} \mapsto \frac{\langle \mathbf{y}, \mathbf{u}\rangle}{\langle \mathbf{u}, \mathbf{u}\rangle}\mathbf{u}$ is linear, for any nonzero $\mathbf{u}$. To verify this, take any $\mathbf{x}$ and $\mathbf{y}$ in the space and any scalar c. Then

$$\frac{\langle \mathbf{x}+\mathbf{y}, \mathbf{u}\rangle}{\langle \mathbf{u}, \mathbf{u}\rangle}\mathbf{u} = \frac{\langle \mathbf{x}, \mathbf{u}\rangle}{\langle \mathbf{u}, \mathbf{u}\rangle}\mathbf{u} + \frac{\langle \mathbf{y}, \mathbf{u}\rangle}{\langle \mathbf{u}, \mathbf{u}\rangle}\mathbf{u}, \quad \text{and} \quad \frac{\langle c\mathbf{x}, \mathbf{u}\rangle}{\langle \mathbf{u}, \mathbf{u}\rangle}\mathbf{u} = \frac{c\langle \mathbf{x}, \mathbf{u}\rangle}{\langle \mathbf{u}, \mathbf{u}\rangle}\mathbf{u} = c\frac{\langle \mathbf{x}, \mathbf{u}\rangle}{\langle \mathbf{u}, \mathbf{u}\rangle}\mathbf{u}$$

Similarly, if $\mathbf{u}_1, \ldots, \mathbf{u}_p$ are any nonzero vectors, then the mapping

$$\mathbf{y} \mapsto \frac{\langle \mathbf{y}, \mathbf{u}_1\rangle}{\langle \mathbf{u}_1, \mathbf{u}_1\rangle}\mathbf{u}_1 + \cdots + \frac{\langle \mathbf{y}, \mathbf{u}_p\rangle}{\langle \mathbf{u}_p, \mathbf{u}_p\rangle}\mathbf{u}_p$$

is a linear transformation. Thus, if $\{\mathbf{u}_1, \ldots, \mathbf{u}_p\}$ is an orthogonal basis for a subspace W, then the mapping $\mathbf{y} \mapsto \text{proj}_W \mathbf{y}$ is a linear transformation. In particular, if W is the vector space of trigonometric polynomials of order at most n, and if f and g are in $C[0,2\pi]$, then

$$\text{proj}_W(f + g) = \text{proj}_W f + \text{proj}_W g$$

That is, the nth order Fourier approximation to $f + g$ is the sum of the nth order Fourier approximations to f and to g. Can you use the linearity of the mapping $f \mapsto \text{proj}_W f$ and the final result of Example 4, to produce (with practically no work) the answer to Exercise 9? [*Hint*: The nth order Fourier approximation to a constant function is the function itself.]

CHAPTER 6 SUPPLEMENTARY EXERCISES

1. Justifications for the answers to the True/False questions are given below. Other justifications are possible.

   a. False. The length of the zero vector is zero.

   b. True. See the box before Example 2 in Section 6.1.

   c. True, by definition of distance.

   d. False. $\|(-1)\mathbf{v}\| \neq (-1)\|\mathbf{v}\|$ for nonzero **v**.

   e. False. Orthogonal *nonzero* vectors are linearly independent.

   f. True. If $\mathbf{x}\cdot\mathbf{u} = 0$ and $\mathbf{x}\cdot\mathbf{v} = 0$, then $\mathbf{x}\cdot(\mathbf{u} - \mathbf{v}) = \mathbf{x}\cdot\mathbf{u} - \mathbf{x}\cdot\mathbf{v} = 0$.

   g. True. From the calculations on pages 373, $\|\mathbf{u} + \mathbf{v}\|^2 = \|\mathbf{u}\|^2 + \|\mathbf{v}\|^2 + 2\mathbf{u}\cdot\mathbf{v}$. Thus $\|\mathbf{u} + \mathbf{v}\|^2 = \|\mathbf{u}\|^2 + \|\mathbf{v}\|^2$ if and only if $\mathbf{u}\cdot\mathbf{v} = 0$.

   h. True. The same argument as in (g) applies, with **v** replaced by $-\mathbf{v}$.

   i. False. The orthogonal projection of **y** onto **u** is a scalar multiple of **u**, not **y** (except when **y** itself is a multiple of **u**).

   j. True, because the orthogonal projection of **y** (onto W) is in W.

   k. True. Given **w**, let $H = (\text{Span}\{\mathbf{w}\})^{\perp}$. Certainly **0** is in H. If **u**, **v** are in H, then for any scalars c, d, $(c\mathbf{u} + d\mathbf{v})\cdot\mathbf{w} = (c\mathbf{u})\cdot\mathbf{w} + (d\mathbf{v})\cdot\mathbf{w} = c(\mathbf{u}\cdot\mathbf{w}) + d(\mathbf{v}\cdot\mathbf{w}) = 0$. So $c\mathbf{u} + d\mathbf{v}$ is in H. Thus H is a subspace of $\mathbb{R}^n$.

   l. False. The zero vector is in both W and $W^{\perp}$.

   m. True. If $\mathbf{v}_i\cdot\mathbf{v}_j = 0$, then $(c_i\mathbf{v}_i)\cdot(c_j\mathbf{v}_j) = c_ic_j(\mathbf{v}_i\cdot\mathbf{v}_j) = c_ic_j(0) = 0$.

   n. False. The statement is true only for a *square* matrix. See Theorem 10 in Section 6.3.

   o. False. An orthogonal matrix is square and has ortho*normal* columns.

   p. True. See Exercises 27 and 28 in Section 6.2.

   q. True, by the Pythagorean Theorem and the Orthogonal Decomposition Theorem.

   r. False. A least-squares solution is a vector $\hat{\mathbf{x}}$ (not $A\hat{\mathbf{x}}$) such that the vector $A\hat{\mathbf{x}}$ is the closest point to **b** in Col A.

   s. False. By definition, the normal equations are $A^TA\hat{\mathbf{x}} = A^T\mathbf{b}$. When A^TA is invertible, $\hat{\mathbf{x}}$ is the *solution* of the normal equations.

Warning: Many students miss parts (r) and (s) of Exercise 1.

2. For $p = 1$ and $\mathbf{x} = c_1\mathbf{v}_1$, $\|\mathbf{x}\|^2 = |c_1|^2\|\mathbf{x}\|^2 = |c_1|^2$. The *Hint* in the text shows that the stated equality holds for $p = 2$. Suppose that the equality holds for $p = k$, with $k \geq 2$, and let $\{\mathbf{v}_1, \ldots, \mathbf{v}_{k+1}\}$ be an orthonormal set. Suppose

$$\mathbf{x} = c_1\mathbf{v}_1 + \cdots + c_{k+1}\mathbf{v}_{k+1} = \mathbf{u} + c_{k+1}\mathbf{v}_{k+1}$$

where $\mathbf{u} = c_1\mathbf{v}_1 + \cdots + c_k\mathbf{v}_k$. Since $\mathbf{v}_{k+1}\cdot\mathbf{v}_j = 0$ for $1 \leq j \leq k$, $\mathbf{v}_{k+1}$ is orthogonal to $\mathbf{u}$. By the Pythagorean theorem,

$$\|\mathbf{x}\|^2 = \|\mathbf{u}\|^2 + \|c_{k+1}\mathbf{v}_{k+1}\|^2 = \|\mathbf{u}\|^2 + |c_{k+1}|^2\|\mathbf{v}_{k+1}\|^2 = \|\mathbf{u}\|^2 + |c_{k+1}|^2$$

and by the assumed equality for $p = k$

$$\|\mathbf{u}\|^2 = |c_1|^2 + \cdots + |c_k|^2$$

Hence $\|\mathbf{x}\|^2 = |c_1|^2 + \cdots + |c_{k+1}|^2$, which shows that the equality for $p = k + 1$ follows from the equality for $p = k$. By the principle of induction, the equality is true for all integers $p \geq 2$.

7. If $Q = I - 2\mathbf{u}\mathbf{u}^T$, then Q is square and $Q^T = Q$, because $(\mathbf{u}\mathbf{u}^T)^T = \mathbf{u}^{TT}\mathbf{u}^T = \mathbf{u}\mathbf{u}^T$. By Exercise 13(c) in the Supplementary Exercises for Chapter 2, $Q^2 = I$. Thus, $Q^TQ = QQ = I$, which shows that Q is an orthogonal matrix, by Theorem 6 in Section 6.2.

10. Let $\mathbf{x} = \begin{bmatrix} x \\ y \\ z \end{bmatrix}$, $\mathbf{b} = \begin{bmatrix} a \\ b \\ c \end{bmatrix}$, $\mathbf{v} = \begin{bmatrix} 1 \\ -2 \\ 5 \end{bmatrix}$, and $A = \begin{bmatrix} \mathbf{v}^T \\ \mathbf{v}^T \\ \mathbf{v}^T \end{bmatrix} = \begin{bmatrix} 1 & -2 & 5 \\ 1 & -2 & 5 \\ 1 & -2 & 5 \end{bmatrix}$. Then the given set of equations is $A\mathbf{x} = \mathbf{b}$, and the set of all least-squares solutions coincides with the set of solutions of the normal equations $A^TA\mathbf{x} = A^T\mathbf{b}$. Now

$$A^TA\mathbf{x} = \begin{bmatrix} \mathbf{v} & \mathbf{v} & \mathbf{v} \end{bmatrix}\begin{bmatrix} \mathbf{v}^T \\ \mathbf{v}^T \\ \mathbf{v}^T \end{bmatrix}\mathbf{x} = (\mathbf{v}\mathbf{v}^T + \mathbf{v}\mathbf{v}^T + \mathbf{v}\mathbf{v}^T)\mathbf{x}$$ Column-row expansion in Section 2.4

$$= 3\mathbf{v}\mathbf{v}^T\mathbf{x} = 3\mathbf{v}(\mathbf{v}^T\mathbf{x}) = 3(x - 2y + 5z)\mathbf{v}$$

and

$$A^T\mathbf{b} = \begin{bmatrix} \mathbf{v} & \mathbf{v} & \mathbf{v} \end{bmatrix}\begin{bmatrix} a \\ b \\ c \end{bmatrix} = a\mathbf{v} + b\mathbf{v} + c\mathbf{v} = (a + b + c)\mathbf{v}$$

Thus the normal equations reduce to $3(x - 2y + 5z)\mathbf{v} = (a + b + c)\mathbf{v}$, which is equivalent to $x - 2y + 5z = (a + b + c)/3$, since $\mathbf{v} \neq \mathbf{0}$. The points on this plane are the least-squares solutions of the original system (which itself describes a set of planes, all parallel to the least-squares solution plane).

CHAPTER 6 GLOSSARY CHECKLIST

Check your knowledge by attempting to write definitions of the terms below. Then compare your work with the definitions given in the text's Glossary. Ask your instructor which definitions, if any, might appear on a test.

angle (between nonzero vectors **u** and **v** in $\mathbb{R}^2$ or $\mathbb{R}^3$): The angle ϑ between the .... Related to the scalar product by: $\mathbf{u}\cdot\mathbf{v}$ = ...

best approximation: The closest point ....

Cauchy-Schwarz inequality: ... for all **u**, **v**.

component of y orthogonal to u (for $\mathbf{u} \neq \mathbf{0}$): The vector ....

design matrix: The matrix X in the linear model ..., where the columns of X are determined in some way by ....

distance between u and v: ..., denoted by dist(**u**,**v**).

Fourier approximation (of order n): The closest point in ... to ....

Fourier coefficients: The weights used to make ....

Fourier series: An infinite series that ... in $C[0,2\pi]$, with the inner product given by a definite integral.

fundamental subspaces (of A): The ... space and ... space of A, and the ... space and ... space of A^T, with Col A^T commonly called the ....

general least-squares problem: Given an $m \times n$ matrix A and a vector **b** in $\mathbb{R}^m$, find ... such that ... for all ....

Gram-Schmidt process: An algorithm for producing ....

inner product: The scalar ..., usually written as $\mathbf{u}\cdot\mathbf{v}$, where **u** and **v** are vectors in $\mathbb{R}^n$ viewed as .... Also called the ... of **u** and **v**. In general, a function on a vector space that assigns to each pair of vectors **u** and **v** a number ..., subject to certain axioms.

inner product space: A vector space on which is defined ....

least-squares line: The line ... that minimizes ... in the equation ....

least-squares solution (of $A\mathbf{x} = \mathbf{b}$): A vector ... such that ....

length (of **v**): The scalar $\|\mathbf{x}\|$ = ...; also called the ... of **v**.

linear model (in statistics): Any equation of the form ..., where X and **y** are known and $\boldsymbol{\beta}$ is to be chosen to minimize ....

mean-deviation form (of a vector): A vector whose entries ....

multiple regression: A linear model involving ... variables and ....

normal equations: The system of equations represented by ..., whose solution yields all ... solutions of $A\mathbf{x} = \mathbf{b}$. In statistics, a common notation is ....

normalizing (a vector $\mathbf{v}$): The process of creating a ... vector $\mathbf{u}$ that ....

observation vector: The vector ... in the linear model $\mathbf{y} = X\boldsymbol{\beta} + \boldsymbol{\varepsilon}$, where the entries in ... are the observed values of ....

orthogonal basis: A basis that ....

orthogonal complement (of W): The set $W^{\perp}$ of ....

orthogonal matrix: A ... matrix U such that ....

orthogonal projection of y onto u (or onto the line through $\mathbf{u}$ and the origin, for $\mathbf{u} \neq \mathbf{0}$): The vector $\hat{\mathbf{y}}$ defined by ....

orthogonal projection of y onto W: The unique vector $\hat{\mathbf{y}}$ such that .... Notation: $\hat{\mathbf{y}} = \text{proj}_W \mathbf{y}$.

orthogonal set: A set S of vectors such that ... for ....

orthogonal to W: Orthogonal to every ....

orthonormal basis: A basis that is ....

orthonormal set: An ... set of ....

parameter vector: The unknown vector ... in the linear model ....

regression coefficients: The coefficients ... in the ....

residual vector: The quantity ... that appears in the general linear model: ..., the difference between ... and the ... values (of y).

QR factorization: A factorization of an $m \times n$ matrix A with linearly independent columns, $A = QR$, where Q is an ... matrix whose ..., and R is an ... matrix.

same direction (as a vector $\mathbf{v}$): A vector that is ....

scale (a vector): Multiply a vector by ....

trend analysis: The use of ... to fit data, with the inner product ....

triangle inequality: ....

trigonometric polynomial: A linear combination of ... and ... functions such as ....

unit vector: A vector $\mathbf{v}$ such that ....

weighted least squares: Least-squares problems with a ... inner product such as $\langle \mathbf{x}, \mathbf{y} \rangle = \ldots$.

7 SYMMETRIC MATRICES AND QUADRATIC FORMS

7.1 DIAGONALIZATION OF SYMMETRIC MATRICES

To prepare for this section, review Section 6.2 and the Diagonalization Theorem of Section 5.3, along with Example 3 and Theorem 6 of Section 5.3.

KEY IDEAS

If a symmetric matrix has distinct eigenvalues, as in Example 2, then the ordinary diagonalization process produces a matrix P with *orthogonal* columns, because eigenvectors from different eigenspaces are automatically orthogonal. However, the P you need here must have ortho*normal* columns. Forgetting to normalize the columns of P is the main error students make in this section.

The statements in the spectral theorem, together with the general approach used in Example 3, lead to the following outline for orthogonally diagonalizing any symmetric matrix.

Procedure for Orthogonally Diagonalizing a Symmetric Matrix A

1. *Find the characteristic equation of A.*
2. *For each eigenvalue with multiplicity 1, find an eigenvector and normalize it.*
3. *For each eigenvalue with multiplicity $k \geq 2$, find a basis of k vectors for the eigenspace. Then use the Gram-Schmidt process to construct an orthonormal basis.*
4. *The union of the bases for all the eigenspaces is an orthonormal basis for $\mathbb{R}^n$ (if A is $n \times n$). Use these vectors as the columns of an orthogonal matrix P.*

5. *Construct D from the eigenvalues, in an order corresponding to the columns of P. The number of times an eigenvalue appears on the diagonal of D is equal to the dimension of the corresponding eigenspace.*

6. *Finally,* $A = PDP^{-1} = PDP^T$.

SOLUTIONS TO EXERCISES

1. $A = \begin{bmatrix} 3 & 5 \\ 5 & -7 \end{bmatrix} = A^T$, because the (1,2) and (2,1) entries match. The entries on the main diagonal of A can have any values.

7. $P = \begin{bmatrix} .6 & .8 \\ .8 & -.6 \end{bmatrix} = [\mathbf{p}_1 \quad \mathbf{p}_2]$. To show that P is orthogonal by hand calculations, show that its columns are orthonormal: $\mathbf{p}_1 \cdot \mathbf{p}_2 = .48 - .48 = 0$, $\|\mathbf{p}_1\|^2 = (.6)^2 + (.8)^2 = 1$, and similarly, $\|\mathbf{p}_2\|^2 = 1$. Since P is square, P is an orthogonal matrix.

13. $A = \begin{bmatrix} 3 & 1 \\ 1 & 3 \end{bmatrix}$. Characteristic polynomial: $(3 - \lambda)^2 - 1 = \lambda^2 - 6\lambda + 8 = (\lambda - 4)(\lambda - 2)$. So the eigenvalues are 4 and 2.

<u>For $\lambda = 4$:</u> $[A - 4I \quad \mathbf{0}] = \begin{bmatrix} -1 & 1 & 0 \\ 1 & -1 & 0 \end{bmatrix} \sim \begin{bmatrix} 1 & -1 & 0 \\ 0 & 0 & 0 \end{bmatrix}$, $\begin{array}{l} x_1 = x_2 \\ x_2 \text{ is free} \end{array}$

Take $x_2 = 1$ to get a basis for the eigenspace: $\begin{bmatrix} 1 \\ 1 \end{bmatrix}$. Then normalize to get a unit vector: $\mathbf{u}_1 = \begin{bmatrix} 1/\sqrt{2} \\ 1/\sqrt{2} \end{bmatrix}$. (Don't forget this step.)

<u>For $\lambda = 2$:</u> $[A - 2I \quad \mathbf{0}] = \begin{bmatrix} 1 & 1 & 0 \\ 1 & 1 & 0 \end{bmatrix} \sim \begin{bmatrix} 1 & 1 & 0 \\ 0 & 0 & 0 \end{bmatrix}$, $\begin{array}{l} x_1 = -x_2 \\ x_2 \text{ is free} \end{array}$

Take $x_2 = 1$ to get a basis for the eigenspace: $\begin{bmatrix} -1 \\ 1 \end{bmatrix}$. Then normalize to get a unit vector: $\mathbf{u}_2 = \begin{bmatrix} -1/\sqrt{2} \\ 1/\sqrt{2} \end{bmatrix}$.

Set $P = [\mathbf{u}_1 \quad \mathbf{u}_2] = \begin{bmatrix} 1/\sqrt{2} & -1/\sqrt{2} \\ 1/\sqrt{2} & 1/\sqrt{2} \end{bmatrix}$. The corresponding D is $\begin{bmatrix} 4 & 0 \\ 0 & 2 \end{bmatrix}$.

Study Tip: The fact that eigenvectors for distinct eigenvalues are orthogonal gives you a check on your work. After you find $\mathbf{u}_2$ in Exercise 13, verify that $\mathbf{u}_2 \cdot \mathbf{u}_1 = 0$. When $\mathbf{u}_1$ and $\mathbf{u}_2$ are in $\mathbb{R}^2$, you can easily guess what $\mathbf{u}_2$

must be, once you know $\mathbf{u}_1$. If you do this, you should compute $A\mathbf{u}_2$, to make sure that $\mathbf{u}_2$ is indeed an eigenvector.

19. Be sure to *work* this problem before reading the solution. Use Exercises 13—24 to sharpen your skills. They are critical for the rest of the chapter. Here, $A = \begin{bmatrix} 3 & 2 & 0 \\ 2 & 4 & 2 \\ 0 & 2 & 5 \end{bmatrix}$, and the eigenvalues are given: 7, 4, 1.

<u>For $\lambda = 7$:</u> $[A - 7I \quad \mathbf{0}] = \begin{bmatrix} -4 & 2 & 0 & 0 \\ 2 & -3 & 2 & 0 \\ 0 & 2 & -2 & 0 \end{bmatrix} \sim \begin{bmatrix} 1 & 0 & -1/2 & 0 \\ 0 & 1 & -1 & 0 \\ 0 & 0 & 0 & 0 \end{bmatrix}$

$x_1 = (1/2)x_3$, $x_2 = x_3$, and x_3 is free. Take $x_3 = 2$ to avoid fractions.

A basis for the eigenspace: $\begin{bmatrix} 1 \\ 2 \\ 2 \end{bmatrix}$; a unit eigenvector: $\mathbf{u}_1 = \begin{bmatrix} 1/3 \\ 2/3 \\ 2/3 \end{bmatrix}$.

<u>For $\lambda = 4$:</u> $[A - 4I \quad \mathbf{0}] = \begin{bmatrix} -1 & 2 & 0 & 0 \\ 2 & 0 & 2 & 0 \\ 0 & 2 & 1 & 0 \end{bmatrix} \sim \begin{bmatrix} 1 & 0 & 1 & 0 \\ 0 & 1 & 1/2 & 0 \\ 0 & 0 & 0 & 0 \end{bmatrix}$

$x_1 = -x_3$, $x_2 = -(1/2)x_3$, and x_3 is free. Take $x_3 = 2$ to avoid fractions.

A basis for the eigenspace: $\begin{bmatrix} -2 \\ -1 \\ 2 \end{bmatrix}$; a unit eigenvector: $\mathbf{u}_2 = \begin{bmatrix} -2/3 \\ -1/3 \\ 2/3 \end{bmatrix}$.

(Remember to check mentally that $\mathbf{u}_2 \cdot \mathbf{u}_1 = 0$.)

<u>For $\lambda = 1$:</u> $[A - I \quad \mathbf{0}] = \begin{bmatrix} 2 & 2 & 0 & 0 \\ 2 & 3 & 2 & 0 \\ 0 & 2 & 4 & 0 \end{bmatrix} \sim \begin{bmatrix} 1 & 0 & -2 & 0 \\ 0 & 1 & 2 & 0 \\ 0 & 0 & 0 & 0 \end{bmatrix}$

$x_1 = 2x_3$, $x_2 = -2x_3$, and x_3 is free. Take $x_3 = 1$.

A basis for the eigenspace: $\begin{bmatrix} 2 \\ -2 \\ 1 \end{bmatrix}$; a unit eigenvector: $\mathbf{u}_3 = \begin{bmatrix} 2/3 \\ -2/3 \\ 1/3 \end{bmatrix}$.

(Also, check that $\mathbf{u}_3 \cdot \mathbf{u}_1 = 0$ and $\mathbf{u}_3 \cdot \mathbf{u}_2 = 0$.)

Set $P = [\mathbf{u}_1 \quad \mathbf{u}_2 \quad \mathbf{u}_3] = \begin{bmatrix} 1/3 & -2/3 & 2/3 \\ 2/3 & -1/3 & -2/3 \\ 2/3 & 2/3 & 1/3 \end{bmatrix}$, and $D = \begin{bmatrix} 7 & 0 & 0 \\ 0 & 4 & 0 \\ 0 & 0 & 1 \end{bmatrix}$.

Because the A in this exercise has three distinct eigenvalues, the only freedom you have in choosing P is to rearrange its columns (and change

D accordingly) and to multiply any column of P by -1 (because the eigenvectors must have *unit* length). Notice how the matrix P above differs from the P in the text's answer section.

Study Tips: The matrices in Exercises 21 and 22 have only two distinct eigenvalues, so the dimension of one eigenspace must be at least two. (Why?) You will have to construct an orthonormal basis for the eigenspace. Exercises 23 and 24 are good models for exam questions because they give you information from which you can quickly diagonalize A (orthogonally).

25. a. See Theorem 2 and the paragraph preceding the theorem.

b. See Theorem 1. **c.** See Example 3.

d. See the paragraph following formula (2) and exercise 35.

29. By hypothesis, $A = PDP^{-1}$, where P is orthogonal and D is diagonal. Since A is invertible, 0 is not an eigenvalue and D is invertible. Then

$$A^{-1} = (PDP^{-1})^{-1} = (P^{-1})^{-1}D^{-1}P^{-1} = PD^{-1}P^{-1}$$

Since D^{-1} is diagonal, A^{-1} is orthogonally diagonalizable.

A second argument: By Theorem 2, A is symmetric. Since A is invertible, a property of transposes shows that $(A^{-1})^T = (A^T)^{-1} = A^{-1}$, so A^{-1} is symmetric. By Theorem 2, again, A^{-1} is orthogonally diagonalizable.

31. The solution is in the text.

35. a. The matrix $B = \mathbf{u}\mathbf{u}^T$ is an outer product, or rank 1 matrix. Given $\mathbf{x}$ in $\mathbb{R}^n$, $B\mathbf{x} = (\mathbf{u}\mathbf{u}^T)\mathbf{x} = \mathbf{u}(\mathbf{u}^T\mathbf{x}) = (\mathbf{u}^T\mathbf{x})\mathbf{u}$, because $\mathbf{u}^T\mathbf{x}$ is a scalar. Using dot products, $B\mathbf{x} = (\mathbf{x}\cdot\mathbf{u})\mathbf{u}$. Since $\mathbf{u}$ is a unit vector, this *is* the orthogonal projection of $\mathbf{x}$ onto $\mathbf{u}$. See Section 6.2.

b. B is symmetric, because $B^T = (\mathbf{u}\mathbf{u}^T)^T = \mathbf{u}^{TT}\mathbf{u}^T = \mathbf{u}\mathbf{u}^T = B$. Also, $B^2 = (\mathbf{u}\mathbf{u}^T)(\mathbf{u}\mathbf{u}^T) = \mathbf{u}(\mathbf{u}^T\mathbf{u})\mathbf{u}^T = \mathbf{u}\mathbf{u}^T = B$, because $\mathbf{u}^T\mathbf{u} = 1$.

MATLAB Orthogonal Diagonalization

The command **[P D] = eig(A)** orthogonally diagonalizes any symmetric matrix A, but you miss the opportunity to learn the procedure of this section. Instead, use **eig(A)** for eigenvalues and **nulbasis** to obtain eigenvectors, as in Section 5.3. If you encounter a two-dimensional eigenspace, with a basis $\{\mathbf{v}_1, \mathbf{v}_2\}$, use the command

```
v2 = v2 - (v2'*v1)/(v1'*v1)*v1
```

or

```
v2 = v2 - proj(v2,v1)
```

to make the new eigenvector $\mathbf{v}_2$ orthogonal to $\mathbf{v}_1$. (Review Section 6.2.) The command **proj** was introduced in the MATLAB note for Section 6.4. After you normalize the eigenvectors and create P, check that $P^TP = I$ to verify that P is indeed an orthogonal matrix.

7.2 QUADRATIC FORMS

This section, together with Section 7.1, forms the foundation for the rest of the chapter.

KEY IDEAS

The main point here is to learn how a change of variable, $\mathbf{x} = P\mathbf{u}$, with P an orthogonal matrix, can transform a quadratic form into a new quadratic form with no cross-product terms.

If you study the various classes of quadratic forms (or, equivalently, classes of symmetric matrices), you should learn both the definitions and the characterizations (in Theorem 5) of these classes. Exercise 24 describes another useful way to characterize quadratic forms. (The 2×2 case can be generalized to $n\times n$ matrices.)

SOLUTIONS TO EXERCISES

1. a. $\mathbf{x}^TA\mathbf{x} = [x_1 \quad x_2]\begin{bmatrix} 5 & 1/3 \\ 1/3 & 1 \end{bmatrix}\begin{bmatrix} x_1 \\ x_2 \end{bmatrix} = [x_1 \quad x_2]\begin{bmatrix} 5x_1 + (1/3)x_2 \\ (1/3)x_1 + x_2 \end{bmatrix}$

$= 5x_1^2 + (2/3)x_1x_2 + x_2^2$

b. When $\mathbf{x} = (6,1)$, $\mathbf{x}^TA\mathbf{x} = 5(6)^2 + (2/3)(6)(1) + (1)^2 = 185$

c. When $\mathbf{x} = (1,3)$, $\mathbf{x}^TA\mathbf{x} = 5(1)^2 + (2/3)(1)(3) + (3)^2 = 16$

7. The matrix of the quadratic form is $A = \begin{bmatrix} 1 & 5 \\ 5 & 1 \end{bmatrix}$. The characteristic polynomial is $\lambda^2 - 2\lambda - 24 = (\lambda - 6)(\lambda + 4)$; eigenvalues are 6 and -4.

<u>For $\lambda = 6$:</u> an eigenvector is $\begin{bmatrix} 1 \\ 1 \end{bmatrix}$, normalized: $\mathbf{u}_1 = \frac{1}{\sqrt{2}}\begin{bmatrix} 1 \\ 1 \end{bmatrix}$

For $\lambda = -4$: an eigenvector is $\begin{bmatrix} -1 \\ 1 \end{bmatrix}$, normalized: $\mathbf{u}_2 = \frac{1}{\sqrt{2}}\begin{bmatrix} -1 \\ 1 \end{bmatrix}$

Thus $A = PDP^{-1}$ and $D = P^{-1}AP = P^TAP$ when $P = \frac{1}{\sqrt{2}}\begin{bmatrix} 1 & -1 \\ 1 & 1 \end{bmatrix}$ and $D = \begin{bmatrix} 6 & 0 \\ 0 & -4 \end{bmatrix}$.

The desired change of variable is $\mathbf{x} = P\mathbf{y}$, so that

$$\mathbf{x}^TA\mathbf{x} = (P\mathbf{y})^TA(P\mathbf{y}) = \mathbf{y}^TP^TAP\mathbf{y} = \mathbf{y}^TD\mathbf{y} \qquad (*)$$

$$= 6y_1^2 - 4y_2^2$$

Study Tip: To make the "change of variable" requested in Exercise 7, you should: (1) write the equation $\mathbf{x} = P\mathbf{y}$ and specify P; (2) show the matrix algebra in (*) that produces the new quadratic form; and (3) include the new quadratic form. Find out how much of this information you should supply if a problem like Exercise 7 were to appear on an exam.

13. The matrix of the quadratic form is $A = \begin{bmatrix} 1 & -3 \\ -3 & 9 \end{bmatrix}$. The characteristic polynomial is $\lambda^2 - 10\lambda = \lambda(\lambda - 10)$; the eigenvalues are 10 and 0. Thus the quadratic form is positive semidefinite. To find the change of variable, proceed as in Exercise 7:

For $\lambda = 10$: an eigenvector is $\begin{bmatrix} 1 \\ -3 \end{bmatrix}$, normalized: $\mathbf{u}_1 = \frac{1}{\sqrt{10}}\begin{bmatrix} 1 \\ -3 \end{bmatrix}$

For $\lambda = 0$: an eigenvector is $\begin{bmatrix} 3 \\ 1 \end{bmatrix}$, normalized: $\mathbf{u}_2 = \frac{1}{\sqrt{10}}\begin{bmatrix} 3 \\ 1 \end{bmatrix}$

Take $P = \frac{1}{\sqrt{10}}\begin{bmatrix} 1 & 3 \\ -3 & 1 \end{bmatrix}$ and $D = \begin{bmatrix} 10 & 0 \\ 0 & 0 \end{bmatrix}$. Since P orthogonally diagonalizes A, the desired change of variable is $\mathbf{x} = P\mathbf{y}$, and

$$\mathbf{x}^TA\mathbf{x} = (P\mathbf{y})^TA(P\mathbf{y}) = \mathbf{y}^TP^TAP\mathbf{y} = \mathbf{y}^TD\mathbf{y} = 10y_1^2$$

The new quadratic form is $10y_1^2$.

19. Because 8 is larger than 5, you should make the x_2^2 term as large as possible. The constraint $x_1^2 + x_2^2 = 1$ keeps x_2 from exceeding 1. When $x_1 = 0$ and $x_2 = 1$, the value of the quadratic form is $5(0) + 8(1) = 8$.

21. **a.** See the definition before Example 1.

b. See the paragraph following Example 3.

c. See the Principal Axes Theorem and the Diagonalization Theorem (in Section 5.3).

d. Check the definition. **e.** See Theorem 5.

f. See the Numerical Note after Example 6.

25. The text's answer showed that $\mathbf{x}^T B^T B\mathbf{x} \geq 0$ for all $\mathbf{x}$. To show $B^T B$ is positive definite, suppose that $\mathbf{x}^T B^T B\mathbf{x} = 0$. Then $(B\mathbf{x})^T B\mathbf{x} = 0$, so that $\|B\mathbf{x}\|^2 = 0$ and $B\mathbf{x} = 0$. If B is invertible, then $\mathbf{x} = \mathbf{0}$, which shows that, in this case, the form $\mathbf{x}^T B^T B\mathbf{x}$ is positive definite.

27. The quadratic forms $\mathbf{x}^T A\mathbf{x}$ and $\mathbf{x}^T B\mathbf{x}$ are both positive definite, by Theorem 5, because all eigenvalues of A and B are positive. Then for any nonzero $\mathbf{x}$, $\mathbf{x}^T(A + B)\mathbf{x} = \mathbf{x}^T(A\mathbf{x} + B\mathbf{x}) = \mathbf{x}^T A\mathbf{x} + \mathbf{x}^T B\mathbf{x} > 0$, so the quadratic form $\mathbf{x}^T(A + B)\mathbf{x}$ is positive definite. Also, the matrix $A + B$ is symmetric, because $(A + B)^T = A^T + B^T = A + B$. By Theorem 5, $A + B$ has positive eigenvalues.

Mastering Linear Algebra Concepts: Diagonalization and Quadratic Forms

Since the end of the course draws near, I recommend that you prepare a review sheet that contrasts the Diagonalization Theorem in Section 5.3 with Theorem 2 in Section 7.1. You might begin by copying the statements of these theorems, and then use the following questions to guide your review:

- What special properties does an orthogonal diagonalization have that are not present in all diagonalizations? Consider the eigenvalues, the eigenspaces, the eigenvectors, and the matrix P in PDP^{-1}.

- Suppose A is symmetric, and all eigenspaces are one-dimensional. What differences are there between a general diagonalization and an orthogonal diagonalization of A? What differences are there when A is symmetric and one eigenspace is two-dimensional?

- Why is an orthogonal diagonalization needed to simplify a quadratic form? Make the change of variable $\mathbf{x} = P\mathbf{y}$ in $\mathbf{x}^T A\mathbf{x}$ and show the algebra involved.

- If you studied Section 5.6 or 5.7, compare Figure 4 in Section 5.6 with Figure 3 in Section 5.7. How do the general shapes of the trajectories differ? Why do they differ? Could Figure 5 in Section 5.6 or Figure 5 in Section 5.7 be associated with a symmetric matrix?

7.3 CONSTRAINED OPTIMIZATION

This section is important in its own right, since constrained optimization problems arise in many mathematical problems and applications. The main results of the section are also used in the following two sections.

KEY IDEAS

Theorem 6 gives the main idea. The maximum value of a quadratic form $\mathbf{x}^T A\mathbf{x}$ over the set of all unit vectors can be computed by finding the greatest eigenvalue of A; this maximum value is attained at a corresponding eigenvector. The key step in the proof is to diagonalize A by P, substitute $\mathbf{x} = P\mathbf{y}$, and use the fact that in this case, $\mathbf{x}$ and $\mathbf{y}$ have the same norm.

Example 6 presents a topic that is widely discussed in elementary economics texts.

SOLUTIONS TO EXERCISES

1. We are given an equality of two quadratic forms:

$$5x_1^2 + 6x_2^2 + 7x_3^2 + 4x_1x_2 - 4x_2x_3 = 9y_1^2 + 6y_2^2 + 3y_3^2$$

The matrix of the left quadratic form obviously is $A = \begin{bmatrix} 5 & 2 & 0 \\ 2 & 6 & -2 \\ 0 & -2 & 7 \end{bmatrix}$.

The equality between the two quadratic forms indicates that the eigenvalues of A are 9,6,3. (Proof: The diagonal matrix D of the quadratic form $9y_1^2 + 6y_2^2 + 3y_3^2$ obviously has eigenvalues 9,6,3. Since A is similar to D, A has the same eigenvalues as D.) The standard calculations produce a unit eigenvector for each eigenvalue. Don't forget to normalize each eigenvector.

$$\lambda = 9: \quad \mathbf{u}_1 = \begin{bmatrix} 1/3 \\ 2/3 \\ -2/3 \end{bmatrix}; \qquad \lambda = 6: \quad \mathbf{u}_2 = \begin{bmatrix} 2/3 \\ 1/3 \\ 2/3 \end{bmatrix}; \qquad \lambda = 3: \quad \mathbf{u}_3 = \begin{bmatrix} -2/3 \\ 2/3 \\ 1/3 \end{bmatrix}$$

These eigenvectors are mutually orthogonal because they correspond to distinct eigenvalues. So the desired change of variable is

$$\mathbf{x} = P\mathbf{y}, \text{ where } P = \begin{bmatrix} 1/3 & 2/3 & -2/3 \\ 2/3 & 1/3 & 2/3 \\ -2/3 & 2/3 & 1/3 \end{bmatrix}$$

Study Tip: Review the matrix algebra that leads from $\mathbf{x}^T A\mathbf{x}$ to $\mathbf{y}^T D\mathbf{y}$. Also, be sure you can show that $\|\mathbf{x}\| = \|\mathbf{y}\|$ when P is an orthogonal matrix.

7. The matrix of $Q(\mathbf{x}) = -2x_1^2 - x_2^2 + 4x_1x_2 + 4x_2x_3$ is $A = \begin{bmatrix} -2 & 2 & 0 \\ 2 & -1 & 2 \\ 0 & 2 & 0 \end{bmatrix}$. The hint in the exercise lists 2, -1, and -4 as the eigenvalues. The greatest eigenvalue is 2, not -4, because "greatest" here refers to the eigenvalue that is farthest to the right on the real line. The maximum value of $Q(\mathbf{x})$ (for $\mathbf{x}$ a unit vector) is attained at a unit eigenvector for $\lambda = 2$. Standard calculations produce the eigenvector:

$$\mathbf{v}_1 = \begin{bmatrix} 1/2 \\ 1 \\ 1 \end{bmatrix}, \text{ scaled to } \begin{bmatrix} 1 \\ 2 \\ 2 \end{bmatrix}, \text{ and normalized to } \mathbf{u}_1 = \begin{bmatrix} 1/3 \\ 2/3 \\ 2/3 \end{bmatrix}.$$

Warning: Exercise 7 illustrates the potential error of selecting -4 instead of 2 as the greatest eigenvalue.

12. This exercise can be done by using a theorem, but try to do it by direct computation, using the hint in the text.

13. If $m = M$ and $\mathbf{x}$ is the unit eigenvector $\mathbf{u}_1$, then $\mathbf{x}^TA\mathbf{x} = \mathbf{u}_1^TA\mathbf{u}_1 = \mathbf{u}_1^T(m\mathbf{u}_1) = m$. Otherwise, set $\alpha = (t - m)/(M - m)$. The hint in the text shows that $0 \le \alpha \le 1$ when $m \le t \le M$. For such α, let $\mathbf{x} = \sqrt{1-\alpha}\,\mathbf{u}_n + \sqrt{\alpha}\,\mathbf{u}_1$. Then the vectors $\sqrt{1-\alpha}\,\mathbf{u}_n$ and $\sqrt{\alpha}\,\mathbf{u}_1$ are orthogonal because they are eigenvectors for different eigenvalues (or one of the vectors is $\mathbf{0}$). By the Pythagorean Theorem,

$$\begin{aligned} \mathbf{x}^T\mathbf{x} = \|\mathbf{x}\|^2 &= \|\sqrt{1-\alpha}\,\mathbf{u}_n\|^2 + \|\sqrt{\alpha}\,\mathbf{u}_1\|^2 \\ &= |1 - \alpha|\,\|\mathbf{u}_n\|^2 + |\alpha|\,\|\mathbf{u}_1\|^2 \\ &= (1 - \alpha) + \alpha = 1 \end{aligned}$$

because $\mathbf{u}_n$ and $\mathbf{u}_1$ are unit vectors and $0 \le \alpha \le 1$. Also, using the fact that $\mathbf{u}_n$ and $\mathbf{u}_1$ are orthogonal, we have

$$\begin{aligned} \mathbf{x}^TA\mathbf{x} &= (\sqrt{1-\alpha}\,\mathbf{u}_n + \sqrt{\alpha}\,\mathbf{u}_1)^TA(\sqrt{1-\alpha}\,\mathbf{u}_n + \sqrt{\alpha}\,\mathbf{u}_1) \\ &= (\sqrt{1-\alpha}\,\mathbf{u}_n + \sqrt{\alpha}\,\mathbf{u}_1)^T(m\sqrt{1-\alpha}\,\mathbf{u}_n + M\sqrt{\alpha}\,\mathbf{u}_1) \\ &= |1 - \alpha|\,m\mathbf{u}_n^T\mathbf{u}_n + |\alpha|\,M\mathbf{u}_1^T\mathbf{u}_1 = (1 - \alpha)m + \alpha M = t \end{aligned}$$

Thus the quadratic form $\mathbf{x}^TA\mathbf{x}$ assumes every value between m and M for a suitable unit vector $\mathbf{x}$.

7.4 THE SINGULAR VALUE DECOMPOSITION

This section is the capstone of the text. It completes the story of the linear transformation $\mathbf{x} \mapsto A\mathbf{x}$ for a general $m \times n$ matrix A and, in so doing, gives you an opportunity to review many basic concepts from Chapters 4–7. In addition, this section opens the door into the modern world of applied linear algebra. An understanding of the singular value decomposition is essential for advanced work in science and engineering that requires matrix computations.

KEY IDEAS

The first singular value σ_1 of an $m \times n$ matrix A is the maximum of $\|A\mathbf{x}\|$ over all unit vectors. This maximum value is attained at a unit eigenvector $\mathbf{v}_1$ of A^TA corresponding to the greatest eigenvalue λ_1 of A^TA. The second singular value is the maximum of $\|A\mathbf{x}\|$ over all unit vectors orthogonal to $\mathbf{v}_1$. The following algorithm produces the singular value decomposition for A. (As mentioned in the text, other more reliable methods are used in professional software.)

Procedure for Computing a Singular Value Decomposition

1. Find an orthonormal basis $\{\mathbf{v}_1, \ldots, \mathbf{v}_n\}$ for $\mathbb{R}^n$ consisting of eigenvectors of A^TA, arranged so that the associated eigenvalues satisfy $\lambda_1 \geq \cdots \geq \lambda_r > 0$ and $\lambda_{r+1} = \cdots = 0$, where $r = \text{rank } A$.
2. Construct the $n \times n$ orthogonal matrix $V = [\mathbf{v}_1 \quad \cdots \quad \mathbf{v}_n]$.
3. Let $\sigma_j = \sqrt{\lambda_j}$ $(1 \leq j \leq n)$, and construct the $m \times n$ diagonal matrix Σ whose (j, j)-entry is σ_j $(1 \leq j \leq n)$ and has zeros elsewhere.
4. The set $\{A\mathbf{v}_1, \ldots, A\mathbf{v}_r\}$ is orthogonal and $\sigma_j = \|A\mathbf{v}_j\|$. Compute $\mathbf{u}_j = (1/\sigma_j)A\mathbf{v}_j$ $(1 \leq j \leq r)$.
5. Extend $\{\mathbf{u}_1, \ldots, \mathbf{u}_r\}$ to an orthonormal basis $\{\mathbf{u}_1, \ldots, \mathbf{u}_m\}$ for $\mathbb{R}^m$. Write the $m \times m$ orthogonal matrix $U = [\mathbf{u}_1 \quad \cdots \quad \mathbf{u}_m]$.
6. $A = U\Sigma V^T$.

The diagram on the next page illustrates how the SVD splits the action of A into first multiplication by V^T (which amounts to an orthogonal change of basis in $\mathbb{R}^n$), then a scaling by Σ in the directions of the standard basis vectors $\mathbf{e}_1, \ldots, \mathbf{e}_n$, and finally multiplication by U (an orthogonal change of basis in $\mathbb{R}^m$).

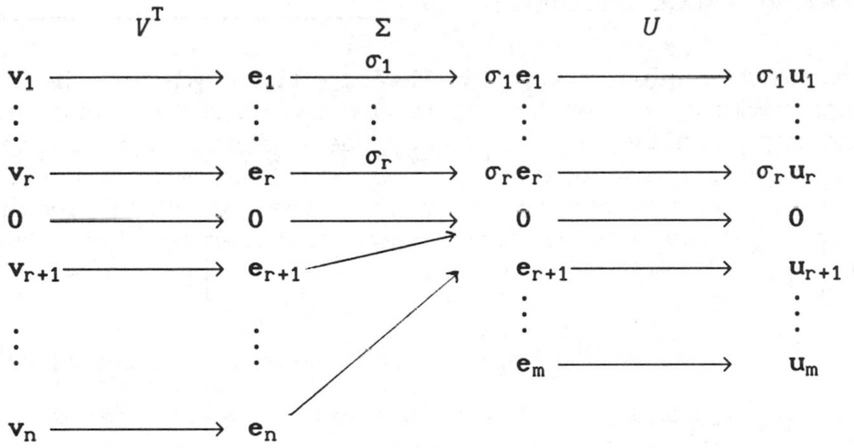

The Singular Value Decomposition: $A = U\Sigma V^T$

Note that the so-called *left* singular vectors of A are the columns of U (because U appears on the left in the factorization of A), even though $\mathbf{u}_1, \ldots, \mathbf{u}_n$ appear on the right side of the diagram above.

SOLUTIONS TO EXERCISES

1. $A = \begin{bmatrix} 1 & 0 \\ 0 & -3 \end{bmatrix}$, $A^TA = \begin{bmatrix} 1 & 0 \\ 0 & 9 \end{bmatrix}$. Eigenvalues and eigenvectors of A^TA are:

$$\lambda_1 = 9: \quad \mathbf{v}_1 = \begin{bmatrix} 0 \\ 1 \end{bmatrix}; \qquad \lambda_2 = 1: \quad \mathbf{v}_2 = \begin{bmatrix} 1 \\ 0 \end{bmatrix}$$

(Remember to arrange the eigenvalues in decreasing order.) Thus

$$V = \begin{bmatrix} 0 & 1 \\ 1 & 0 \end{bmatrix}$$

The singular values are $\sigma_1 = \sqrt{9} = 3$ and $\sigma_2 = 1$. The matrix Σ is the same shape as A, and

$$\Sigma = \begin{bmatrix} \sigma_1 & 0 \\ 0 & \sigma_2 \end{bmatrix} = \begin{bmatrix} 3 & 0 \\ 0 & 1 \end{bmatrix}$$

Next, computc

$$A\mathbf{v}_1 = \begin{bmatrix} 1 & 0 \\ 0 & -3 \end{bmatrix}\begin{bmatrix} 0 \\ 1 \end{bmatrix} = \begin{bmatrix} 0 \\ -3 \end{bmatrix}, \qquad A\mathbf{v}_2 = \begin{bmatrix} 1 & 0 \\ 0 & -3 \end{bmatrix}\begin{bmatrix} 1 \\ 0 \end{bmatrix} = \begin{bmatrix} 1 \\ 0 \end{bmatrix}$$

and normalize:

$$\mathbf{u}_1 = \frac{1}{\sigma_1}A\mathbf{v}_1 = \frac{1}{3}\begin{bmatrix}0\\-3\end{bmatrix} = \begin{bmatrix}0\\-1\end{bmatrix}, \quad \mathbf{u}_2 = \frac{1}{\sigma_2}\begin{bmatrix}1\\0\end{bmatrix} = \begin{bmatrix}1\\0\end{bmatrix}$$

Finally, $\{\mathbf{u}_1, \mathbf{u}_2\}$ is already a basis for $\mathbb{R}^2$, so the basis for $\mathbb{R}^2$ is complete, and

$$U = \begin{bmatrix}0 & 1\\-1 & 0\end{bmatrix}, \text{ and } A = U\Sigma V^T = \begin{bmatrix}0 & 1\\-1 & 0\end{bmatrix}\begin{bmatrix}3 & 0\\0 & 1\end{bmatrix}\begin{bmatrix}0 & 1\\1 & 0\end{bmatrix}$$

This happens to equal V.

7. $A = \begin{bmatrix}2 & -1\\2 & 2\end{bmatrix}$. $A^TA = \begin{bmatrix}8 & 2\\2 & 5\end{bmatrix}$. Find the eigenvalues of A^TA from the characteristic equation.

$$0 = \lambda^2 - 13\lambda + 36 = (\lambda - 9)(\lambda - 4); \quad \lambda_1 = 9, \ \lambda_2 = 4$$

Corresponding unit eigenvectors for A^TA (calculations omitted) are:

$$\lambda_1 = 9: \quad \mathbf{v}_1 = \begin{bmatrix}2/\sqrt{5}\\1/\sqrt{5}\end{bmatrix}; \qquad \lambda_2 = 4: \quad \mathbf{v}_2 = \begin{bmatrix}-1/\sqrt{5}\\2/\sqrt{5}\end{bmatrix}$$

Take

$$V = \begin{bmatrix}2/\sqrt{5} & -1/\sqrt{5}\\1/\sqrt{5} & 2/\sqrt{5}\end{bmatrix}$$

The singular values are $\sigma_1 = \sqrt{9} = 3$ and $\sigma_2 = \sqrt{4} = 2$. The matrix Σ is the same shape as A and $\Sigma = \begin{bmatrix}\sigma_1 & 0\\0 & \sigma_2\end{bmatrix} = \begin{bmatrix}3 & 0\\0 & 2\end{bmatrix}$. Next, compute

$$A\mathbf{v}_1 = \begin{bmatrix}2 & -1\\2 & 2\end{bmatrix}\begin{bmatrix}2/\sqrt{5}\\1/\sqrt{5}\end{bmatrix} = \begin{bmatrix}3/\sqrt{5}\\6/\sqrt{5}\end{bmatrix}, \quad A\mathbf{v}_2 = \begin{bmatrix}2 & -1\\2 & 2\end{bmatrix}\begin{bmatrix}-1/\sqrt{5}\\2/\sqrt{5}\end{bmatrix} = \begin{bmatrix}-4/\sqrt{5}\\2/\sqrt{5}\end{bmatrix}$$

To check your work at this point, verify that $A\mathbf{v}_1$ and $A\mathbf{v}_2$ are orthogonal. (They are.) Then normalize:

$$\mathbf{u}_1 = \frac{1}{\sigma_1}A\mathbf{v}_1 = \frac{1}{3}\begin{bmatrix}3/\sqrt{5}\\6/\sqrt{5}\end{bmatrix} = \begin{bmatrix}1/\sqrt{5}\\2/\sqrt{5}\end{bmatrix}, \quad \mathbf{u}_2 = \frac{1}{\sigma_2} = \frac{1}{2}\begin{bmatrix}-4/\sqrt{5}\\2/\sqrt{5}\end{bmatrix} = \begin{bmatrix}-2/\sqrt{5}\\1/\sqrt{5}\end{bmatrix}$$

Since $\{\mathbf{u}_1, \mathbf{u}_2\}$ is a basis for $\mathbb{R}^2$, take $U = \begin{bmatrix}1/\sqrt{5} & -2/\sqrt{5}\\2/\sqrt{5} & 1/\sqrt{5}\end{bmatrix}$. Thus

$$A = U\Sigma V^T = \begin{bmatrix}1/\sqrt{5} & -2/\sqrt{5}\\2/\sqrt{5} & 1/\sqrt{5}\end{bmatrix}\begin{bmatrix}3 & 0\\0 & 2\end{bmatrix}\begin{bmatrix}2/\sqrt{5} & 1/\sqrt{5}\\-1/\sqrt{5} & 2/\sqrt{5}\end{bmatrix}$$

Use V^T, not V.

Study Tip: Your answer for a singular value decomposition may differ from that given in the text. To check your work, compute AV and $U\Sigma$. If $AV = U\Sigma$, then $A = U\Sigma V^T$ and your answer is correct (provided U and V truly are orthogonal matrices).

13. The matrix A^TA is 3×3. Because the text has not given you practice computing and solving a cubic characteristic equation, the *Hint* suggests that you consider A^T instead of A. (You are free to work on A itself, if you prefer.) Using A^T, compute

$$(A^T)^TA^T = AA^T = \begin{bmatrix} 3 & 2 & 2 \\ 2 & 3 & -2 \end{bmatrix}\begin{bmatrix} 3 & 2 \\ 2 & 3 \\ 2 & -2 \end{bmatrix} = \begin{bmatrix} 17 & 8 \\ 8 & 17 \end{bmatrix}$$

The characteristic equation is

$$0 = \lambda^2 - 34\lambda + 225 = (\lambda - 25)(\lambda - 9); \quad \lambda_1 = 25, \ \lambda_2 = 9$$

The corresponding unit eigenvectors and the matrix V are

$$\lambda_1 = 25: \ \mathbf{v}_1 = \begin{bmatrix} 1/\sqrt{2} \\ 1/\sqrt{2} \end{bmatrix}; \quad \lambda_2 = 9: \ \mathbf{v}_2 = \begin{bmatrix} -1/\sqrt{2} \\ 1/\sqrt{2} \end{bmatrix}; \quad V = \begin{bmatrix} 1/\sqrt{2} & -1/\sqrt{2} \\ 1/\sqrt{2} & 1/\sqrt{2} \end{bmatrix}$$

The singular values are $\sigma_1 = 5$ and $\sigma_2 = 3$. Thus Σ is $\begin{bmatrix} 5 & 0 \\ 0 & 3 \\ 0 & 0 \end{bmatrix}$, the same size as A^T. To get $\mathbf{u}_1$ and $\mathbf{u}_2$, compute $A^T\mathbf{v}_1$ and $A^T\mathbf{v}_2$,

$$A^T[\mathbf{v}_1 \quad \mathbf{v}_2] = \begin{bmatrix} 3 & 2 \\ 2 & 3 \\ 2 & -2 \end{bmatrix}\begin{bmatrix} 1/\sqrt{2} & -1/\sqrt{2} \\ 1/\sqrt{2} & 1/\sqrt{2} \end{bmatrix} = \begin{bmatrix} 5/\sqrt{2} & -1/\sqrt{2} \\ 5/\sqrt{2} & 1/\sqrt{2} \\ 0 & -4/\sqrt{2} \end{bmatrix}$$

and normalize:

$$\mathbf{u}_1 = \begin{bmatrix} 1/\sqrt{2} \\ 1/\sqrt{2} \\ 0 \end{bmatrix}, \quad \mathbf{u}_2 = \begin{bmatrix} -1/\sqrt{18} \\ 1/\sqrt{18} \\ -4/\sqrt{18} \end{bmatrix}$$

We need one more vector, orthogonal to $\mathbf{u}_1$ and $\mathbf{u}_2$. So write the equations $\mathbf{u}_1^T\mathbf{x} = 0$ and $\mathbf{u}_2^T\mathbf{x} = 0$ and solve for $\mathbf{x}$. Simpler equations are

$$\begin{aligned} \sqrt{2}\mathbf{u}_1^T\mathbf{x} &= 0 \\ \sqrt{18}\mathbf{u}_2^T\mathbf{x} &= 0 \end{aligned} \quad \text{or} \quad \begin{aligned} x_1 + x_2 \phantom{{}- 4x_3} &= 0 \\ -x_1 + x_2 - 4x_3 &= 0 \end{aligned}$$

The solution is $x_1 = -2x_3$, $x_2 = 2x_3$, x_3 free. A suitable unit vector is

$$\mathbf{u}_3 = \begin{bmatrix} -2/3 \\ 2/3 \\ 1/3 \end{bmatrix}$$

Thus an SVD of A^T is

$$A^T = [\mathbf{u}_1 \quad \mathbf{u}_2 \quad \mathbf{u}_3] \begin{bmatrix} 5 & 0 \\ 0 & 3 \\ 0 & 0 \end{bmatrix} [\mathbf{v}_1 \quad \mathbf{v}_2]^T$$

So an SVD of A appears by taking transposes:

$$A = \begin{bmatrix} 1/\sqrt{2} & -1/\sqrt{2} \\ 1/\sqrt{2} & 1/\sqrt{2} \end{bmatrix} \begin{bmatrix} 5 & 0 & 0 \\ 0 & 3 & 0 \end{bmatrix} \begin{bmatrix} 1/\sqrt{2} & 1/\sqrt{2} & 0 \\ -1/\sqrt{18} & 1/\sqrt{18} & -4/\sqrt{18} \\ -2/3 & 2/3 & 1/3 \end{bmatrix}$$

This *is* an SVD because the outside matrices are orthogonal matrices, and the center matrix is a diagonal matrix of the proper type. Another way to find $\mathbf{u}_3$ is to realize that $\mathbf{u}_1$ and $\mathbf{u}_2$ form an orthonormal basis for Col A^T = Row A. The remaining $\mathbf{u}_3$ must be a basis for $(\text{Row } A)^\perp$ = Nul A.

19. Let $A = U\Sigma V^T$. Then

$$A^TA = (U\Sigma V^T)^T U\Sigma V^T = V\Sigma^T U^T U\Sigma V^T$$

$$= V(\Sigma^T\Sigma)V^{-1} \qquad \text{Because U and V are orthogonal}$$

If $\sigma_1, \ldots, \sigma_r$ are the nonzero diagonal entries in Σ, then $\Sigma^T\Sigma$ is diagonal, with diagonal entries $\sigma_1^2, \ldots, \sigma_r^2$, and possibly some zeros. Thus V diagonalizes A^TA. By the Diagonalization Theorem in Section 5.3, the columns of V are eigenvectors of A^TA, and $\sigma_1^2, \ldots, \sigma_r^2$ are the nonzero eigenvalues of A^TA. Hence $\sigma_1, \ldots, \sigma_r$ are the nonzero singular values of A. A similar calculation of AA^T shows that the columns of U are eigenvectors of AA^T.

23. From the proof of Theorem 10, $U\Sigma = [\sigma_1\mathbf{u}_1 \ \cdots \ \sigma_r\mathbf{u}_r \quad \mathbf{0} \ \cdots \ \mathbf{0}]$. The column-row expansion of a matrix product shows that

$$A = (U\Sigma)V^T = (U\Sigma)\begin{bmatrix} \mathbf{v}_1^T \\ \vdots \\ \mathbf{v}_n^T \end{bmatrix} = \sigma_1\mathbf{u}_1\mathbf{v}_1^T + \cdots + \sigma_r\mathbf{u}_r\mathbf{v}_r^T$$

This expansion generalizes the spectral decomposition in Section 7.1.

Study Tip: In Exercise 23, the *left* singular vectors are the columns $\mathbf{u}_1, \ldots, \mathbf{u}_m$ of the *left* factor U in $U\Sigma V^T$, but the *right* singular vectors are the columns of V, *not* V^T.

25. Consider the SVD for the standard matrix of T, say, $A = U\Sigma V^T$. Let $\mathcal{B} = \{\mathbf{v}_1, \ldots, \mathbf{v}_n\}$ and $\mathcal{C} = \{\mathbf{u}_1, \ldots, \mathbf{u}_m\}$ be bases constructed from the columns of V and U, respectively. Observe that, since the columns of V are orthonormal, $V^T\mathbf{v}_j = \mathbf{e}_j$, where $\mathbf{e}_j$ is the jth column of the $n \times n$ identity matrix. To find the matrix of T relative to $\mathcal{B}$ and $\mathcal{C}$, compute

$$T(\mathbf{v}_j) = A\mathbf{v}_j = U\Sigma V^T\mathbf{v}_j = U\Sigma\mathbf{e}_j = U\sigma_j\mathbf{e}_j = \sigma_j U\mathbf{e}_j = \sigma_j\mathbf{u}_j$$

So $[T(\mathbf{v}_j)]_{\mathcal{C}} = \sigma_j\mathbf{e}_j$. Formula (4) in the discussion at the beginning of Section 5.4 shows that the "diagonal" matrix Σ is the matrix of T relative to $\mathcal{B}$ and $\mathcal{C}$.

MATLAB The Singular Value Decomposition

The command **[P D] = eig(A'*A)** produces an orthogonal matrix P of eigenvectors and a diagonal matrix D of eigenvalues of A^TA, but the eigenvalues in D may not be in decreasing order. In such a case, you will have to rearrange things to form V and Σ (denoted below by S).

For instance, if P is 3×3, the command

```
V = P(:,[1 3 2])
```

interchanges columns 2 and 3 of P to form V. The commands

```
S = zeros(size(A)); S(2,2) = sqrt(D(3,3))
```

produce a zero matrix for "Σ" the same size as A and place the square root of the (3,3)-entry of D into the (2,2)-entry of S. Other diagonal entries for S can be entered similarly. To form U for the SVD, normalize the nonzero columns of $A*V$. If U needs more columns, use the method of Example 4.

This construction helps you to think about properties of the factorization. In practical work, however, you should use the much faster and more numerically reliable command **[U,S,V] = svd(A)** .

7.5 APPLICATIONS TO IMAGE PROCESSING AND STATISTICS

If you find remote sensing or image processing interesting, or if you plan to use multivariate statistics later in your career, then you will want to study this section thoroughly. You may have difficulty finding an elementary explanation of this material elsewhere. The idea for the application to image processing came from a student in my linear algebra class—a geography major who was taking an undergraduate course in remote sensing. The book by Lillesand and Kiefer, referenced in the text, was one of the texts for her course.

KEY IDEAS

The **first principal component** of the data in the matrix of observations is a unit eigenvector $\mathbf{u}_1$ corresponding to the largest eigenvalue of the covariance matrix S. If $\mathbf{u}_1 = (c_1, \ldots, c_p)$, then the entries in $\mathbf{u}_1$ are weights in a linear combination of the original variables, $x_1, \ldots, x_p$, that creates a new variable y_1 (sometimes called a composite score or *index*):

$$y_1 = \mathbf{u}_1^T\mathbf{X} = c_1x_1 + \cdots + c_px_p$$

The variance of the values of this index is the largest possible among all indices whose coefficients $c_1, \ldots, c_p$ form a unit vector. (The variance of y_1 is the largest eigenvalue of S.) The **second principal component** is the unit eigenvector corresponding to the second largest eigenvalue of S. The entries in the second principal component determine the index with greatest variance among all possible indices (from a unit vector) that are uncorrelated (in a statistical sense) with y_1. Additional principal components are defined similarly.

Checkpoints: (1) If the variables x_1 and x_3 are uncorrelated, what can you say about the covariance matrix S? (2) What is the covariance matrix of the new variables $y_1, \ldots, y_p$ formed from the principal components of S?

SOLUTIONS TO EXERCISES

1. The matrix of observations is $X = \begin{bmatrix} 19 & 22 & 6 & 3 & 2 & 20 \\ 12 & 6 & 9 & 15 & 13 & 5 \end{bmatrix}$, and the sample mean $\mathbf{M}$ is $\begin{bmatrix} 12 \\ 10 \end{bmatrix}$. Subtract $\mathbf{M}$ from each column of X to obtain

$$B = \begin{bmatrix} 7 & 10 & -6 & -9 & -10 & 8 \\ 2 & -4 & -1 & 5 & 3 & -5 \end{bmatrix}$$

The sample covariance matrix is

$$S = \frac{1}{N-1}BB^T = \frac{1}{5}\begin{bmatrix} 7 & 10 & -6 & -9 & -10 & 8 \\ 2 & -4 & -1 & 5 & 3 & -5 \end{bmatrix}\begin{bmatrix} 7 & 2 \\ 10 & -4 \\ -6 & -1 \\ -9 & 5 \\ -10 & 3 \\ 8 & -5 \end{bmatrix}$$

$$= \frac{1}{5}\begin{bmatrix} 430 & -135 \\ -135 & 80 \end{bmatrix} = \begin{bmatrix} 86 & -27 \\ -27 & 16 \end{bmatrix}$$

Usually, S contains decimals.

Study Tip: Note that the formula for the sample mean involves division by N, but for statistical reasons, the covariance matrix formula involves division by $N-1$.

7. Let x_1, x_2 denote the variables for the two-dimensional data in Exercise 1. The characteristic equation of the covariance matrix S from Exercise 1 is $\lambda^2 - 102\lambda + 647 = 0$. By the quadratic formula, the roots of this equation are $\lambda_1 = 95.20$ and $\lambda_2 = 6.80$ (to two decimal places). The first principal component of the data is a unit eigenvector corresponding to λ_1, which turns out to be $(-.95, .32)$, or $(.95, -.32)$. The two possible choices for the new variable are $y_1 = -.95x_1 + .32x_2$ and $y_1 = .95x_1 - .32x_2$. The variance of y_1 is 95.20, while the total variance is $95.20 + 6.80 = 102$. Since $95.20/102 = .933$, the new variable y_1 explains about 93.3% of the variance in the data.

11. **a.** The solution in the text shows that the $\mathbf{Y}_k$ are in mean-deviation form, where $\mathbf{Y}_k = P^T\mathbf{X}_k$ for some $p \times p$ matrix P.

b. By part (a), the covariance matrix of $\mathbf{Y}_1, \ldots, \mathbf{Y}_N$ is

$$\frac{1}{N-1}[\mathbf{Y}_1 \ \cdots \ \mathbf{Y}_N][\mathbf{Y}_1 \ \cdots \ \mathbf{Y}_N]^T$$

$$= \frac{1}{N-1}P^T[\mathbf{X}_1 \ \cdots \ \mathbf{X}_N]\left(P^T[\mathbf{X}_1 \ \cdots \ \mathbf{X}_N]\right)^T$$

$$= P^T\left(\frac{1}{N-1}[\mathbf{X}_1 \ \cdots \ \mathbf{X}_N][\mathbf{X}_1 \ \cdots \ \mathbf{X}_N]^T\right)P$$

$$= P^TSP$$

because $\mathbf{X}_1, \ldots, \mathbf{X}_N$ are in mean-deviation form.

13. Let $\mathbf{M}$ be the sample mean of the data, and for $k = 1,\ldots,N$, write $\hat{\mathbf{X}}_k$ for $\mathbf{X}_k - \mathbf{M}$. Let $B = [\hat{\mathbf{X}}_1 \ \cdots \ \hat{\mathbf{X}}_N]$, the matrix of observations in mean deviation form. By the column-row expansion of BB^T, the sample covariance matrix is

$$S = \frac{1}{N-1}BB^T = \frac{1}{N-1}[\hat{\mathbf{X}}_1 \ \cdots \ \hat{\mathbf{X}}_N]\begin{bmatrix}\hat{\mathbf{X}}_1^T \\ \vdots \\ \hat{\mathbf{X}}_N^T\end{bmatrix}$$

$$= \frac{1}{N-1}\sum_{1}^{N} \hat{\mathbf{X}}_k\hat{\mathbf{X}}_k^T = \frac{1}{N-1}\sum_{1}^{N}(\mathbf{X}_k - \mathbf{M})(\mathbf{X}_k - \mathbf{M})^T$$

Answers to Checkpoints: (1) The (1,3)-entry and (3,1)-entry of S are zero. (2) The covariance matrix of $y_1,\ldots,y_p$ is the diagonal matrix formed from the eigenvalues of S. This matrix is diagonal because the new variables are pairwise uncorrelated.

MATLAB Computing Principal Components

The command **mean(X')** produces a row vector whose jth entry lists the average of the jth row of X, and **diag(mean(X'))** creates a diagonal matrix whose diagonal entries are the row averages of X. (Be careful not to use **mean(X)** , which lists the averages of the *columns* of X.) Finally, the command **diag(mean(X'))*ones(size(X))** creates a matrix the size of X, whose columns are all the same, each one listing the row averages of X. To convert the data in X into mean-deviation form, use

```
B = X - diag(mean(X'))*ones(size(X))
```

The sample covariance matrix is produced by

```
S = B*B'/(N-1)
```

The principal component data is produced by

```
[U,D,V] = svd(B'/sqrt(N-1))
```

The columns of V are the principal components of the data, and the diagonal entries of **D^2** list the variances of the new variates.

CHAPTER 7 SUPPLEMENTARY EXERCISES

1. Justifications for the answers to the True/False questions are given below. Other justifications are possible.

a. True, by Theorem 2 in Section 7.1.

b. False. Counterexample: $A = \begin{bmatrix} 0 & -1 \\ 1 & 0 \end{bmatrix}$.

c. True. $\|A\mathbf{x}\|^2 = (A\mathbf{x})^T(A\mathbf{x}) = \mathbf{x}^TA^TA\mathbf{x} = \mathbf{x}^T\mathbf{x} = \|\mathbf{x}\|^2$, so $\|A\mathbf{x}\| = \|\mathbf{x}\|$, all $\mathbf{x}$.

d. False. The principal axes of $\mathbf{x}^TA\mathbf{x}$ are the columns of any *orthogonal* matrix P that diagonalizes A. Note: When A has an eigenvalue whose eigenspace has dimension greater than 1 (for example, when $A = I$), then the principal axes are not uniquely determined.

e. False. Counterexample: $P = \begin{bmatrix} 1 & -1 \\ 1 & 1 \end{bmatrix}$. If P is an $n \times n$ matrix with ortho*normal* columns, then $P^T = P^{-1}$.

f. False. See Exercise 11 in Section 7.2.

g. False. Counterexample: If $A = \begin{bmatrix} 2 & 0 \\ 0 & -3 \end{bmatrix}$ and $\mathbf{x} = \begin{bmatrix} 1 \\ 0 \end{bmatrix}$, then $\mathbf{x}^TA\mathbf{x} = 2 > 0$, but $\mathbf{x}^TA\mathbf{x}$ is an indefinite quadratic form.

h. True, by the Principal Axes Theorem, since the matrix of any quadratic form can be taken to be symmetric.

i. False. See Example 3 in Section 7.3.

j. False. The maximum value must computed over the set of *unit* vectors; otherwise $\mathbf{x}^TA\mathbf{x}$ can be made as large as desired.

k. False. Any orthogonal change of variable $\mathbf{x} = P\mathbf{y}$ changes a positive definite form $\mathbf{x}^TA\mathbf{x}$ into another positive definite form, because the the matrix of the new quadratic form is $P^{-1}AP$, which is similar to A and therefore has the same eigenvalues as A.

l. False. The term "definite eigenvalue" is undefined.

m. True. If $\mathbf{x} = P\mathbf{y}$, then $\mathbf{x}^TA\mathbf{x} = (P\mathbf{y})^TA(P\mathbf{y}) = \mathbf{x}^TP^TAP\mathbf{y} = \mathbf{x}^TP^{-1}AP\mathbf{y}$.

n. False. Counterexample: Let $U = \begin{bmatrix} 1 & -1 \\ 1 & 1 \end{bmatrix}$. The columns of U must be ortho*normal* to make $UU^T\mathbf{x}$ the orthogonal projection of $\mathbf{x}$ onto Col U.

o. True. See Examples 1 and 2 in Section 7.4.

p. True. See Theorem 10 in Section 7.4.

q. False. The singular values of $A = \begin{bmatrix} 2 & 0 \\ 0 & 1 \end{bmatrix}$ are 2 and 1, but the singular values of $A^TA = \begin{bmatrix} 4 & 0 \\ 0 & 1 \end{bmatrix}$ are 4 and 1.

6. Since A is symmetric, there is an orthonormal eigenvector basis $\{\mathbf{u}_1, \ldots, \mathbf{u}_n\}$ for $\mathbb{R}^n$. Let $r = \text{rank } A$. We may assume that $0 < r < n$, because otherwise the decomposition of Exercise 4(b) is either $\mathbf{y} = \mathbf{0} + \mathbf{y}$ or $\mathbf{y} = \mathbf{y} + \mathbf{0}$, and we are finished.

Note that $\dim \text{Nul } A = n - r$, so 0 is an eigenvalue with multiplicity $n - r$, and we can assume that $\mathbf{u}_1, \ldots, \mathbf{u}_r$ are the unit eigenvectors corresponding to the remaining r nonzero eigenvalues. (See the solution to Exercise 3.) By Exercise 5, $\mathbf{u}_1, \ldots, \mathbf{u}_r$ are in Col A. Also, $\mathbf{u}_{r+1}, \ldots, \mathbf{u}_n$ are in Nul A, because these vectors are eigenvectors for $\lambda = 0$. For any $\mathbf{y}$ in $\mathbb{R}^n$, there are scalars c_j such that

$$\mathbf{y} = \underbrace{c_1\mathbf{u}_1 + \cdots + c_r\mathbf{u}_r}_{\hat{\mathbf{y}}} + \underbrace{c_{r+1}\mathbf{u}_{r+1} + \cdots + c_n\mathbf{u}_n}_{\mathbf{z}}$$

The vector $\hat{\mathbf{y}}$, shown above, is in Col A and the vector $\mathbf{z}$ is in Nul A.

7. If $A = R^TR$, where R is invertible, then A is positive definite, by Exercise 25 in Section 7.2. Conversely, suppose that A is positive definite. Then by Exercise 26 in Section 7.2, $A = B^TB$ for some positive definite matrix B. Since the eigenvalues of B are positive, 0 is not an eigenvalue and so B is invertible. In particular, the columns of B are linearly independent. By Theorem 12 in Section 6.4, $B = QR$ for some $n \times n$ matrix Q with orthonormal columns and some upper triangular matrix R with positive elements on its diagonal. Since Q is square, $Q^TQ = I$. So

$$A = B^TB = (QR)^T(QR) = R^TQ^TQR = R^TR$$

and R has the required properties.

8. Suppose that A is positive definite, and let A have a Cholesky factorization, $A = R^TR$, with R upper triangular and positive entries on its diagonal. Let D be the diagonal matrix whose diagonal entries are the entries on the diagonal of R. Since right-multiplication by a diagonal matrix scales the columns of the matrix on its left, the matrix $L = R^TD^{-1}$ is lower triangular with 1's on its diagonal. If $U = DR$, then $A = R^TD^{-1}DR = LU$.

11. Start with an SVD decomposition, $A = U\Sigma V^T$. Since U is orthogonal, $U^TU = I$, and so $A = U\Sigma U^T UV^T = PQ$, where $P = U\Sigma U^T = U\Sigma U^{-1}$ and $Q = UV^T$. The matrix P is symmetric, because Σ is symmetric, and P has nonnegative eigenvalues because it is similar to Σ (which is diagonal with nonnegative entries). Thus P is positive semidefinite. The matrix Q is orthogonal because it is the product of orthogonal matrices.

CHAPTER 7 GLOSSARY CHECKLIST

Check your knowledge by attempting to write definitions of the terms below. Then compare your work with the definitions given in the text's Glossary. Ask your instructor which definitions, if any, might appear on a test.

condition number (of A): The quotient σ_1/σ_n, where ....

covariance (of variables x_i and x_j, for $i \neq j$): The entry ... in the covariance matrix S for a matrix of observations, where x_i and x_j vary over the ... coordinates, respectively of the observation vectors.

covariance matrix (or **sample covariance matrix**): The $p \times p$ matrix S defined by $S = \ldots$, where B is a $p \times N$ matrix of observations ....

indefinite matrix: A symmetric matrix A such that ....

indefinite quadratic form: A quadratic form Q such that $Q(\mathbf{x})$ ....

left singular vectors (of A): The columns of ... in the singular value decomposition $A = \ldots$.

matrix of observations: A $p \times N$ matrix whose columns are ..., each column listing p measurements made on ....

mean-deviation form (of a matrix of observations): A matrix whose ... vectors are ....

Moore-Penrose inverse: *See* pseudoinverse.

negative definite matrix: A symmetric matrix A such that ....

negative definite quadratic form: A quadratic form Q such that $Q(\mathbf{x})$ ....

negative semidefinite matrix: A symmetric matrix A such that ....

negative semidefinite quadratic form: A quadratic form Q such that ....

orthogonally diagonalizable: A matrix A that admits a factorization, $A = PDP^{-1}$, with P ... and D ....

positive definite matrix: A symmetric matrix A such that ....

positive definite quadratic form: A quadratic form Q such that $Q(\mathbf{x})$ ....

positive semidefinite matrix: A symmetric matrix A such that ....

positive semidefinite quadratic form: A quadratic form Q such that ....

principal axes (of a quadratic form $\mathbf{x}^T A\mathbf{x}$): The orthonormal columns of an orthogonal matrix P such that ....

principal components (of the data in a matrix of observations B): The ... eigenvectors of a sample covariance matrix S for B, with the eigenvectors arranged so that the corresponding ....

projection matrix (or **orthogonal projection matrix**): A symmetric matrix B such that .... A simple example is $B =$ ....

pseudoinverse (of A): The matrix ..., when UDV^T is a reduced singular value decomposition of A.

quadratic form: A function Q defined for $\mathbf{x}$ in $\mathbb{R}^n$ by $Q(\mathbf{x}) =$ ..., where A is an $n\times n$ ... matrix A (called the matrix of the quadratic form).

reduced singular value decomposition: A factorization $A =$ ..., for an $m\times n$ matrix A of rank r, where U is $\_\_\times\_\_$ with orthonormal columns, D is $\_\_\times\_\_$ with ..., and V is $\_\_\times\_\_$ with orthonormal columns.

right singular vectors (of A): The columns of ... in the singular value decomposition $A =$ ....

row sum: The sum of the entries ....

sample mean: The average M of a set of vectors, $\mathbf{X}_1, \ldots, \mathbf{X}_N$, given by $\mathbf{M} =$ ....

singular value decomposition (of an $m\times n$ matrix A): $A =$ ..., where U is an $\_\_\times\_\_$ ... matrix, V is an $\_\_\times\_\_$ ... matrix, and Σ is an $\_\_\times\_\_$ matrix with ....

singular values (of A): The ... of the eigenvalues of ..., arranged ....

spectral decomposition (of A): A representation $A =$ ... , where ....

symmetric matrix: A matrix A such that ....

total variance: The ... of the covariance matrix S of a matrix of ....

uncorrelated variables: Any two variables x_i and x_j (with $i \neq j$) that range over the ith and jth coordinates of the observation vectors in an observation matrix, such that ....

variance of a variable x_j: The diagonal entry ... in the ... matrix S for a matrix of observations, where x_j varies over the jth coordinates of the ....

INDEX OF MATLAB COMMANDS AND OPERATIONS

Notes for the TI-85 Graphics Calculator

ENTERING MATRICES. There are several ways to enter matrices and vectors into the TI-85 calculator, two such methods are: direct entry as an expression, and through the use of the matrix (vector, respectively) editor found in the [MATRX] ([VECTR], respectively) menu. Below are examples that illustrate the creation of matrices and vectors using these two methods. The entries in a matrix or vector can be either real or complex (pairs of real numbers). Some special matrices and vectors are created with other commands found in the [MATRX] ([VECTR], respectively) menu, we illustrate these in the appropriate notes. For other entry methods and additional information consult the TI-85 Graphics Calculator Guidebook.

Direct Entry. Begin a matrix with a square bracket, each row of the matrix must be inside a matching pair of square brackets, and the entries in the rows are separated by commas. Figure 1 below shows how you can enter the matrix $\begin{bmatrix} 1 & 2 & 3 \\ 4 & 5 & 6 \\ 7 & 8 & 9 \end{bmatrix}$ directly and assign to it the name MAT1 (case sensitive)—the symbol → appears after pressing the sto▸ key, the closing ']]' is not necessary at the end of the command)—press ENTER to display the matrix MAT1 on the screen. For complex numbers, follow the same procedure, only enter the entries as pairs of real numbers.

```
[[1,2,3][4,5,6][7,8,9
]]→MAT1
                [[1 2 3]
                 [4 5 6]
                 [7 8 9]]
█
```

Figure 1 MAT1 entered directly.

The Matrix Editor. Press 2nd (yellow key), [MATRX] (7 key), to see EDIT in the menu. The editor allows you to create a matrix by specifying a variable name, a size—row size and column size—and the entries of the matrix. The editor, Figure 2, is accessed by pressing the F2 menu key. Select from the matrix names shown or enter a new name for the matrix. For example to enter the matrix $Q = \begin{bmatrix} -1 & 0 & 4 \\ 5 & 1 & 3 \end{bmatrix}$, type Q and press ENTER, your screen should similar to Figure 3.

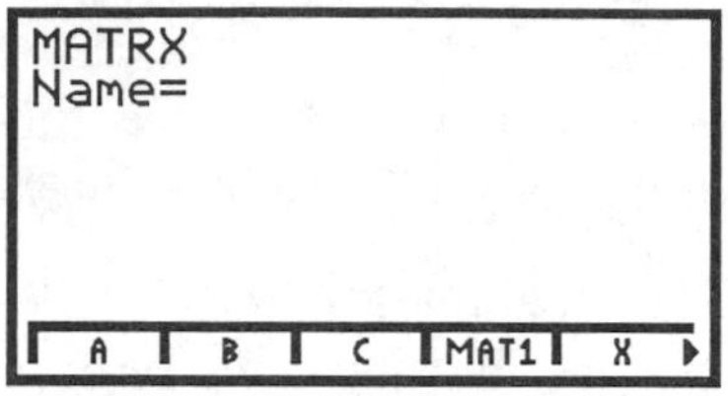

Figure 2 The matrix editor.

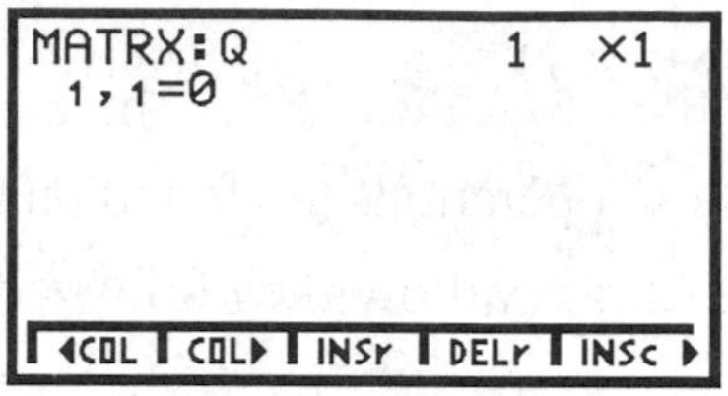

Figure 3 The matrix editor, ready to receive Q.

The cursor is blinking at the top right, ready for you to enter the number of rows in the matrix, press 2 and ENTER; to enter the number of columns, press 3 and ENTER. Type in the entries as prompted by rows, or use the cursor arrows to move up or down among columns; to move among rows, use the F1 or F2 menu keys. To exit the editor press the EXIT key. To view the matrix Q on the screen, press ALPHA (blue key), [Q] (4 key) and ENTER.

You can create vectors in a similar fashion. If using direct entry, a vector is entered as a row vector, that is, as [1, 2, 3, 4]. If using the vector editor, the procedure is similar to the matrix editor described above.

WARNING: Vectors and matrices are different objects; operations in the [MATRX] menu are *only* for matrices, and operations in the [VECTR] menu are *only* for vectors. Vectors can also be entered as n×1 matrices, in which case it will be impossible to use the operations in the [VECTR] menu on these. In the following notes the term vector is *never* used for an n×1 matrix. The programs MtoV, and VtoM convert an n×1 matrix into a vector, and a vector into an n×1 matrix, respectively. For the TI-85 there is no such thing as a row vector.

STUDY GUIDE NOTES

Section 1.1

Row Operations

In this exercise set, the data for each exercise are stored in a matrix M. Row operations on M are performed by the following commands:

mRAdd(m, M, s, r)	Replaces row r of M by (row r) + m × (row s)
rSwap(M, r, s)	Interchanges rows r and s of M
multR(m, M, r)	Multiplies row r of M by a nonzero scalar m

rAdd(M, s, r) Replaces row r of M by row r + row s

These operations arc found on the second page of the [MATRX] [OPS] menu. (Press the [2nd] (yellow) key followed by [7], to bring up the [MATRX] menu, Figure 4. The row operations are found when you press the [F4] menu key followed by the [MORE] key, Figure 5.) To use the row operations efficiently it is convenient to store matrices in memory. The name of any matrix in your calculator memory can be inserted in place of M; the letters r, m, and s stand for numbers you choose. Press [ENTER] after each command.

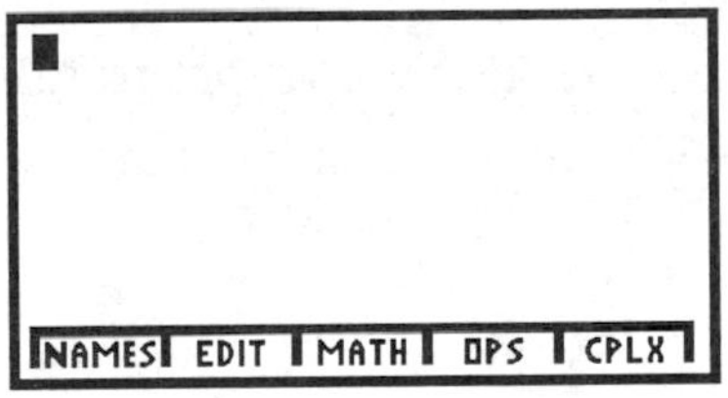

Figure 4 The matrix menu

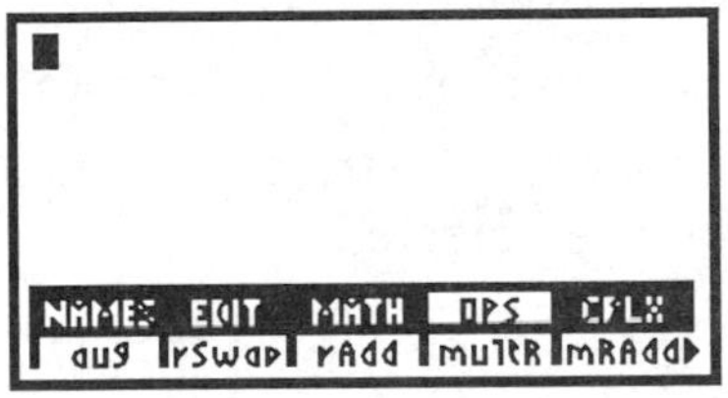

Figure 5 The matrix menu showing row operations

If you enter one of these commands, say, **rSwap(M, 1, 3)**, then the new matrix, produced from M, is stored in the matrix "ANS" (for "answer"). If instead you enter **rSwap(M, 1, 3)→M1**, then the answer is stored in a new matrix M1. If the new operation is **mRAdd(5, M1, 1, 2)→M2**, then the result of changing M1 is placed in M2, and so on.

The advantage of giving a new name to each new matrix is that you can easily go back a step if you don't like what you just did to a matrix. If, instead you enter **mRAdd(5, M, 1, 2)→M**, then the result is placed back in M and the "old" M is lost. Of course the "reverse" operation **mRAdd(-5, M, 1, 2)→M**, will bring back the old M.

Note: For the simple problems in this section and the next, the multiple m you need in the command **mRAdd(m, M, s, r)** will usually be a small integer or fraction that you can compute in your head. In general, m may not be so easy to compute mentally. The next two paragraphs describe how to handle such a case.

The entry in row r and column c of a matrix M is denoted by M(r, c). If the number stored in M(r, c) is displayed with a decimal point, then the displayed values may be accurate to only about five digits. In this case, use the symbol M(r, c) instead of the displayed value in calculations.

For instance, if you want to use the entry M(s, c) to change M(r, c) to 0, enter the commands

-M(r, c)/M(s, c)→m	The multiple of row s to be added to row r
mRAdd(m, M, s, r)→M	Adds m times row s to row r

or you can just use the command **mRAdd(-M(r, c)/M(s, c), M, s, r)→M**.

Section 1.3

Constructing a Matrix

The **aug** operator concatenates two matrices or a matrix and a vector. (see the second page of the [MATRX] [OPS] menu. Press [2nd] [MATRX] [F4] [MORE] to view the appropriate command, then [F1] to select the it, Figure 6. Fill in with the desired arguments and press [ENTER].) The command **aug(aug(a1, a2), aug(a3, b))→M** can create the matrix $[\mathbf{a}_1 \quad \mathbf{a}_2 \quad \mathbf{a}_3 \quad \mathbf{b}]$ and save it as M; but, if you entered **a1, a2, a3** and **b** as vectors, you must convert **a1** and **a3** to n×1 matrices (use the [VtoM] program). The same matrix is created by the command **aug(A, b)→M**—assuming that the matrix A and the vector **b**, have been created.

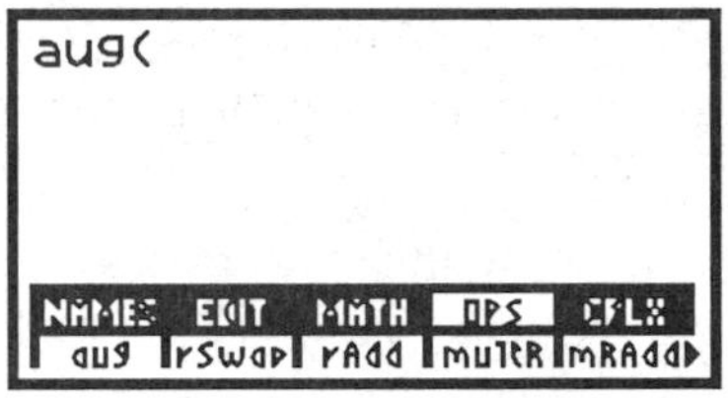

Figure 6 The **aug(** command.

Exercises 11–14 and 25–28 can be solved using the commands **mRAdd**, **rSwap**, and (occasionally) **multR**, described in section 1.1.

Section 1.4

[GAUS] AND [BGAUS]

To solve A**x** = **b**, row reduce the matrix [A **b**]**→M**. The command **[5, 3, ⁻7]→x** creates a vector **x** with entries 5, 3, –7. Matrix-vector multiplication is **A*x**.

To speed up row reduction of [A **b**]**→M**, the program [GAUS] will use the leading entry in row r of M as a pivot, and use row replacements to create zeros in the pivot column below this pivot entry. The result is stored in the matrix U. You can assign it to some other variable, such as M itself.

For the backward phase of row reduction, use the [BGAUS] program, which selects the leading entry in row r as the pivot, and creates zeros in the column *above* the pivot. Use **multR** to create leading 1's in the pivot positions.

To run a program enter [PRGM] [NAMES], use the [MORE] key as many times as necessary until the name of the program appears in the menu. Press the desired menu key

(F1 through F5) then ENTER to begin execution of the program. Both GAUS and BGAUS will prompt the user for the matrix and the row.

Section 1.5

The command **{m, n}→dim Z** creates an m×n matrix of zeros and stores it in Z, if Z does not already exist (if Z already exists, the command will reformat Z). The same procedure applies to vectors. The '{' and '}' symbols are found in the [LIST] menu, and the **dim** command is in the [MATRX] OPS menu. When solving the equation A**x** = **0** create an augmented matrix **aug(A, Z)→M**, and assign it to the variable M, or execute the commands

```
A→M
{m, n+1}→dim M              m and n are the number of rows and columns of A
```

to reformat the matrix A as an m×(n + 1) matrix with a column of zeros appended and save it as M. Then use GAUS, BGAUS and **multR** to row reduce M completely.

Section 1.9

Generating a Sequence

Enter and store the matrix M and the vector $\mathbf{x}_0$ as **x**. Then use the command **M*x→x** repeatedly to generate the sequence $\mathbf{x}_1, \mathbf{x}_2, \ldots$. You only enter the command once. After that, press 2nd [ENTRY] to recall **M*x→x** and press ENTER.

In Exercise 11 you need 6 decimal places to get four significant figures in M(1, 2). The TI-85 displays up to 12 digits; press 2nd [MODE], use the cursor keys to select Normal and Float, then EXIT.

Numbers are entered into the TI-85 without commas. The number 600,000 in TI-85 scientific notation is **6E5**. A small number such as .00000012 is **1.2E-7**.

Section 2.1

Matrix Notation and Operations

To create a matrix, begin with a square bracket, enter the data row-by-row, with a comma between entries, each row of the matrix must begin and end with square brackets. For instance, the command

```
[[1, 2, 3][4, 5, 6]]→A          Use brackets around the data
```

Creates a 2×3 matrix A. If A is m×n , then **dim A** (found in the [MATRX] OPS menu) is the list {m n}. The (i, j)-entry in the matrix A is A(i, j). A(i) is the ith row of the matrix A, *given as a vector*. To extract the jth column of A, use the program COL, the program will prompt for a matrix, and column number, it will return the jth column *as a vector*. A(i, j, r, s) is a submatrix of A starting at row i, column j, and ending at row r, column s. Examples:

A(1, 3, m, 3)	is column 3 of A, *as a matrix*
A(2)	is row 2 of A, *as a vector*

To specify columns 3, 4, and 5 of A, you can use

A(1, 3, m, 5)	m is the number of rows of A

The TI-85 uses the +, −, and × keys to denote matrix addition, subtraction and multiplication, respectively. If A is square and k is a positive integer, A ^ k denotes the kth power of A (A x^2 is equivalent to A^2). The transpose of A is **A^T** (the ^{T} operator is in the [MATRX] MATH menu). Note: when A has complex entries, the (i, j)-entry of **A^T** is the complex conjugate of the (i, j)-entry of A.

If **u** and **v** are two vectors of the same size, then **dot(u, v)** is their inner product (the **dot** operator is in the [VECTR] MATH menu). If the vectors are entered as n×1 matrices the inner product is **v^T*u**. The outer product of two real vectors **u** and **v** is computed as **u*v^T** and you must enter the vectors as n×1 matrices.

Section 2.2

The n×n identity matrix is **ident n** (the **ident** command is in the [MATRX] OPS menu). If A is n×n , then **aug(A, ident n)→M** creates the augmented matrix M = [A | I]. Use the GAUS and BGAUS programs, and the **multR** operation to reduce [A | I].

There are other commands that can be used to row reduce matrices, invert matrices, and solve equations A**x** = **b**, but they are not discussed here because they will not help you to learn the concepts in this section.

Section 2.3

Inverse

Determining whether a specific numerical matrix is invertible is not always a simple matter. A fast and fairly reliable method is to enter the matrix A and press [2nd] A [x^{-1}], which computes the inverse of A. If the matrix in question is not invertible, the TI-85 will display the message "SINGULAR MAT".

Section 2.4

Partitioned Matrices

The TI-85 uses partitioned matrix notation. For example, if A, B, C, D, E, and F are matrices of appropriate sizes, then the command

aug(aug(aug(A, B), C)ᵀ, aug(aug(D, E), F)ᵀ)ᵀ→M

creates a larger matrix of the form $M = \begin{bmatrix} A & B & C \\ D & E & F \end{bmatrix}$. Once M is formed, there is no record of the partition that was used to create M. For instance, although B was the (1, 2)-block used to form M, the number M(1, 2) is the same as the (1, 2)-entry of A.

Section 2.5

LU Factorization

Row reduction of A using the [GAUS] program will produce the intermediate matrices needed for an LU factorization of A. You can try this on the matrix in Example 2. The matrices in (5) on page 136 in the text are produced by running the commands

[GAUS]	A	1	Returns a matrix U with 0's below the first pivot
[GAUS]	U	2	Returns a matrix U with 0's below pivots 1 and 2
[GAUS]	U	3	Returns an echelon form, U

You can copy the information from your screen onto your paper, and divide by the pivot entries to produce L as in the text. For most text exercises, the pivots are integers and so are displayed accurately.

The TI-85 command **LU(A, L, U, P)** produces a permuted LU factorization for some square matrices A, but it does not handle the general case.

Section 2.6

Jacobi and Gauss-Seidel Methods

The $\boxed{\text{MdiaV}}$ program produces a vector (not an n×1 matrix) that lists the diagonal entries of a given matrix. To create an n×n diagonal matrix that has the same diagonal entries as the matrix A use the $\boxed{\text{MdiaM}}$ program. Both programs prompt you for the matrix A.

The Jacobi and Gauss-Seidel methods use the same commands, except for the construction of M. Enter **M − A→N**, and make sure the initial data is in the vector **x**. To produce the "next" **x**, you must solve the equation M* $\mathbf{x}^{(1)}$ = N* $\mathbf{x}$ + **b** for $\mathbf{x}^{(1)}$. The correct way to do this is to use the TI-85's inverse key ($\boxed{\text{2nd}}$ [x^{-1}]): compute

$\mathbf{x}^{(1)} = M^{-1}*(N*\mathbf{x} + \mathbf{b})$. However, instead of assigning the solution to $\mathbf{x}^{(1)}$, you should put the solution back into **x**, so you can repeat the operation *with exactly the same symbols*, that is, the command

$$\mathbf{M^{-1}*(N*x + b)→x}$$

is used over and over, creating the sequence $\mathbf{x}^{(k)}$. These vectors can be recorded on paper as they appear, you may need to scroll down to view all of the entries. You only enter the command once. After that, use the $\boxed{\text{2nd}}$ [ENTRY] $\boxed{\text{ENTER}}$ keys to recall and execute the command.

If you wish to monitor how close the approximations are to each other, use the following sequence of commands (repeatedly):

M^{-1}*(N*x + b)→t	t is temporary storage for the new **x**
t − x→change	Compare t with the old **x**
t→x	Update **x** with new values

Section 2.9

rref and ref

The command **rref A** (the number of columns in A must be greater then the number of rows in A), in the [MATRX] $\boxed{\text{OPS}}$ menu, produces the r̲educed r̲ow e̲chelon f̲orm of A. From that you can write a basis for Col A or write the homogeneous equations that describe Nul A. (Don't forget that A is a coefficient matrix, not an augmented matrix.)

Your calculator also has the command **ref A** (the number of columns in A must be greater then the number of rows in A), in the [MATRX] $\boxed{\text{OPS}}$ menu, that produces a r̲ow e̲chelon f̲orm of A. This form gives you the pivot positions, but you will not be able to write down the homogeneous equations that describe Nul A.

You can use **ref A** to check the rank of A, but roundoff error of an extremely small pivot entry can produce an incorrect echelon form.

Section 3.2

Computing Determinants

To compute det A, repeatedly use the [GAUS] program and the **rSwap(matrix, row1, row2)** command to reduce A to a matrix U which is an echelon form of A. Keep track of how many times you swap rows. Then except for a ±1, the determinant is obtained by first using the program [MdiaV] to extract the diagonal of U followed by the [PRDCT] program to compute the product of the diagonal. You can, of course, use the [det] key (found on the first page of the [MATRX] [MATH] menu) to check your work, but the longer sequence of commands helps you to think about the *process* of computing det A.

Section 4.3

rref

The command **rref A** (the number of columns in A must be greater then the number of rows in A), in the [MATRX] [OPS] menu, produces the reduced row echelon form of A. From that you can write down a basis for Col A or write the homogeneous equations that describe Nul A. (Don't forget that A is a coefficient matrix, not an augmented matrix.)

In some cases, roundoff error or an extremely small pivot entry can cause **rref A** to produce an incorrect echelon form.

Section 4.4

If the equation A**x** = **b** has a unique solution and A is a square matrix, the TI-85 will automatically produce **x** if you use the command

A⁻¹*b→x

In this section, the equation will probably have the form P**u** = **x**, with **u** the $\mathcal{B}$-coordinate vector of **x**, and the command will be **P⁻¹*x→u**.

Section 4.6

ref and rank

In this course, you can use either **ref A** or **rref A** to check the rank of A. Or use the GAUS program.

Section 4.7

The **rref A** command (the number of columns in A must be greater then the number of rows in A) will completely reduce a matrix such as $[c_1 \quad c_2 \quad b_1 \quad b_2]$ to the desired form. See the note on section 1.3 to on how to construct this matrix.

Section 4.8

To find the roots of a polynomial, use the 2nd [POLY] sequence, then input the coefficients as prompted, and press the SOLVE menu key. The roots of the polynomial are displayed on the screen. Complex roots appear as pairs of real numbers. Refer to your TI-85 Graphics Calculator Guidebook (p. 14–8).

Section 4.9

The TI-85 note for Section 1.9 contains information that is useful for homework here.

Section 5.1

Finding Eigenvectors

The program NULB will simplify your homework by automatically producing a basis for an eigenspace. For example, if A is a 3×3 matrix with an eigenvalue 7, first store A, then use the keystrokes

A - 7* 2nd [MATRX] OPS ident 3→C

to produce the matrix C = A – 7I. Then run the NULB program; at the prompt for a matrix, enter C. The output is a matrix B whose columns form a basis for the eigenspace of A corresponding to $\lambda = 7$. In general **ident k** (in the [MATRX] OPS menu) produces the k×k identity matrix and NULB produces a matrix whose columns form a basis for Nul C (the same basis you would get if you started with **rref C** and made the calculations by hand).

Remarks:

1. The program NULB uses the command **rref** which requires that the number of columns of the input matrix be greater than the number of rows.

   F5 MORE F1 ENTER which shows the last answer in fractions, if possible.

2. If the numbers in the basis matrix B are messy, try the sequence 2nd [MATH] F5 MORE F1 ENTER which shows the last answer in fractions, if possible.

3. Although your TI-85 has more powerful commands for eigenvector calculations, you should not use them yet, because you need to learn basic concepts about eigenvalues and eigenvectors.

Section 5.2

You can use the CHAR program to check your answers in Exercises 9–14. Note that if A is n×n, running this program produces a vector listing the coefficients of the characteristic polynomial of A, in order of decreasing powers of λ, beginning with λ^n. If the polynomial is of odd degree, the coefficients are multiplied by −1, to make +1 the coefficient of λ^n; this corresponds to finding the determinant of λI – A. (The program works for matrices up to size 3×3.)

Section 5.3

eigVl and eigVc

The command **eigVl A→ev** (see the first page of the [MATRX] MATH menu) produces a list, containing the eigenvalues of the matrix A. To learn the diagonalization procedure, you should use the method of Section 5.1 to produce the eigenvectors. For instance if A is 5×5, create the matrix

```
A - ev(1)*ident 5→B
```

then use B as input for the NULB program to create a matrix whose column(s) give a basis for the eigenspace corresponding to the first eigenvalue of A listed in **ev**.

In later work you can automate the diagonalization process. The command `eigVc→P` (eigVc is in the same menu as eigVl) produces a matrix P such that AP = PD, where D is a diagonal matrix you can create from the list of eigenvalues, with the eigenvalues in the diagonal. If A happens to be diagonalizable, then P will be invertible. In any case, P is likely to be quite different from what you construct for your homework. The columns of P may be scaled. (2nd [MATH] F5 MORE F1 ENTER shows the last answer in fractions, if possible.)

Section 5.5

Complex Eigenvalues

The eigVl and eigVc keys (mentioned in Section 5.3) also work for matrices with complex eigenvalues. In this case the resulting list of eigenvalues or the matrix containing the eigenvectors as columns have some complex entries.

For any matrix V the real and imag keys in the [MATRX] CPLX menu produce the real and imaginary parts of the entries in V, displayed as matrices of the same size as V.

Section 5.6

Plotting Trajectories

Given a vector **x**, the command `A*x→x` will compute the "next" point on the trajectory. Use the 2nd [ENTRY] keys to repeat the command over and over.

The following program creates a "trajectory" matrix whose columns are the points **x**, A**x**, A^2**x**, ..., A^{15}**x**. (Change 15 to any number you wish.) The program TRAJT below, assumes that the matrix A and the vector **x** have been stored, then use the program VtoM to convert the vector **x** into an n×1 matrix. The program creates a trajectory matrix T.

TRAJT

`x→T`	**x** is the first column of T
`For(K, 1, 15, 1)`	Repeat the next two lines 15 times
`    A*x→x`	Compute the next point on the trajectory
`    aug(T, x)→T`	Store the new point in T
`End`	

If you want to plot the points of the matrix T then enter the commands

`vc▸li T(1)→X`	Convert the first row of T into a list X
`vc▸li T(2)→Y`	Convert the second row of T into a list Y

The command **vc▸li** is in the second page of the [VECTR] [OPS] menu. To plot the data go to the [STAT] menu and select [EDIT]. Make sure that the lists that appear in the editor are the X and Y lists that you created, select the [DRAW] menu ([2nd] [F3]), select the [CLDRW] option (this clears other plots that may have been created before), and then select the [SCAT] option. You may want to change the viewing window so that your plot is visible. If you have data from another trajectory stored in two other lists, you can plot both trajectories on the same graph, by plotting first one and then the other. Do not select [CLDRW] between your plots. See the TI-85 Graphics Calculator Guidebook for more information about programming and plotting with the TI-85.

Section 5.8

Power Method and Inverse Power Method

Set your calculator to display as many decimal places as possible. The algorithms below assume that A has a strictly dominant eigenvalue, and the initial vector is **x**, with largest entry 1 (in magnitude). (If your initial vector is called $\mathbf{x}_0$, rename it by entering **x0→x**.)

The Power Method When the following steps are executed over and over, the values of **x** approach (in many cases) an eigenvector for a strictly dominant eigenvalue (the program [VAMX] prompts for a vector and returns the entry in the vector with the maximum absolute value and stores it in **mu**.)

`A*x→y`		(1)
[VAMX]	Returns **mu** = estimate for the eigenvalue	(2)
`y/mu→x`	Estimate for the eigenvector	(3)

As these commands are repeated, the numbers that appear are the μ_k that approach the dominant eigenvalue. You can program your TI-85 to perform this sequence a certain number of times by using a loop structure. See the TI-85 Graphics Calculator Guidebook for more information about programming with the TI-85.

The Inverse Power Method Store the initial estimate for the eigenvalue in the variable **a**, and enter the command **A - a*ident n→C**, where n is the number of columns of A. Then enter the commands

C⁻¹*x→y	Solves the equation (A – aI)**y** = **x**	(1)
[VAMX]		
[2nd] [ANS] [2nd] [x⁻¹] + a [sto▸] nu	**nu** = estimate for the eigenvalue	(2)
y/nu→x	Estimate for the eigenvector	(3)

As these commands are repeated, lines (2) and (3) produce the sequences { ν_k } and { x_k } described in the text.

Section 6.1

The inner product of two vectors is found with the [dot] command in the [VECTR] [MATH] menu. The [norm] command in the same menu produces the length of a vector. See the TI-85 note for Section 2.1.

Section 6.2

In exercises 1–9 and 17–22, the fastest way (counting keystrokes) with the TI-85 to test a set such as { $\mathbf{u}_1$, $\mathbf{u}_2$, $\mathbf{u}_3$ } for orthogonality is to use a matrix U = [$\mathbf{u}_1$ $\mathbf{u}_2$ $\mathbf{u}_3$]. See the proof of Theorem 6.

For vectors **y** and **u**, the orthogonal projection of **y** onto **u**, is:

```
(dot(y, u)/dot(u, u))*u
```

There is a program [PROJ] that prompts for the vectors and will produce the orthogonal projection of **y** onto **u**. It is important that you understand the formula before you use the short cut to compute the projection.

Section 6.3

Orthogonal Projections

The orthogonal projection of **y** onto a single vector was described in the TI-85 note for Section 6.2. The orthogonal projection onto the set spanned by an orthogonal set of nonzero vectors is the sum of the one-dimensional projections. Another way to construct this projection is to normalize the orthogonal vectors, (use the [unitV] key in the [VECTR] [MATH] menu) place them in the columns of a matrix U, and use Theorem 10. For instance if $\{\mathbf{y}_1, \mathbf{y}_2, \mathbf{y}_3\}$ is an orthogonal set of nonzero vectors, then the matrix

$$U = \left[\frac{\mathbf{y}_1}{\text{norm}(\mathbf{y}_1)} \quad \frac{\mathbf{y}_2}{\text{norm}(\mathbf{y}_2)} \quad \frac{\mathbf{y}_3}{\text{norm}(\mathbf{y}_3)} \right]$$ can be created with the sequence:

`unitV y1→u1`	u1 is a unit vector in the direction of y1
[VtoM] `u1`	Convert u1 to an n×1 matrix
[2nd] `[ANS]→U`	Store the resulting matrix in U
`unitV y2→u2`	u2 is a unit vector in the direction of y2
`aug(U, u2)→U`	Append u2 to U
`unitV y3→u3`	u3 is a unit vector in the direction of y3
`aug(U, u3)→U`	Append u3 to U.

The resulting matrix U has orthonormal columns, and `U*Uᵀ*y` produces the orthogonal projection of y onto the subspace spanned by $\{\mathbf{y}_1, \mathbf{y}_2, \mathbf{y}_3\}$.

Section 6.4

The Gram-Schmidt Process

If A has only two columns, then the Gram-Schmidt process can be implemented using the following commands:

`Aᵀ→B`	B is the transpose of A
`B(1)→v1`	v1 is the first vector
`B(2)→x2`	x2 is the second column of A (a vector)
`x2-(dot(x2, v1)/dot(v1, v1))*v1→v2`	v2 is the second vector

If A has three columns, then add the commands:

```
B(3)→x3
x3-(dot(x3, v1)/dot(v1, v1))*v1-(dot(x3, v2)/dot(v2, v2))*v2→v3
```

You can also use the [PROJV] program, which computes the projection of a vector **x** onto the subspace spanned by a set of vectors given as columns of a matrix V. For example,

`Aᵀ→B`	B is the transpose of A
`A(1, 1, m, 1)→v1`	v1 is the first column of A (as a matrix)
`B(2)→x2`	x2 is the second column of A (as a vector)

PROJV x2 V	Compute the projection of the x2 onto V = [**v1**]
x2 - 2nd [ANS]→v2	v2 is the second vector in the process
aug(V, v2)→V	V = [**v1 v2**]
B(3)→x3	x3 is the third column of A (as a vector)
PROJV x3 V	Compute the projection of the x3 onto V = [**v1 v2**]
x3 - 2nd [ANS]→v3	v3 is the third vector in the process

Note: in this example, **v1** is an n×1 matrix, and **v2** and **v3** are vectors, this is because the **aug** operator requires at least one matrix, and the PROJV program takes a vector and a matrix as data.

To check your work or save time, you can use the GS and GSO programs to perform the Gram-Schmidt process on the columns of a given matrix. The GS program will produce a matrix whose columns are orthogonal, while the GSO program will produce a matrix with orthonormal columns. Thus to find Q for a matrix A, use the GSO program.

Section 6.5

For Exercises 15 and 16, see the Numerical Note on page 410 in the text.

For Exercise 26, the command

aug(A1ᵀ, A2ᵀ)ᵀ

creates a (partitioned) matrix whose top block is A1 and the bottom block is A2. This command works as long as A1 and A2 have the same number of columns.

Section 7.1

Orthogonal Diagonalization

The eigVl and eigVc keys orthogonally diagonalize any symmetric matrix A, but you miss the opportunity to learn the procedure of this section. Use eigVl to find the eigenvalues of the matrix and the NULB program to obtain eigenvectors, as in Section 5.3. If you encounter a two-dimensional eigenspace with a basis { $\mathbf{v}_1$, $\mathbf{v}_2$ }, use the command 2nd [ANS]→V to save the matrix of eigenvectors in V. Then use the sequence

```
U(1, 1, m, 1)→v1                    v1 is the first eigenvector (as a matrix)
U(1, 2, m, 2)→v2                    v2 is the second eigenvector (as a matrix)
v2-((v2ᵀ*v1)/(v1ᵀ*v1))*v1→v2
```

or the sequence

```
Uᵀ→U                                U is a matrix that has the eigenvectors as rows
U(1)→v1                             v1 is the first eigenvector (as a vector)
U(2)→v2                             v2 is the second eigenvector (as a vector)
v2-(dot(v2, v1)/dot(v1, v1))*v1→v2
```

or the sequence

```
Uᵀ→U                                U is a matrix that has the eigenvectors as rows
U(1)→v1                             v1 is the first eigenvector (as a vector)
U(2)→v2                             v2 is the second eigenvector (as a vector)
[PROJ] v2 v1                        Returns the projection of v2 onto v1 (v1 and v2 vectors)
[2nd] [ANS]→x                       Store the answer in x
v2-x→v2                             v2 is the new eigenvector
```

to make the new eigenvector **v2** orthogonal to **v1**. (Review Section 6.2.) After you normalize the eigenvectors and create P, check that $P^TP = I$ to verify that P is indeed an orthogonal matrix.

Section 7.4

The Singular Value Decomposition

The commands **eigVl Aᵀ A→eG** and **eigVc AᵀA→P** produce the list of eigenvalues and an orthogonal matrix P of eigenvectors of the matrix A^TA, but the eigenvalues may not be in decreasing order. In such a case, you will have to rearrange things to form V and Σ (denoted below by S).

For instance, if P is 3×3 the commands

```
Pᵀ→M
rSwap(M, 2, 3)→M
Mᵀ→P
```

interchange columns 2 and 3 of P to form V. The commands

```
A→S: Fill (0, S)
2nd [√]eG(3)→S(2, 2)
```

produce a zero matrix for "Σ" the same size as A and place the square root of the 3rd element in the list of eigenvalues into the (2, 2)-entry of S. (The [Fill] key is in the [MATRX] [OPS] menu, its purpose is to fill a matrix with a specified value). Other diagonal entries for S can be entered similarly. To form U for the SVD, normalize the nonzero columns of A*V. If U needs more columns, use the method of Example 4.

Section 7.5

Computing Principal Components

The [RMEAN] program takes a stored matrix X and produces a vector whose entries list the averages of the rows of X, use the [VdiaM] program with this vector as input to create a diagonal matrix whose diagonal entries are the row averages of X. Multiply this matrix on the right by a matrix of all ones to create a matrix A which is the size of X, whose columns are all the same: each column of A lists the row averages of X. To create a matrix of all ones of the appropriate size, make a copy of the matrix X, say Ones, and use the [Fill] key in the [MATRX] [OPS] menu. (**Fill (1, Ones)** fills the matrix Ones with ones.)

To convert the data in X into mean-deviation form, use

```
X - A→B
```

The sample covariance matrix S is produced by

```
(1/(N-1))*B*Bᵀ→S
```

Notes for the HP-46G Graphics Calculator

Section 1.1 – Systems of Linear Equations

HP-48G Row Operations

Row operations on a matrix A are performed by the following keys, which are found in the `MATH` `MATR` `ROW` menu: `RSWP` swaps rows, `RCI` scales a row by a non-zero constant, and `RCIJ` performs a row replacement operation. The `RSWP` key is on the second page of the menu. These keys are used as follows; you may also find the User's Guide (p. 14-19) helpful.

`RSWP`: With the matrix on level 1, enter the numbers of the rows you want swapped. For example, 1 `ENTER` 2 `ENTER` `RSWP` will swap rows 1 and 2.

`RCI`: With the matrix on level 1, enter the constant c by which you want to multiply row i. Next enter the row number i. For example, 5 `ENTER` 1 `ENTER` `RCI` will multiply row 1 by 5.

`RCIJ`: With the matrix on level 1, first enter the constant c by which you want to multiply row number i. Next enter the row number i of the row you wish to multiply, then enter the row number j of the row to which you want to add c times row i. For example, 5 `ENTER` 1 `ENTER` 2 `ENTER` `RCIJ` will add 5 times row 1 to row 2.

The new matrix will now be on level 1 of the stack; if you wish to keep a copy of it for later use, simply press `ENTER`; a copy of it will then appear on level 2. If you then perform a row operation that you don't like for some reason, simply use `DROP` (the purple backspace key) to remove it from the stack. The old matrix on level 2 will now move down to level 1, ready for your next operation. Make sure to recopy it using `ENTER` before proceeding. More permanent storage can be achieved using the `STO` key; see the User's Guide (p. 5-11) for more information.

Note: For the simple problems in this section and the next, the multiple c you will need in the `RCI` and `RCIJ` commands will usually be a small integer or fraction that you can compute in your head. In general, c may not be so easy to compute mentally. The paragraphs that follow describe a simple way to write c in terms of the entries in A.

The (i, j) entry in A is denoted by the algebraic expression `'A(i,j)'`. To use this expression in calculations, it must be surrounded by single quotes, and the matrix A must be stored as a variable in your current directory. You can use this expression to help you row reduce matrices.

For instance, if you want to scale row i of A to change the value of `A(i,k)` to 1, you can enter A, then `'A(i,k)'`, then press `1/x` and `EVAL`. The proper scaling factor should now be on level 1. Finally, enter the row number i and press `RCI`.

If you want to use a pivot entry `A(i,j)` to change `A(k,j)` to 0, you can enter A, then `'A(k,j)'` `'A(i,j)'` `÷` `+/−` `EVAL` to produce the proper factor. Then enter i, then k, then press `RCIJ`.

Section 1.3 – Vector Equations

HP-48G Constructing a Matrix

To create the matrix $A = \begin{bmatrix} \mathbf{a}_1 & \mathbf{a}_2 & \mathbf{a}_3 & \mathbf{b} \end{bmatrix}$, do the following steps. First enter $\mathbf{a}_1$ *as a vector*. This can be confusing – you must enter it as a row. Open one pair of brackets with the [[]] (purple [×]) key, then enter the entries of the vector from top to bottom as the cursor proceeds from left to right. Separate the entries with a [SPC]; when you have completed typing the entries, press [ENTER]. Enter $\mathbf{a}_2$, $\mathbf{a}_3$, and $\mathbf{b}$ onto the stack in similar fashion. You then enter the number of columns (4 in this case), and press [COL→], which is found in the [MTH] [MATR] [COL] menu.

To append the column vector $\mathbf{b}$ to the matrix A, thus forming $\begin{bmatrix} A & \mathbf{b} \end{bmatrix}$, first place A on the stack, then enter $\mathbf{b}$ as described above. You then enter the number of the column in A which you wish $\mathbf{b}$ to become, and press the [COL+] key in the [MTH] [MATR] [COL] menu. Consult your User's Guide (pp.14-3, 14-5) for more details.

Exercises 11-14 and 25-28 can be solved using the [RSWP], [RCI], and [RCIJ] keys described in the HP-48G Note for Section 1.1.

Section 1.4 – The Matrix Equation $A\mathbf{x} = \mathbf{b}$

HP-48G [GAUS] and [BGAU]

To solve $A\mathbf{x} = \mathbf{b}$, row reduce the matrix $\begin{bmatrix} A & \mathbf{b} \end{bmatrix}$, which you can create by the method outlined in the HP-48G Note for Section 1.3. Recall that you enter column vectors as rows; thus the vector

$$\mathbf{x} = \begin{bmatrix} 1 \\ 2 \\ 3 \end{bmatrix} \text{ is entered as } \begin{bmatrix} 1 & 2 & 3 \end{bmatrix}.$$

To multiply a matrix A by a vector $\mathbf{x}$ place A on the stack, then place $\mathbf{x}$ on the stack and press [×]. The number of entries in $\mathbf{x}$ must match the number of columns in A. You should interpret the result as a column vector.

To speed up row reduction of the augmented matrix $M = [\ A \mid \mathbf{b}\ ]$, the [GAUS] key in your [TBOX] directory may be used. Place the matrix M on the stack, then enter the number of the row you wish to use. The [GAUS] program will now use the leading entry in the given row of M as a pivot, and use row replacements to create zeroes in the pivot column below this pivot entry. The result is returned to the stack. For the backward phase of row reduction, use the [BGAU] key which is also in your [TBOX] directory. The key works exactly as the [GAUS] key, except that the program creates zeroes in the pivot column *above* the pivot entry. You may then use the [RCI] key to create 1's in the pivot positions. The [TBOX] directory which contains the [GAUS] and [BGAU] programs is a subdirectory of the directory [LALG]. The [LALG] directory is available from your instructor.

Section 1.5 – Solution Sets of Linear Systems

HP-48G To create an $m \times n$ matrix of zeroes, enter the list { m n }, then 0, then press the [CON] key in the [MTH] [MATR] [MAKE] menu. To create a vector containing m zeroes, proceed as above, except use the list { m }. When solving the equation $A\mathbf{x} = \mathbf{0}$, where A is an $m \times n$ matrix, you can create the matrix $\begin{bmatrix} A & \mathbf{0} \end{bmatrix}$ by entering A, then creating a vector of m zeroes by the above method. Finally use the [COL+] key in the [MTH] [MATR] [COL] menu (described in the HP-48G Note for Section 1.3) to append the vector onto the matrix. You can then use the [GAUS], [BGAU] and [RCI] keys to row reduce $\begin{bmatrix} A & \mathbf{0} \end{bmatrix}$ completely.

Section 1.9 – Linear Models in Business, Science, and Engineering

HP-48G Generating a Sequence

To generate the sequence $\mathbf{x}_1, \mathbf{x}_2, \ldots$, enter and store the matrix M. You can then enter the vector $\mathbf{x}_0$ onto the stack, and press [ENTER] to copy it. Press [VAR] if necessary to produce a list of your variables, then press the menu key labelled [M]. The series of commands [SWAP] [×] [ENTER] will compute $\mathbf{x}_1$ and copy it onto the stack. Repeating this process will yield the sequence of vectors in order on your stack; the final vector in the sequence will appear twice. The stack will extend upwards as long as the calculator has memory to hold it; you shouldn't worry about exhausting your calculator's memory with the exercises in this section.

Numbers are entered into the HP-48G without commas. The number 6,000,000,000,000 in HP-48G scientific notation is 6.E12. A small number such as .0000000000012 is 1.2E-12.

Section 2.1 – Matrix Operations

HP-48G Matrix Notation and Operations

To create a matrix, you may use the MatrixWriter; see Chapter 8 of your User's Guide for more information. You may also use the command line to enter a matrix. For example, the keystrokes

[[]] [[]] 1 [SPC] 2 [SPC] 3 [▷] 4 [SPC] 5 [SPC] 6 [ENTER]

will create the 2×3 matrix

$$\begin{bmatrix} 1 & 2 & 3 \\ 4 & 5 & 6 \end{bmatrix}.$$

If A is an $m \times n$ matrix, you can find its size by placing A on level 1 of the stack and pressing the [SIZE] key in the [PRG] [LIST] [ELEM] menu. The list { m n } will be returned to the stack. As was noted in Section 1.5, the (i, j) element in the matrix A is 'A(i,j)'.

The HP-48G uses the [+], [−], and [×] keys to denote matrix addition, subtraction, and multiplication, respectively. Note that the [y^x] key will not operate on matrices, but the

[x^2] key will. You can produce the transpose of a matrix by using the [TRN] key in the [MTH] [MATR] [MAKE] menu. To compute the inner product of two vectors, place them at levels 1 and 2 of the stack and use the [DOT] key in [MTH] [VECTR] menu. In order to take the outer product $\mathbf{uv}^T$ of two vectors, you must enter them as $n \times 1$ matrices, and use the matrix commands [TRN] and [$\times$]. For complex vectors or matrices, consult your User's Guide.

Section 2.2 – The Inverse of a Matrix

HP-48G To produce the $n \times n$ identity matrix, enter n and press the [IDN] key in the [MTH] [MATR] [MAKE] menu. You may augment the matrix A with an identity matrix by means of the [COL+] command mentioned in Section 1.3. Use the [GAUS], [BGAU] and [RCI] keys to row reduce $\begin{bmatrix} A & I \end{bmatrix}$.

There are other keys that can be used to row reduce matrices, invert matrices, and solve equations $A\mathbf{x} = \mathbf{b}$, but they are not discussed here because they will not help you learn the concepts in this section.

Section 2.3 – Characterizations of Invertible Matrices

HP-48G [$1/x$] **and** [RANK]

Determining whether a specific numerical matrix is invertible is not always a simple matter. A fast and fairly reliable method is to enter the matrix onto the stack and press [$1/x$], which computes the inverse of the matrix. An error message is given if the calculator finds that the matrix is not invertible.

Another method to test invertibility is to enter the matrix and press the [RANK] key located on the second page of the [MTH] [MATR] [NORM] menu. Section 4.6 discusses rank and shows that an $n \times n$ matrix is invertible if and only if rank $A = n$. The HP-48G [RANK] program requires more calculations than [$1/x$], but is more reliable when the matrix is nearly singular.

Section 2.4 – Partitioned Matrices

HP-48G Partitioned Matrices

You may use the [COL+] key in the [MTH] [MATR] [COL] menu and the [ROW+] key in the [MTH] [MATR] [ROW] menu to append matrices to each other, thus creating partitioned matrices. Consult your User's Guide (p. 14-5) for more details.

Section 2.5 – Matrix Factorizations

HP-48G LU Factorization and the [$\div$] **Key**

Row reduction of A using the [GAUS] key described in the HP-48G Note for Section 2.2 will produce the intermediate matrices needed for an LU factorization of A. You can try this on the matrix in Example 2 of Section 2.5. The matrices in equation (5) on page 136 of the text are produced by placing A on the stack and keying

1	GAUS	This produces a matrix with 0's below the first pivot
2	GAUS	This produces a matrix with 0's below pivots 1 and 2
3	GAUS	This produces the echelon form of the matrix

You can copy the information from your screen onto your paper, and divide by the pivot entries to produce L as in the text. (For most text exercises, the pivots are integers and so are displayed accurately.) The LU key in the MTH MATR FACTR menu produces the ingredients for a permuted LU factorization of a square matrix A, but does not handle the general case. The calculator produces three matrices on the stack. On level 1 you will find a matrix P, on level 2 an upper triangular matrix U, and on level 3 a lower triangular matrix L. Notice that in this case the ones lie on the diagonal of U, not L. These three matrices satisfy the identity $PA = LU$, or $A = P^{-1}LU$. The matrix $P^{-1}L$ is a permuted lower triangular matrix, so the factorization $A = (P^{-1}L)U$ is a permuted LU factorization of A. As the algorithm used by the calculator differs from that in your text, you should not expect your permuted LU factorization to agree with that of the calculator.

When A is invertible, the best way to solve $A\mathbf{x} = \mathbf{b}$ is to use the ÷ key. Enter $\mathbf{b}$ onto the stack (as a vector), then enter A. Pressing ÷ now will cause $\mathbf{x}$ to be produced, again as a vector. The HP-48G performs a permuted LU factorization on A and uses the matrices P, L, and U to find $A^{-1} = U^{-1}L^{-1}P$, then $\mathbf{x} = A^{-1}\mathbf{b}$. The ÷ operation uses 15-digit internal precision, which provides for a more accurate answer than would be obtained by calculating A^{-1} by using the $1/x$ operation. The $1/x$ operation also uses a permuted LU decomposition, but does not carry as great an internal precision.

Section 2.6 – Iterative Solutions of Linear Systems

HP-48G Jacobi and Gauss-Seidel Methods

The →DIAG key in the second page of the MTH MATR menu produces a vector that lists the diagonal entries of a given matrix. To create an $n \times n$ diagonal matrix that has the same diagonal entries as the matrix A, place A on the stack and press →DIAG, then enter the number n, and press the DIAG→ key, which is also on the second page of the MTH MATR menu.

The Jacobi and the Gauss-Seidel methods use the same commands, except for the construction of M. Store M, N, and $\mathbf{b}$, and place your initial data vector $\mathbf{x}_0$ on the stack. To produce the "next" $\mathbf{x}$, you must solve the equation $M\mathbf{x} = N\mathbf{x}_0 + \mathbf{b}$ for $\mathbf{x}$. The correct way to do this is to use the HP-48G's division key: compute $N\mathbf{x}_0 + \mathbf{b}$ and place it on the stack, then enter M and press ÷. This operation will produce $\mathbf{x}$ on level 1 of the stack. If you repeat this process over and over, you will create the sequence $\mathbf{x}^{(k)}$. The vectors may be recorded on paper as you calculate them.

If you wish to monitor how close the approximations are to each other, you can store the approximations on the stack and subtract them when appropriate.

Section 2.9 – Vectors in R^n

HP-48G   RREF and RANK

By now, the row reduction algorithm should be second nature, so now it's time to cut to the chase. Applying the RREF key in the MTH MATR FACTR menu to a matrix A produces the r̲educed r̲ow e̲chelon f̲orm of A. From that you will immediately be able to write a basis for Col A and to write the homogeneous equations that describe Nul A. Don't forget that A is a coefficient matrix, not an augmented matrix.

You can use the RREF key to check the rank of A, but roundoff error or small pivot entries can produce an incorrect reduced row echelon form. A more reliable strategy is to use RANK. The RANK key is located on the second page of the MTH MATR NORM menu. By default the HP-48G sets all "tiny" elements in a matrix to 0. This helps avoid problems with roundoff error in calculations, but can also generate unexpected results. See the User's Guide (pp. 14-9, 14-20, D-5) for more information.

Section 3.2 – Properties of Determinants

HP-48G   Computing Determinants

To compute det A, place A on the stack and then repeatedly use the GAUS and RSWP keys as needed to reduce A to a matrix U which is in echelon form. (See the HP-48G Note for Section 2.2.) Keep track of how many times you swap rows. Then except for a ± 1, the determinant of A can be found by placing U on the stack and executing the keystrokes →DIAG PROD. The →DIAG key is found on the second page of the MTH MATR menu; the PROD key is found in the TBOX directory. The →DIAG key extracts the diagonal entries from U and places them in a vector, and the PROD key computes the product of those entries. You can, of course, use the DET key (found on the second page of the MTH MATR NORM menu) to check your work, but the longer sequence of commands helps you to think about the *process* of computing det A.

Section 4.3 – Linearly Independent Sets; Bases

HP-48G   RREF and COS

Applying the RREF key in the MTH MATR FACTR menu to a matrix A produces the r̲educed r̲ow e̲chelon f̲orm of A. From that you will immediately be able to write a basis for Col$A$ and to write the homogeneous equations that describe Nul$A$. Don't forget that A is a coefficient matrix, not an augmented matrix.

For Exercise 34: To form the necessary coefficient matrix in this case, you can first produce each column then use the COL→ key (See the HP-48G Note for Section 1.3 or the User's Guide p. 14-3). To produce each column, you will want to apply the function $\cos^k(t)$ to each element in a vector. To do this, enter the vector on the stack **enclosed not by brackets, but by set braces ({ and })**. This is an example of what the HP-48G calls a list. You may operate on lists just as you do on numbers, so pressing COS k y^x will apply the function $\cos^k(t)$ to each element in the list. To change this list into a vector, press [OBJ→] [→ARR]. These keys are found in the [PRG] [TYPE] menu. For more on lists, see Chapter 17 of the User's Guide.

Section 4.4 – Coordinate Systems

HP-48G The Divsion Key [÷]

If the equation $A\mathbf{x} = \mathbf{b}$ has a unique solution and A is a square matrix, you may calculate the solution $\mathbf{x}$ by entering first $\mathbf{b}$ then A and pressing [÷]. In this section, the equation will probably have the form $P\mathbf{u} = \mathbf{x}$.

This command actually computes $A^{-1}\mathbf{b}$, but uses 15-digit internal precision. This provides a more precise result than computing $A^{-1}\mathbf{b}$ by inverting A and multiplying. If A is not invertible, the HP-48G will give you an "Infinite Result" error. When A is either not invertible or not square, but $A\mathbf{x} = \mathbf{b}$ still has a solution, you may find a solution by entering $\mathbf{b}$ and A onto the stack, then pressing the [LSQ] key in the [MTH] [MATR] menu. This action produces a least-squares solution to the system (see Section 6.5).

Section 4.6 – Rank

HP-48G [RREF] and [RANK]

In this course you may use either the [RREF] or the [RANK] key to check the rank of a matrix. In practical work the [RANK] key should be used, since this key uses a more reliable algorithm. This algorithm is based on the singular value decomposition (see Section 7.4).

Section 4.7 – Change of Basis

HP-48G The [RREF] key will completely row reduce the matrix $\begin{bmatrix} \mathbf{c}_1 & \mathbf{c}_2 & \mathbf{b}_1 & \mathbf{b}_2 \end{bmatrix}$ to the desired form.

Section 4.8 – Applications to Difference Equations

HP-48G To solve a polynomial, use the `Solve poly` option in the [SOLVE] application. A vector of roots will be returned to the stack. Refer to your User's Guide (p. 18-10) for more information.

Section 4.9 – Applications to Markov Chains

HP-48G The HP-48G Note for Section 1.9 contains information that is useful for homework here.

Section 5.2 – Eigenvectors and Eigenvalues

HP-48G Finding Eigenvectors

Your TBOX directory has a program NULB that will simplify your homework by automatically producing a basis for an eigenspace. For example, if A is a 3×3 matrix with an eigenvalue 7, first place A on level 1 of the stack, then use the keystrokes

3 IDN 7 × −

to produce the matrix $A - 7I$. Then pressing the NULB key will produce a set of vectors on the stack which forms a basis for the eigenspace for A corresponding to $\lambda = 7$. In general, k IDN produces the $k \times k$ identity matrix, and C NULB produces a set of vectors which is a basis for Nul C.

Although your HP-48G has more powerful commands for eigenvalue calculations, you should not use them yet, because you need to learn basic concepts about eigenvalues and eigenvectors.

Section 5.2 – The Characteristic Equation

HP-48G You can use the CHAR key in the TBOX directory to check your answers in Exercises 9-14. Note that if A is $n \times n$, pressing this key with A at level 1 produces a vector listing the coefficients of the characteristic polynomial of A, in order of decreasing powers of λ, beginning with λ^n. If the polynomial is of odd degree, the coefficients are multiplied by -1, to make $+1$ the coefficient of λ^n. This corresponds to finding the determinant of $\lambda I - A$.

Section 5.3 – Diagonalization

HP-48G EGVL and EGV

The EGVL key on the second page of the MTH MATR menu produces a vector containing the eigenvalues of the matrix on level 1 of the stack, which we will call A. To learn the diagonalization procedure, you should use the method of Section 6.1 to produce the eigenvectors. For example, if A is 5×5 with eigenvalue λ, and is at level 1 of the stack, the series of operations

5 IDN λ × − NULB

will produce a basis for the eigenspace corresponding to the eigenvalue λ.

In later work, you can automate the diagonalization process. The EGV key in the MTH MATR menu also produces a vector containing the eigenvalues of A; in addition, it produces a matrix P on level 2 of the stack such that $AP = PD$, where D is a diagonal matrix created from the vector of eigenvalues. If A happens to be diagonalizable, then P will be invertible. In any case P is likely to be quite different from what you construct for your homework.

Section 5.5 – Complex Eigenvalues

HP-48G Complex Eigenvalues

The EGVL and EGV keys mentioned in the HP-48G Note for Section 5.3 also work for matrices with complex eigenvalues.

For a matrix on level 1 of the stack, the RE and IM keys in the CMPL menu produce the real and the imaginary parts of the entries in a matrix, displayed as matrices the same size as the given matrix. The CMPL menu is located on the second page of the MTH menu.

Section 5.6 – Applications to Dynamical Systems

HP-48G Plotting Trajectories

Given a vector **x**, you may compute the product $A\mathbf{x}$ by placing **x** on the stack, then placing A on the stack, pressing SWAP, then ×. The product vector will be left on the stack, and you may repeat the above procedure over and over if you wish.

The following program creates a "trajectory" matrix whose rows are the points **x**, $A\mathbf{x}$, $A^2\mathbf{x}$, …, $A^{15}\mathbf{x}$. (Change 15 to any number you wish.) This program asssumes that the matrix A is on level 2 of the stack and the vector **v** is on level 3 of the stack when the program is run from level 1 of the stack.

```
«→ V A                          Inputs data.
    « V DUP                     Places v on stack.
        1 15 START              This loop repeats the next line 15 times.
            A SWAP * DUP        Computes the next point on the trajectory.
        NEXT                    End of the loop.
        DROP DEPTH ROW→         Assembles points into the matrix.
    »
»
```

If you place this program on level 1 of the stack and press ENTER, the result is the trajectory matrix. If you intend to use the program more than once, store it as a variable. If you want to plot the points in the output matrix, store the matrix under some name and enter the PLOT utility. Choose the plot type "Scatter" in the window labelled TYPE and choose your matrix in the window labelled ΣDAT. You can also elect to change the corners of your viewing window at this point, if you so desire. Pressing the ERASE and DRAW menu keys will produce a graph of the trajectory. If you have the data for another trajectory stored in another matrix, you can plot both trajectories on the same graph by plotting first one and then the other. Do not press ERASE in between your plots. See Chapters 22, 23, and 29 in your User's Guide for more information about programming and plotting with the HP-48G.

Section 5.8 – Iterative Estimates for Eigenvalues

HP-48G Power Method and Inverse Power Method

The algorithms below assume that A has a strictly dominant eigenvalue, and the initial vector is $\mathbf{x}$, with largest entry 1 (in magnitude). Store A, then place A and your initial vector $\mathbf{x}$ on the stack. You proceed as follows, using the [VAMX] key in your [TBOX] directory.

The Power Method When the following keystrokes are repeated over and over, the resulting vectors approach (in many cases) an eigenvector for a strictly dominant eigenvalue:

[×]		(1)
[VAMX]	estimate for eigenvalue at level 1	(2)
[÷]	estimate for the eigenvector	(3)

In (2), the program [VAMX] finds the entry of largest absolute value in the vector $A\mathbf{x}$. To repeat the process, recall A to the stack and press [SWAP]. As these commands are repeated, the numbers that appear at level 1 after you press [VAMX] are the μ_k that approach the dominant eigenvalue. You could program your HP-48G to perform this algorithm a certain number of times by using a loop structure (see the HP-48G Note for Section 5.6).

The Inverse Power Method Store the initial estimate of the eigenvalue in the variable B, then perform the following keystrokes, where n is the number of columns in A:

[A] [B] n [IDN] [×] [−].

Store the resulting matrix as C. Place your initial vector $\mathbf{x}$ on the stack, followed by C. You then enter the keystrokes

[÷]		(1)
[VAMX]		(2)
[ENTER] [1/x] [B] [+]	estimate for eigenvalue at level 1	(3)
[DROP] [÷]	estimate for the eigenvector	(4)

You may now enter C onto the stack and repeat the keystrokes. As you repeat these keystrokes, the numbers at level 1 after step 3 form the sequence referred to as $\{\nu_k\}$ in the text; the vectors at level 1 after step 4 form the sequence $\{\mathbf{x}_k\}$.

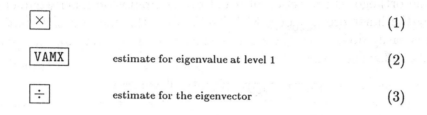

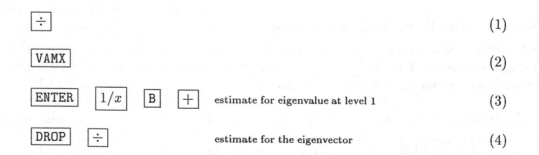

Section 6.1 – Inner Product, Length, and Orthogonality

HP-48G The inner product of two vectors may be found using the [DOT] key in the [MTH] [VECTR] menu. The [ABS] key in the same menu produces the length of a vector. See the HP-48G Note for Section 2.1.

Section 6.2 – Orthogonal Sets

HP-48G In Exercises 1-9 and 17-22, the quickest way to test a set such as $\{\mathbf{u}_1, \mathbf{u}_2, \mathbf{u}_3\}$ for orthogonality is to use the matrix $\begin{bmatrix} \mathbf{u}_1 & \mathbf{u}_2 & \mathbf{u}_3 \end{bmatrix}$. See the proof of Theorem 6.

To find the orthogonal projection of **y** onto **u**, compute the inner product of **y** and **u** and divide by the inner product of **u** with itself. Take this number and multiply it by **u**. This process is easily done using the stack; you should enter 3 copies of **u** and one copy of **y**, then use the following keystrokes:

[DOT] [SWAP] [ABS] [x^2] [÷] [×].

There is a key (labelled [PROJ] in your [TBOX] directory) which will produce the orthogonal projection of **y** onto **u** when given **y** on level 2 of the stack and **u** on level 1. It is important that you understand the formula before using this shortcut.

Section 6.3 – Orthogonal Projections

HP-48G Orthogonal Projections

The orthogonal projection of **y** onto a single vector was described in the HP-48G Note for Section 6.2. The orthogonal projection onto the set spanned by an orthogonal set of vectors is the sum of the one-dimensional projections. Another way to construct this projection is to normalize the orthogonal vectors, place them in the columns of a matrix U, and use Theorem 10. That is, the desired projection is $UU^T\mathbf{y}$.

Section 6.4 – The Gram-Schmidt Process

HP-48G The Gram-Schmidt Process

If A has only two columns, then the Gram-Schmidt process can be implemented using the following keystrokes. The [GCOL] key in your [TBOX] directory will be used to get columns from A. Store the matrix under the variable A.

[A] 1 [GCOL] 'V1' [STO]

[A] 2 [GCOL] [ENTER] [V1] [DOT] [V1] [ENTER] [DOT] [÷] [V1] [×] [−]

'V2' [STO]

If A has three columns, add the keystrokes

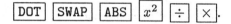

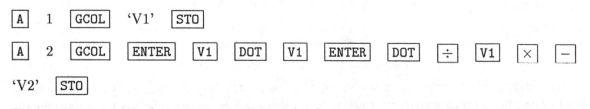

[A] 3 [GCOL] [ENTER] [ENTER] [V1] [DOT] [V1] [ENTER] [DOT] [÷] [V1]

[×] [SWAP] [V2] [DOT] [V2] [ENTER] [DOT] [÷] [V2] [×] [+] [−]

'V3' [STO]

You should use these keystrokes for awhile, to learn the general procedure. After that, you can use the [PROJ] key in your [TBOX] directory, which computes the projection of a vector $\mathbf{x}$ onto the subspace spanned by a set of vectors. To use the [PROJ] program, place $\mathbf{x}$ on the stack, then enter the set of vectors one by one onto the stack. If your set of vectors is the set of columns of a matrix, you may enter the matrix then press [→COL] [DROP] to enter the vectors. Pressing the [PROJ] key will produce the projection of the first vector entered onto the span of the remaining vectors.

To implement the Gram-Schmidt process on a matrix A with three columns using [PROJ], you would use the following keystrokes:

[A] 1 [GCOL] 'V1' [STO]

[A] 2 [GCOL] [V1] [PROJ] [A] 2 [GCOL] [SWAP] [−] 'V2' [STO]

[A] 3 [GCOL] [V1] [V2] [PROJ] [A] 3 [GCOL] [SWAP] [−] 'V3' [STO]

The set of vectors need not be orthogonal for the [PROJ] program to work, but if they are, the resulting vector will usually agree with those computed via Theorem 10 in Section 6.3 to ten or more decimal places.

To check your work or save time, you can use the [GS] and [GS.O] keys in the [TBOX] directory to perform the Gram-Schmidt process on the columns of a given matrix. The [GS] key will produce a matrix whose columns are orthogonal, while the [GS.O] key will produce a matrix with orthonormal columns. Thus to find Q for a matrix A, use the [GS.O] key.

Your calculator computes a permuted QR factorization of a matrix A with the [QR] key in the [MTH] [MATR] [FACTR] menu. This command will produce matrices Q, R, and P at levels 3, 2, and 1 respectively. The matrix Q is orthogonal, R is upper triangular, and P is again a permutation matrix (See the HP-48G Note for Section 2.5) such that $AP = QR$.

Section 6.5 – Least-Squares Problems

HP-48G The [÷] and [LSQ] Keys

For Exercises 15 and 16, see the Numerical Note in Section 6.5 of the text. You can solve the equation $A\mathbf{x} = \mathbf{b}$ using the QR factorization by solving $R\hat{\mathbf{x}} = Q^T\mathbf{b}$. Entering $Q^T\mathbf{b}$ and R onto the stack and then pressing the [÷] key will solve the system. You could also enter $\mathbf{b}$ and A onto the stack directly and press [LSQ]. This key is located in the [MTH] [MATR] menu. Compute the solutions to Exercises 15 and 16 using [LSQ], and compare your answers to those which you obtained using the QR factorization and [÷].

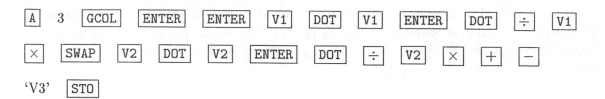

For Exercise 26, you can create a (partitioned) matrix whose top block is $A1$ and bottom block is $A2$ by placing $A1$ and $A2$ on the stack, then entering the number of rows in $A1$ plus 1. Now pressing the [ROW+] key in the [MTH] [MATR] [ROW] menu will produce the desired matrix.

Section 7.1 – Diagonalization of Symmetric Matrices

HP-48G Orthogonal Diagonalization

You can use the [EGVL] and [NULB] keys to orthogonally diagonalize a matrix by the procedure of this section. You can obtain eigenvectors as in Section 5.3; if you encounter a two-dimensional eigenspace with a basis $\{\mathbf{u}_1, \mathbf{u}_2\}$, replace $\mathbf{u}_2$ with a new eigenvector $\mathbf{v}_2$ orthogonal to $\mathbf{u}_1$. (Review Section 6.2.) You can use the [PROJ] key introduced in the HP-48G Note for Section 6.4 to help with this calculation. After you normalize these vectors and create P, you can check that P is orthogonal by confirming that $P^TP = I$.

Section 7.4 – The Singular Value Decomposition

HP-48G The Singular Value Decomposition

Applying the [EGV] operation on the matrix A^TA will produce a matrix of eigenvectors and a vector of eigenvalues for A^TA, but there are two problems. First, the matrix of eigenvectors may not be orthogonal. The procedure in the HP-48G Note for Section 7.1 can help you to produce an orthogonal matrix of eigenvectors. Second, the eigenvalues in the vector may not be in decreasing order. In such a case you will have to rearrange things things to form V and Σ. The [CSWP] key in the [MTH] [MATR] [COL] menu allows you to swap columns just as [RSWP] allows you to swap rows. To form U for the singular value decomposition, normalize the nonzero columns of AV. If U needs more columns, use the method of Example 4.

After you thoroughly understand the singular value decomposition, you will want to use the much faster and more numerically reliable [SVD] key. This key is found in the [MTH] [MATR] [FACTR] menu. If you place the matrix A on level 1 of the stack and press [SVD], the result will be the matrix U on level 3, the matrix V on level 2, and a vector of singular values on level 1. You can create the matrix Σ from this vector by using the [DIAG→] key (described in the HP-48G Note for Section 2.6), then appending rows and/or columns of zeros if necessary.

Section 7.5 – Applications to Image Processing and Statistics

HP-48G Computing Principal Components

The RMEAN menu key (abbreviated [RMEA]) in the [TBOX] directory takes a matrix X on level 1 of the stack and produces a vector whose entries list the averages of the rows of X. With this vector you may create a diagonal matrix whose diagonal entries are the row averages of X by using the [DIAG→] key in the [MTH] [MATR] menu. See the HP-48G Note for Section 2.6 for more information. Finally multiplying this diagonal matrix on the right by a matrix of all ones creates a matrix A which is the size of X, whose columns are all the same: each column of A lists the row averages of X. To create a matrix of all ones of the appropriate size, place a copy of X on the stack, enter a 1, then press the CON key in the MTH MATR MAKE menu.

To convert the data in X into mean-deviation form, find the matrix A above, then use

$$B = X - A.$$

The sample covariance matrix S is produced by the formula

$$S = \frac{1}{N-1} BB^T,$$

where N is the number of columns in B.

The principal component data you need is produced by using the SVD key in the MTH MATR FACTR menu. If you place the matrix

$$\frac{B^T}{\sqrt{N-1}}$$

on level 1 of the stack and press SVD, the result will be the matrix U on level 3, the matrix V on level 2, and a vector of singular values on level 1. The columns of V are the principal components of the data, and the squares of the singular values list the variances of the new variates.